Code of Federal Regulations

CODE OF FEDERAL REGULATIONS

T0199933

Title 7
Agriculture

Part 1950 to 1999

Revised as of January 1, 2019

Containing a codification of documents
of general applicability and future effect

As of January 1, 2019

Published by the Office of the Federal Register
National Archives and Records Administration
as a Special Edition of the Federal Register

Table of Contents

Title 7:

SUBTITLE B—REGULATIONS OF THE DEPARTMENT OF AGRICULTURE (CONTINUED)

Finding Aids:

Many agencies have begun publishing numerous OMB control numbers as amendments to existing regulations in the CFR. These OMB numbers are placed as close as possible to the applicable recordkeeping or reporting requirements.

PAST PROVISIONS OF THE CODE

Provisions of the Code that are no longer in force and effect as of the revision date stated on the cover of each volume are not carried. Code users may find the text of provisions in effect on any given date in the past by using the appropriate List of CFR Sections Affected (LSA). For the convenience of the reader, a "List of CFR Sections Affected" is published at the end of each CFR volume. For changes to the Code prior to the LSA listings at the end of the volume, consult previous annual editions of the LSA. For changes to the Code prior to 2001, consult the List of CFR Sections Affected compilations, published for 1949-1963, 1964-1972, 1973-1985, and 1986-2000.

"[RESERVED]" TERMINOLOGY

The term "[Reserved]" is used as a place holder within the Code of Federal Regulations. An agency may add regulatory information at a "[Reserved]" location at any time. Occasionally "[Reserved]" is used editorially to indicate that a portion of the CFR was left vacant and not accidentally dropped due to a printing or computer error.

INCORPORATION BY REFERENCE

What is incorporation by reference? Incorporation by reference was established by statute and allows Federal agencies to meet the requirement to publish regulations in the Federal Register by referring to materials already published elsewhere. For an incorporation to be valid, the Director of the Federal Register must approve it. The legal effect of incorporation by reference is that the material is treated as if it were published in full in the Federal Register (5 U.S.C. 552(a)). This material, like any other properly issued regulation, has the force of law.

What is a proper incorporation by reference? The Director of the Federal Register will approve an incorporation by reference only when the requirements of 1 CFR part 51 are met. Some of the elements on which approval is based are:

(a) The incorporation will substantially reduce the volume of material published in the Federal Register.

(b) The matter incorporated is in fact available to the extent necessary to afford fairness and uniformity in the administrative process.

(c) The incorporating document is drafted and submitted for publication in accordance with 1 CFR part 51.

What if the material incorporated by reference cannot be found? If you have any problem locating or obtaining a copy of material listed as an approved incorporation by reference, please contact the agency that issued the regulation containing that incorporation. If, after contacting the agency, you find the material is not available, please notify the Director of the Federal Register, National Archives and Records Administration, 8601 Adelphi Road, College Park, MD 20740-6001, or call 202-741-6010.

CFR INDEXES AND TABULAR GUIDES

A subject index to the Code of Federal Regulations is contained in a separate volume, revised annually as of January 1, entitled CFR INDEX AND FINDING AIDS. This volume contains the Parallel Table of Authorities and Rules. A list of CFR titles, chapters, subchapters, and parts and an alphabetical list of agencies publishing in the CFR are also included in this volume.

An index to the text of "Title 3—The President" is carried within that volume.

The Federal Register Index is issued monthly in cumulative form. This index is based on a consolidation of the "Contents" entries in the daily Federal Register.

A List of CFR Sections Affected (LSA) is published monthly, keyed to the revision dates of the 50 CFR titles.

REPUBLICATION OF MATERIAL

There are no restrictions on the republication of material appearing in the Code of Federal Regulations.

INQUIRIES

For a legal interpretation or explanation of any regulation in this volume, contact the issuing agency. The issuing agency's name appears at the top of odd-numbered pages.

For inquiries concerning CFR reference assistance, call 202–741–6000 or write to the Director, Office of the Federal Register, National Archives and Records Administration, 8601 Adelphi Road, College Park, MD 20740-6001 or e-mail *fedreg.info@nara.gov.*

THIS TITLE

Title 7—AGRICULTURE is composed of fifteen volumes. The parts in these volumes are arranged in the following order: Parts 1–26, 27–52, 53–209, 210–299, 300–399, 400–699, 700–899, 900–999, 1000–1199, 1200–1599, 1600–1759, 1760–1939, 1940–1949, 1950–1999, and part 2000 to end. The contents of these volumes represent all current regulations codified under this title of the CFR as of January 1, 2019.

The Food and Nutrition Service current regulations in the volume containing parts 210–299 include the Child Nutrition Programs and the Food Stamp Program. The regulations of the Federal Crop Insurance Corporation are found in the volume containing parts 400–699.

All marketing agreements and orders for fruits, vegetables and nuts appear in the one volume containing parts 900–999. All marketing agreements and orders for milk appear in the volume containing parts 1000–1199.

For this volume, Susannah C. Hurley was Chief Editor. The Code of Federal Regulations publication program is under the direction of John Hyrum Martinez, assisted by Stephen J. Frattini.

Title 7—Agriculture

(This book contains parts 1950 to 1999)

SUBTITLE B—REGULATIONS OF THE DEPARTMENT OF AGRICULTURE (CONTINUED)

Subtitle B—Regulations of the Department of Agriculture (Continued)

CHAPTER XVIII—RURAL HOUSING SERVICE, RURAL BUSINESS-COOPERATIVE SERVICE, RURAL UTILITIES SERVICE, AND FARM SERVICE AGENCY, DEPARTMENT OF AGRICULTURE (CONTINUED)

EDITORIAL NOTE: Nomenclature changes to chapter XVIII appear at 61 FR 1109, Jan. 16, 1996, and 61 FR 2899, Jan. 30, 1996.

SUBCHAPTER H—PROGRAM REGULATIONS (CONTINUED)

SUBCHAPTER H—PROGRAM REGULATIONS (CONTINUED)

PART 1950—GENERAL

Subparts A-B [Reserved]

Subpart C—Servicing Accounts of Borrowers Entering the Armed Forces

Subparts A-B [Reserved]

Subpart C—Servicing Accounts of Borrowers Entering the Armed Forces

AUTHORITY: 5 U.S.C. 301; 7 U.S.C. 1989; and 42 U.S.C. 1480.

§ 1950.101 Purpose.

Borrowers with accounts serviced by the Rural Development who have entered or who are entering military service will require special treatment. This subpart prescribes the authorities, policies, and routines for servicing such cases in addition to those contained in other Rural Development regulations. This subpart is inapplicable to Farm Service Agency, Farm Loan Programs.

[45 FR 43152, June 26, 1980, as amended at 72 FR 64122, Nov. 15, 2007; 80 FR 9890, Feb. 24, 2015]

§ 1950.102 General.

(a) Rural Development will do everything possible to assist borrowers entering the armed forces to adjust their affairs in contemplation of military service. It is not the policy of Rural Development to renew, postpone, or modify annual installments due under a promissory note because of the borrower's entry into the armed services. However, under the Soldiers' and Sailors' Civil Relief Act of 1940, the property of a borrower in the armed forces cannot validly be seized or sold by foreclosure or otherwise during the borrower's tenure of service, or for three months thereafter, except (1) pursuant to an agreement entered into by the borrower after having been accepted for service, or (2) by order of the Court. Any person causing an invalid sale to be made is guilty of a misdemeanor. Regardless of the foregoing, the long-time interest of the borrower can best be served by prompt and satisfactory arrangements for the use and protection, or disposition, of the security property in accordance with the policies expressed herein. Upon request, OGC will inform the State Director with respect to relief which may be secured by a borrower under the Soldiers' and Sailors' Civil Relief Act of 1940.

(b) In connection with Multiple Housing loans to individuals, references to County Supervisor and County Office in this subpart will be read as District Director and District Office.

[50 FR 45763, Nov. 1, 1985, as amended at 80 FR 9890, Feb. 24, 2015]

§ 1950.103 Borrower owing Rural Development loans which are secured by chattels.

(a) *Policy.* (1) Borrowers who owe loans *other than* Farm Ownership (FO), Operating (OL), Soil and Water (SW), Recreation (RL), Emergency (EM), Economic Emergency (EE), Economic Opportunity (EO), Special Livestock (SL), Softwood Timber (ST) loans, and/or Rural Housing loans for farm service buildings (RHF). When information is received that a borrower is entering the armed forces, the County Supervisor will be responsible for contacting the borrower immediately for the purpose of reaching an understanding concerning the actions to take in connection with the government loan indebtedness. The borrower will be permitted to retain the chattel security if arrangements can be worked out which are satisfactory to the borrower and Rural Development. However, because of the nature of chattel security, the borrower will be informed of the usual depreciation of such property and will be encouraged to sell the property and apply the proceeds to the loan(s). In most cases, the interests of both the borrower and the Government can best be served by arranging for a voluntary sale of the security. A borrower retaining security will be expected to make payments on the loan(s) equal to the scheduled payments.

(2) *Borrowers who owe FO, SW, RL, OL, EE, EM, SL, EO, and/or RHF loans.* If the borrower is delinquent in accordance with subpart S of part 1951 of this chapter, or otherwise in default, the County Supervisor will send exhibit A and the appropriate attachments, as outlined in subpart S of part 1951 of this chapter. If the borrower is not delinquent, the County Supervisor will explain the options set out in paragraph (b) of this section.

(b) *Methods of handling.* In carrying out the above policy, the cases of borrowers entering the armed forces will be handled in accordance with one of the following methods:

(1) Voluntary sale of security. This will be accomplished in accordance with § 1962.41 of subpart A of part 1962 of this chapter. Any necessary forms will be signed:

(i) Before being accepted for service in the armed forces, if the sale is to be completed before the borrower is accepted for service, or

(ii) After being accepted for service, if the sale cannot be completed before the borrower is so accepted. For this purpose, an individual will be considered as accepted for service after being ordered to report for induction, or, if in the enlisted reserve, after being ordered to report for service in the armed forces.

(2) Assumption of indebtedness. This will be accomplished in accordance with § 1962.34 of subpart A of part 1962 of this chapter.

(3) Arrangements with third persons. When the borrower arranges with a relative or other reliable person to maintain the security in a satisfactory manner and to make scheduled payments, the State Director is authorized to approve the arrangement. In such a case, the borrower will be required to execute a power of attorney, prepared or approved by OGC, authorizing an attorney-in-fact to act for the borrower during the latter's absence.

(4) Possible legal actions. If the borrower fails or refuses to cooperate in the servicing of the loan indebtedness secured by chattels in accordance with one of the methods set forth in this section, the borrower's case folder will be forwarded to the State Director for referral to OGC for legal advice as to the steps to be taken in protecting the Government's interest.

(c) *Statements of accounts and transfers.* Borrowers entering the armed forces will be requested to designate mailing addresses for the delivery of statements of account. Any changes in these addresses will be processed on Form RD 450-10, "Advice of Borrower's Change of Address, Name, Case Number, or Loan Number" with appropriate explanations. Under this procedure, a statement of account may be mailed to a location other than where the account is maintained and serviced. This is a deviation from the established procedure. These cases will not be transferred unless the security, when retained by the borrower in accordance with paragraph (b)(3) of this section, is moved into another County Office territory. Then the transfer will be processed through the use of Form RD 450-5, "Application to Move Security Property and Verification of Address," and Form RD 450-10 with appropriate explanations. In cases when assumption agreements have been executed, statements of account will be mailed to the assuming borrower. Cases involving assumption agreements will be transferred when the assuming borrower moves from one County Office territory to another.

[45 FR 43152, June 26, 1980, as amended at 50 FR 45763, Nov. 1, 1985; 52 FR 26133, July 13, 1987; 55 FR 40646, Oct. 4, 1990; 80 FR 9890, Feb. 24, 2015]

§ **1950.104 Borrower owing Rural Development loans which are secured by real estate.**

County Supervisors, to the greatest extent possible, should keep themselves informed of the plans of borrowers with Rural Development loans secured by real estate who may enter the armed forces. They should encourage any borrower who is definitely entering the armed forces to consult with them before the borrower's military service begins concerning the most advantageous arrangements that can be made regarding the security. County Supervisors will assist these borrowers in working out mutually satisfactory arrangements. Borrowers who owe FO, SW, RL, OL, EE, EM, SL, EO, ST, and/or RHF loans and who are delinquent or otherwise in default must be sent exhibit A and the appropriate attachments, as outlined in subpart S of part 1951 of this chapter. The County Supervisor will follow the directions in subpart A of part 1965 of this chapter for liquidating real estate security. FO, SW, RL, OL, EE, EM, SL, EO, ST and/or RHF borrowers who are not delinquent will have their accounts handled as set out in the following paragraphs.

(a) *Power of attorney.* Borrowers entering the armed forces who retain ownership of the security should be encouraged to execute a power of attorney authorizing the person of their choice to take any actions necessary to insure proper use and maintenance of the security, payment of insurance and taxes, and repayment of the loan. No Rural Development employee will act as attorney-in-fact for a borrower. The State Director will consult with OGC concerning any limitations upon the use of a power of attorney under local law and the circumstances under which the power of attorney should be exercised. In general, either spouse may act as attorney-in-fact for the other spouse, but, in a few States, a spouse cannot exercise the power of attorney in connection with a sale or encumbrance of the homestead. In a majority of States, a power of attorney is revoked by the death of a person granting the power, but, in some States, the power of attorney executed by a person in the armed services remains valid until actual notice is received of the death of the person granting the power. A power of attorney should not be used in conveying title to the farm except in those States where the power is good until actual notice of death. The State Director will request OGC to prepare a satisfactory form of power of attorney which may be duplicated in the State Office and furnished to County Supervisors with a State supplement concerning its use.

(b) *Borrower retains ownership of the security.* When a borrower retains ownership of the security, Rural Development will assist in making arrangements for the use of the security which will protect the interests of both the Government and the borrower.

(1) *Leasing.* It will be more satisfactory if the security is leased under a written lease in accordance with equitable leasing policies and applicable Rural Development procedures. The borrower should make arrangements for the rental income to be used for regular payments on the loan in order to avoid the accumulation of unpaid interest. The borrower also should make arrangements for the payment of taxes and insurance and maintenance of the security to avoid having these charges paid by the Government and then charged to the account. It would be desirable to provide that the lease will continue for the duration of the borrower's military service unless either party gives written notice of earlier cancellation of the lease.

(2) *Operation by family.* When a borrower wishes to have the farm occupied and operated by family members or relatives without a written lease, the County Supervisor should advise the borrower as to whether or not the proposed arrangements will be in the best interests of the borrower and the Government. When the farm is to be operated by relatives, the hazards and disadvantages to the borrower and the Government which are inherent in unwritten contracts will be discussed, and every effort will be made to induce the borrower to enter into formal contractual arrangements whenever possible to do so.

(c) *Borrower does not retain ownership of the security.* The security may be transferred to another approved applicant or sold in accordance with applicable procedure.

(d) *Borrower abandons the security or fails to make satisfactory arrangements.* This paragraph does *not* apply to borrowers with FO, SW, RL, OL, EE, EM, SL, EO, ST and/or RHF loans. Those borrowers should be sent exhibit A and the appropriate attachments as outlined in subpart S of part 1951 of this chapter. When a borrower abandons the security or fails to make satisfactory arrangements for maintenance of the security and payment of taxes, insurance, and installments on the loan, the

County Supervisor will send a complete report on the case to the State Director. The report will include all the information that can be obtained regarding the borrower's plans for the security and any evidence to indicate that abandonment has, in fact, taken place. In these instances, it must be recognized that the borrower may have entered into verbal arrangements for the care of the security without properly advising the County Supervisor. Whether such cases may be construed to be in violation of the provisions of the mortgage, so as to support foreclosure by order of the Court under the provisions of the Soldiers' and Sailors' Civil Relief Act of 1940, will need to be determined on an individual case basis by the State Director and OGC. Clear-cut abandonment cases or instances in which the borrower fails to take action to transfer or sell the property, while evidencing no interest in it or desire to retain it, will be processed in accordance with applicable procedures.

(e) *Statement of account.* Borrowers entering the armed forces who retain ownership of the security will be requested to designate mailing addresses for the delivery of statements of account. Any changes in addresses will be processed on Form RD 450–10 with appropriate explanations.

[45 FR 43152, June 26, 1980, as amended at 50 FR 45764, Nov. 1, 1985; 52 FR 26134, July 13, 1987; 55 FR 40646, Oct. 4, 1990; 80 FR 9890, Feb. 24, 2015]

§ 1950.105 Interest rate.

(a) The Soldiers and Sailors Relief Act requires that the effective interest rate charged a borrower who enters active military duty after a loan is closed will not exceed 6 percent. This applies only to full-time active military duty and does not include military reserve status or National Guard participation.

(b) As soon as the County Supervisor verifies that a borrower is on active duty, the County Supervisor will send the borrower a letter which states that the interest rate on the borrower's Rural Development loans will not exceed 6 percent. At the same time, the County Supervisor will send the Finance Office a memorandum which states that the borrower is on active duty and that interest of not more

than 6 percent should accrue on the borrower's loans, effective as of the date of the memorandum or as of the date of the last payment, whichever is later, until further notice. If a borrower's interest rate on any loan is less than 6 percent, the loan will continue to accrue interest at the lower rate. The assistance under this section may not be retroactively applied.

(c) As soon as the County Supervisor verifies that a borrower is no longer on active duty, the County Supervisor will send the Finance Office a memorandum advising them to terminate the 6 percent interest rate. The rate will revert to the note rate (or the payment assistance rate), effective with the next scheduled payment. The 6 percent interest rate will not be cancelled retroactively.

(d) Additional directions for handling Single Family Housing Loans are contained in 7 CFR part 3550.

[52 FR 26134, July 13, 1987, as amended at 60 FR 55122, Oct. 27, 1995; 67 FR 78329, Dec. 24, 2002; 80 FR 9890, Feb. 24, 2015]

PART 1951—SERVICING AND COLLECTIONS

Subpart A—Account Servicing Policies

Subparts P–Q [Reserved]

Subpart R—Rural Development Loan Servicing

AUTHORITY: 5 U.S.C. 301; 7 U.S.C 1932 note; 7 U.S.C. 1989; 31 U.S.C. 3716; 42 U.S.C. 1480.

EDITORIAL NOTES: 1. Some of the exhibits referenced in this part 1951 are not published in the Code of Federal Regulations. Exhibits are available in any Rural Development office.

2. Nomenclature changes to part 1951 appear at 80 FR 9890–9894, Feb. 24, 2015.

Subpart A—Account Servicing Policies

SOURCE: 50 FR 45764, Nov. 1, 1985, unless otherwise noted.

§ 1951.1 Purpose.

This subpart sets forth the policies and procedures to use in servicingaccounts. This subpart also applies to Rural Rental Housing Loan (RRH), Rural Cooperative Housing Loan (RCH), Labor Housing Loan (LH), Rural Housing Site Loan (RHS), and Site Option Loan (SO) accounts not covered under the Predetermined Amortization Schedule System (PASS). Loans on PASS will be administered under 7 CFR part 3560, subpart I. Cases involving unauthorized assistance will be serviced under Subparts L and N of this part. Cases involving graduation of borrowers to other sources of credit will be serviced under Subpart F of this part. This subpart does not apply to Water and Waste Programs of the Rural Utilities Service, Watershed loans, or Resource Conservation and Development loans, which are serviced under part 1782 of this title. In addition, this subpart is inapplicable to Farm Service Agency, Farm Loan Programs.

[52 FR 26134, July 13, 1987, as amended at 69 FR 69105, Nov. 26, 2004; 72 FR 55017, Sept. 28, 2007; 72 FR 64122, Nov. 15, 2007]

§ 1951.2 Policy.

Borrowers are expected to pay their debts to the Agency in accordance with their agreements and ability to pay. They will be encouraged to pay ahead of schedule, consistent with sound financial management. When borrowers have acted in good faith and have exercised due diligence in an effort to pay their indebtedness but cannot pay on schedule because of circumstances beyond their control, servicing actions will be consistent with the best interests of the borrower and the Government. It is the policy of this agency to service borrower loan account without regard to race, color, religion, sex, marital status, national origin, age, physical or mental handicap (borrower must possess the capacity to enter into a legal contract for services).

§ 1951.3 Authorities and responsibilities.

County Supervisors and District Directors are responsible for servicing all

Agency accounts serviced by the County and District Offices as prescribed by this subpart under the general guidance and supervision of District Directors and State Office personnel. Full use will be made of the County Office Management System in account servicing. For the purposes of this Subpart, all references to "County Supervisor" shall be construed to mean "District Director" for all loans serviced by the District Office.

§§ 1951.4–1951.6 [Reserved]

§ 1951.7 Accounts of borrowers.

(a) *Accounts of active borrowers.* The foundation for proper and timely debt payment is sound farm and home planning or budgeting, including plans for debt payment, supplemented by effective followup management assistance. Account servicing, therefore, must begin with initial planning and must be an integral part of analysis and subsequent planning, as well as follow-up management assistance.

(b) *Accounts of collection-only borrowers.* (1) Collection-only borrowers are expected to pay debts to the Agency in accordance with their ability to pay. Efforts to collect such debts, including use of collection letters and account servicing visits, must be coordinated with other program activities. If these borrowers are unable to pay in full, appropriate debt settlement policies should be promptly applied.

(2) Envelopes addressed to collection-only borrowers will bear the legend "DO NOT FORWARD." When an envelope is returned indicating the borrower has moved, appropriate steps will be taken to determine the borrower's correct address.

(3) Regular County Office employees are generally expected to service the collection-only caseload when it is of moderate size. State Directors may assign additional employees to County Offices having large collection-only caseloads when necessary to service such cases to a prompt conclusion. State Directors may inform the National Office of the need for employing special collection personnel in urban areas having large collection-only caseloads when employees are not available to assign to such areas.

(4) The following actions will be taken in servicing accounts owed by collection-only borrowers:

(i) District Directors will review, yearly, all collection-only cases in each County Office with the County Supervisor as early in *each* fiscal year as possible. They will jointly agree on the actions to take and will complete Form RD 451-27, "Review of Collection-Only Accounts."

(ii) District Directors will establish with County Supervisors a systematic plan for collecting the accounts or initiating appropriate debt settlement actions during the year.

(iii) County Supervisors will include in their monthly calendars plans for servicing these accounts.

(iv) On visits to County Offices, District Directors will review the progress being made by County Supervisors to insure that goals will be reached.

(v) For collection-only accounts in District Offices, the State Director will review the accounts as required in paragraphs (b)(4)(i) through (b)(4)(iv) of this section and the District Director will service the account.

(c) *Notifying borrowers of payments.* County Supervisors will notify borrowers of the dates and amounts of payments that have been agreed on for all types of accounts. Form RD, "Reminder of Payment to be Made," or similar form approved by the State Director, will be used. The form will not contain any language indicating that an account is delinquent. These notices will be timed to reach borrowers immediately before the receipt of the income from which the payments should be made or before the installment due date on the note, as appropriate, and may include other pertinent information such as a reference to agreements reached during the year and sources of income from which the payment was planned. Such notices need not be sent when frequent payments are scheduled and the borrower customarily makes the payments when due.

(d) *Subsequent servicing.* (1) When a Farmer Program borrower fails to make a payment as agreed, the County Supervisor will notify the borrower in accordance with subpart S of part 1951 of this chapter.

(2) When a borrower other than a Farmer Program borrower fails to make a payment as agreed, the County Supervisor will contact the borrower to discuss the reasons why the payment was not made and to develop specific plans, for making the payment. Form RD, "Notice of Payment Due," may be used to notify borrowers who make payments directly to the Finance Office that their payment has not been received. Form RD 450–13, "Request for Assignment of Income From Trust Property," may be used when other methods of loan collection fail and debt repayment is possible from trust income. In the event the borrower refuses to make the payment when income is available, or if it is determined that income will not be available to make the payment within a reasonable length of time and will not be available to make future payments, action will be taken to protect the Government's interest in accordance with applicable regulations. Followup actions of subsequent servicing will be noted on appropriate Management System Cards.

(e) *Maintaining records of accounts in County Offices.* Records of the accounts of Agency borrowers will be maintained in the County Office as provided in RD Instruction 1905–A (available in any Agency office).

(f) *Inquiry for Multiple Family Housing (MFH) loans.* Inquiry for all RRH, RCH, LH, RHS and SO loans and grants will be made through field terminals using procedures in the "MFH Users Procedures" manual or by contacting the MFH Unit in the Finance Office.

(g) *Inquiry for other than Multiple Family Housing (MFH) loans.* Inquiry for these loan programs will be made through field terminals using procedures in the "Automated Discrepancy Processing System (ADPS)" manuals.

(h) *Loan Summary Statements.* Upon request of a borrower, Rural Development issues a loan summary statement that shows the account activity for each loan made or insured under the Consolidated Farm and Rural Development Act. The field office will post on the bulletin board a notice informing the borrower of the availability of the loan summary statement. See Exhibit A for a sample of the required notice.

(1) The loan summary statement period is from January 1 through December 31. The Finance Office forwards a copy of Form RD 1951–9, "Annual Statement of Loan Account," to field offices to be retained in borrower files as a permanent record of borrower activity for the year.

(2) Quarterly Forms RD 1951–9 are retained in the Finance Office on microfiche. These quarterly statements reflect cumulative data from the beginning of the current year through the end of the most recent quarter. If a borrower requests a loan summary statement with data through the most recent quarter, county supervisors may request copies of these quarterly or annual statements by sending Form RD 1951–57, "Request for Loan Summary Statement," to the Finance Office.

(3) When a loan summary statement is requested by the borrower, the field office will copy the applicable annual or quarterly Forms RD 1951–9. A copy(ies) of Form RD 1951–9 and a copy of the promissory note showing borrower installments will constitute the loan summary statement provided to the borrower.

[50 FR 45764, Nov. 1, 1985, as amended at 52 FR 11457, Apr. 9, 1987; 53 FR 35716, Sept. 14, 1988; 54 FR 10269, Mar. 13, 1989]

§ 1951.8 Types of payments.

(a) *Regular payments.* Regular payments are all payments other than extra payments and refunds. Usually, regular payments are derived from farm income, as defined § 1962.4 of subpart A of part 1962 of this chapter. Regular payments also include payments derived from sources such as Agricultural Stabilization and Conservation Service payments (other than those referred to in paragraph (b) of this section), off-farm income, inheritances, life insurance, mineral royalties and income from mineral leases (see § 1965.17 (c) of subpart A of part 1965 of this chapter), including income from leases or bonuses. Regular payments in the case of a Section 502 RH loan to an applicant involved in a mutual self-help project will include loan funds advanced for the payment of any part of the first and second installments. All payments to the lock box facility(s) by direct payment borrowers are considered regular payments.

(b) *Extra payments*. Extra payments are payments derived from:

(1) Sale of chattels other than chattels which will be sold to produce farm income or real estate security, including rental or lease of real estate security of a depreciating or depleting nature.

(2) Refinancing of the real estate debt.

(3) Cash proceeds of real property insurance as provided in subpart A of part 1806 of this chapter (RD Instruction 426.1).

(4) A sale of real estate not mortgaged to the Government, pursuant to a condition of loan approval.

(5) Agricultural Conservation Program payments as provided in subpart A of part 1941 of this chapter.

(6) Transactions of a similar nature which reduce the value of security other than chattels which will be sold to produce farm income.

(c) *Refunds*. Refunds are payments derived from the return of unused loan or grant funds, except that the term "refunds" as used in Form 1940–17, "Promissory Note," will be construed to mean the return of funds advanced for capital goods, when a loan is made for operating purposes.

[50 FR 45764, Nov. 1, 1985, as amended at 51 FR 4137, Feb. 3, 1986; 53 FR 35717, Sept. 14, 1988; 58 FR 52646, Oct. 12, 1993]

§1951.9 Distribution of payments when a borrower owes more than one type of Agency loan.

"Distribution" means dividing a payment into parts according to the rules set out in this section. This section only applies after the County Supervisor determines the amount of proceeds that will be released for other purposes in accordance with the annual plan (Form RD 431–2, "Farm and Home Plan") and Form RD 1962–1, "Agreement for the Use of Proceeds/Release of Chattel Security."

(a) *Distribution of regular payments.* (1) When a borrower owes more than one type of Agency loan, regular payments received from *each crop year's income* will be distributed in accordance with the following priorities:

(i) *First*, to an amount equal to any advances made by RD for the crop year's living and operating expenses. If no advances were made, distribute the payment according to paragraph (a)(1)(ii) of this section. If the amount of the payment was greater than the amount of any advances, the excess should be distributed according to paragraph (a)(1)(ii) of this section.

(ii) *Second*, to Agency loans in proportion to the approximate amounts due on each for the year. In determining the amounts due for the year, deduct an amount equal to any advances for the year's living and operating expenses. If the amount of the payment exceeds the amount of any advances *plus* the amount due on each loan for the year, the excess should be distributed according to paragraph (a)(1)(iii) of this section.

(iii) *Third*, to Agency loans in proportion to the delinquencies existing on each. If the amount of the payment exceeds the amount of any advances *plus* the amount due on each loan for the year *plus* any delinquencies, the excess should be distributed according to paragraph (a)(1)(iv) of this section.

(iv) *Fourth*, as advance payments on Agency loans. In making such distribution consider the principal balance outstanding on each loan, the security position of the liens securing each loan, the borrower's request, and related circumstances.

(2) When the County Supervisor determines it is reasonable to expect that the income which will be available for payment on Agency debts will be sufficient to pay the installments scheduled for the year under the first and second priorities, collections may be distributed so as to avoid unnecessary delinquencies, and regular payments derived from rental or lease of real estate security after approval of foreclosure or voluntary conveyance will be distributed to the real estate lien of the highest priority.

(3) Payments will be distributed differently than the priorities provided in this section if accounts are out of balance or a different distribution is needed to protect the government's interest.

(4) Any income received from the sale of softwood timber on marginal land

converted to the production of softwood timber must be applied on the ST loan(s).

(b) *Distribution of extra payments.* Extra payments will be distributed first to the Agency loan having highest priority of lien on the security from which the payment was derived. When the payment is in excess of the unpaid balance of the Rural Development lien having the highest priority, the balance of such payment will be distributed to the Rural Development loan having the next highest priority.

(c) *Application of payments.* After the decision is reached as to the amount of each payment that is to be distributed to the different loan types, application of the payment will be governed by §§ 1951.10 or 1951.11 of this subpart as appropriate.

[50 FR 45764, Nov. 1, 1985, as amended at 52 FR 26134, July 13, 1987; 53 FR 35717, Sept. 14, 1988]

§ 1951.10 **Application of payments on production type loan accounts.**

Employees receiving payments on OL, EO, SW codes "24," EM for subtitle B purposes, EE operating-type, and other production-type loan accounts will select, in accordance with the provisions of this section, the account(s) to which such payment will be applied. All payments on OL and EM loans approved on or before December 31, 1971, will be credited first to any administrative costs, then to noncapitalized interest, then to the amount of accrued deferred interest, and then to principal. All payments on all other loans including OL and EM loans approved after December 31, 1971, will be credited first to any administrative costs, then to noncapitalized interest, then to the amount of accrued deferred interest, then to interest accrued to the date of the payment and then to principal, in accordance with the terms of the note. This section only applies after the County Supervisor determines the amount of proceeds that will be released for other purposes in accordance with the annual plan (Form RD 431-2) and Form RD 1962-1.

(a) *Rules for selection of accounts.* The following rules will govern the selection of accounts and installments to which payments will be applied. As used in this section, "recoverable costs" are those which the loan agreement documents say the borrower is primarily responsible for paying and which the government can charge to the borrower's account.

(1) Payments from farm income or from assignments of income will be applied first to accounts with small balances, including recoverable costs, to remove such accounts from the records. Any balance will be applied on debts secured by the lien in the following order:

(i) To amounts due or falling due on loans made in connection with the current year's operations, except:

(A) When funds loaned for the purchase of capital goods were used to meet the current year's operating expenses, payments will be applied first to the final unpaid installments to the extent of the loan funds so used. These payments will be treated as extra payments.

(B) When installments on loans previously made fall due before the installment on the loan for the current year's operations or when such loans are delinquent and it is anticipated that sufficient income will be received to meet the installment on the current year's operations when due, collections may be applied first to installments on loans made in previous years.

(ii) To accounts having the oldest delinquencies, or if no delinquencies, to the oldest unpaid account, except that the amount available for payment on OL and EM loan accounts will be prorated between the two accounts on the basis of:

(A) The delinquent amount owed on each, or

(B) The total amount owed on each if there are no delinquencies.

(2) Non-farm income and payments derived from the sale of real estate security, will be applied to the earliest account secured by the earliest lien covering such security. The amount to be applied to principal will be applied to the final unpaid installment(s).

(3) On partial refunds of loan advances, the amount to be applied to the principal will be applied to the final unpaid installment on the note which evidences such advance; however, a refund of an advance for current farm

and home expenses repayable within the year may be applied to the principal on the first unpaid installment on such note as a regular payment.

(4) Total refunds of loan advances will be applied to the notes which evidence such advances.

(5) In applying payments from sources other than those in paragraphs (a)(2), (3), and (4) of this section the borrower has the right to select the loan account or accounts on which such payments will be applied. In the absence of the borrower's selection, such payments generally will be applied in the following order:

(i) To accounts with small balances, including recoverable costs.

(ii) To accounts with the oldest unsecured note(s).

(iii) To accounts with the oldest delinquencies.

(iv) To accounts with the oldest secured note or notes.

(6) Employees receiving collections are authorized to make exceptions to paragraphs (a)(1), (2), and (6) of this section when it is necessary to apply a part of a payment to delinquent accounts to prevent the Federal Statute of Limitations from being asserted as a defense in suits on Agency claims.

(b) *Payments in full.* Errors of a significant amount in computation or collection will be called to the attention of the collection official by the Finance Office. The borrower's note will not be returned until the balance on the loan account is paid in full. Claims by or on behalf of the borrowers that the amounts owed have been computed incorrectly will be referred to the Finance Office.

[50 FR 45764, Nov. 1, 1985, as amended at 53 FR 35717, Sept. 14, 1988; 54 FR 46844, Nov. 8, 1989; 57 FR 18680, Apr. 30, 1992]

·**§1951.11 Application of payments on real estate accounts.**

(a) *Regular payments.* If a borrower owes more than one type of real estate loan, or has received initial and subsequent real estate loans on which separate accounts are maintained, payments on such accounts should be applied so as to maintain the note accounts approximately in balance at the end of the year with respect to installments due on the notes, other charges, and delinquencies.

(b) *Refunds and extra payments.* (1) Refunds will be applied to the note representing the loan from which the advance was made.

(2) Extra payments will be applied to the note secured by the earliest mortgage on the property from which the extra payment was obtained.

(3) Funds remaining from an RH grant or a combination loan and grant, after completion of development, will be refunded. If the borrower received a combination loan and grant, the remaining funds up to the amount of the grant are considered to be grant funds.

(c) *County Office actions.* (1) The collecting official will complete Form RD 451-1, "Acknowledgment of Cash Payment," in accordance with the FMI when cash or money orders are received as a payment.

(2) The collection official will complete Form RD, "Schedule of Remittances," in accordance with the FMI.

(d) *Finance Office handling.* (1) Regular payment will be handled as follows.

(i) Payments will be applied first to satisfy any administrative costs such as a charge for an uncollectible check. (The amounts of any such charges are available from any Rural Development office.)

(ii) Amounts paid on direct loan accounts will be credited to the borrower's account as of the date of Form RD 451-2 or for direct payments the date payment is received in the Finance Office, and will be applied first to a portion of any interest which accrues during the deferral period, second to interest accrued to the date received and third to principal, in accordance with the terms of the note.

(iii) Amounts paid on insured loan accounts will be credited to the borrower's account as of the date of Form RD 451-2 or for direct payments the date payment is received in the Finance Office, and will be applied in the following order:

(A) Advances from the insurance funds as shown on the latest *Form RD 389-404*, "Analysis of Accounts Maturing." (If the collection is intended for final payment of the loan, or to pay the insurance account in connection with an assumption agreement, the collection will be applied first to the interest accrued on the advance to the date of the payment.)

(B) Principal advanced from the insurance fund.

(C) Unamortized costs.

(D) Amount due for amortized costs for taxes and insurance.

(E) Unpaid loan insurance charges, including the current year's charge, when applicable.

(F) First to a portion of any interest which accrues during the deferral period, second to accrued interest to the date of the payment on the note account and then to the principal balance of the note account in accordance with the terms of the note.

(2) Extra payments and refunds will be credited to the borrower's note account as of the date of Form RD 451-2 and will be applied first to a portion of any interest which accrues during the deferral period, second to interest accrued to the date of the receipt and third to principal in accordance with the terms of the note. The amount to be applied to principal will be applied to the final unpaid installment(s). Extra payments and refunds will not affect the schedule status of a borrower except indirectly in connection with the amortization of a direct loan.

(3) The Finance Office will remit final payments promptly to lenders. Other collections (regular, extra, and refunds) applied to a borrower's insured note will be accumulated until the annual installment due date, and will be remitted along with any advances from the insurance fund to the lender within 30 days after the installment due date. All payments to a lender will be credited first to interest to the date of the Treasury check and then to principal. Since the application of a payment to a borrower's account with the Government and the Government's account with a lender is of a different effective date, the balance owed by a borrower to the government and by the Government to a lender ordinarily will not be the same.

[50 FR 45764, Nov. 1, 1985, as amended at 54 FR 46845, Nov. 8, 1989]

§ 1951.12 Changes in the application of loan payments.

(a) *Authority to change payments.* County Supervisors and Assistant County Supervisors are hereby authorized to approve requests for changes in the application of payments between loan accounts when payments have been applied in error and such requests conform to the policies expressed in this Subpart. However, no change will be made if the payment applied in error resulted in the payment in full of any Agency loan and the canceled note or notes have been returned to the borrower.

(b) Form RD 1951-7, "Request for Change in Application." Requests for changes in application of payments will be made on Form RD 1951-7. For requests which County Supervisors or Assistant County Supervisors are authorized to approve, the County Supervisor or Assistant County Supervisor will sign the original of Form RD 1951-7 and forward it to the Finance Office. The Finance Office will send Form RD 451-26 to the County Office when the change is made on Finance Office records.

(c) *Changes by the Finance Office in application of remittances.* (1) When reapplication of collection is made by the Finance Office Form RD 451-8, "Journal Voucher for Loan Account Adjustments," will be prepared. Form RD 451-26 will be forwarded to the County Office to show the reapplication.

(2) When necessary, the Finance Office will correct Form RD 451-2 as prepared by the County Office.

[50 FR 45764, Nov. 1, 1985, as amended at 54 FR 18883, May 3, 1989]

§ 1951.13 Overpayments and refunds.

(a) The Finance Office will mail any overpayment refund check to the County Supervisor, who will verify that the refund is due before delivering the check.

(b) Borrower requests for overpayment refunds must be in writing. Borrowers will be discouraged from requesting refunds when the County Office records show that a refund is not due, however, the County Supervisor will forward any request to the Finance Office. Finance Office computations will control in determining the amount of any refund.

(c) Underpayments or overpayments of less than $10 will not be collected or refunded (except as provided in paragraph (b) of this section) since the expense of processing the action would be more than the amount involved.

§ 1951.14 Recoverable and nonrecoverable cost charges.

(a) The County Supervisor will:

(1) Prepare vouchers for recoverable and nonrecoverable cost charges according to the applicable instruction for the type of advance being made. ("Recoverable costs" is defined in § 1951.10(a) of this subpart).

(2) If a recoverable cost, show on the voucher the fund code to which the advance is to be charged.

(3) If the cost item relates to security for more than one type of account, show the code for the loan secured by the earliest promissory note (if lien secures more than one note).

(b) The Finance Office will forward Form RD 451–26, to the County Office when the recoverable cost charge is processed.

§ 1951.15 Return of paid-in-full or satisfied notes to borrower.

(a) *Notes not held in County Office.* When the original of the note is not held in the County Office the County Supervisor will request the Finance Office to acquire and forward the note to the County Office.

(b) *Return of notes after collection.* When a note (or loan-type account) evidencing an OL, EM, EE, EO, special livestock (SL), SW loan coded "24", or other production-type loan has been satisfied by payment in full, the County Supervisor will examine the borrower's records in the County Office and determine that the account has been satisfied before delivering the note to the borrower (See § 1962.27 of subpart A of part 1962 on the satisfaction of chattel security instruments). The note(s) will be returned to the borrower immediately except that:

(1) When the final payment is made in a form other than currency and coin, Treasury check, cashier's check, cer-

tified check, Postal or bank money order, bank draft, or a check issued by a responsible lending institution or a responsible title insurance or title and trust company, the note or notes will not be surrendered until 30 days after the date of final payment, and

(2) When notes are needed in making marginal releases or satisfactions or security instruments, the notes will be held until the instruments are satisfied.

(c) *Surrender of notes to effect collection.* (1) County Supervisors are authorized to surrender notes to borrowers when final payment of the amount due is made in the form of currency and coin, Treasury check, cashier's check, certified check, Postal or bank money order, bank draft, or a check issued by a responsible lending institution or a responsible title insurance or title and trust company.

(2) The amount due on the note(s) to be surrendered will be confirmed with the Finance Office. County Supervisors will request the original note(s) from the Finance Office if it is not in the County Office.

(d) *Return of notes reduced to judgment.* Notes which have been reduced to judgment are a part of the court records and ordinarily cannot be withdrawn and returned to the borrower even after satisfaction of the judgment. Therefore, no effort will be made to obtain and return such notes except on the written request of the judgment debtor or debtor's attorney. Such requests will be referred to the Office of the General Counsel (OGC).

(e) *Debt settlement case.* See subparts B or C of part 1956 of this chapter for the handling of notes in debt settlement cases.

(f) *Lost notes.* (1) All promissory notes dated on or after 11–1–73 are held in the County Office. A few notes (with the exception of OL notes) are still held by investors. If a note dated prior to 11–1–73 cannot be located in the County Office and it is needed for servicing the case, the County Supervisor will write a memorandum to the Finance Office explaining why the note is needed. The request should give the name and case number of the borrower, date and original amount of the loan, type of loan and loan code.

(2) If a promissory note is lost in the County Office and it is needed for servicing a case, the State Director may authorize the County Supervisor to execute an appropriate affidavit regarding the lost note. The form of such an affidavit will be provided by OGC.

[50 FR 45764, Nov. 1, 1985, as amended at 51 FR 45432, Dec. 18, 1986; 53 FR 13100, Apr. 21, 1988; 56 FR 10147, Mar. 11, 1991]

§ 1951.16 Other servicing actions on real estate type loan accounts.

(a) *Installment on note and other charges*—(1) *Direct loan accounts.* For a borrower with a direct loan, the term "installation on note and other charges," as used in this Subpart, will be the sum of the following:

(i) Annual installment for the year as provided in the promissory note(s).

(ii) Any recoverable cost charges paid for the borrower during the year. ("Recoverable costs" is defined in § 1951.10(a) of this Subpart.)

(2) *Insured loan accounts.* "Loan insurance charge" means a separate insurance charge applying to FO and SW insured loans evidenced by promissory note forms bearing a form date before January 8, 1959. For all insured loans evidenced by note forms bearing a form date of January 8, 1959, or later, the insurance charge is called "annual charge" and is included in the interest position of the annual installment in the note. For a borrower with an insured loan, the term "Installment on note and other charge" means the sum of the following:

(i) Annual installment for the year as provided in the promissory note.

(ii) Amounts owed the Agricultural Credit Insurance Fund. These amounts are covered by the general term "Insurance Account" and consist of the following:

(A) Unpaid loan insurance charges from prior years.

(B) Loan insurance charge for the current year. The loan insurance charge is computed on the basis of the amount of the unpaid principal obligation as of the installment due date and is due and payable on or before the next installment due date.

(C) Any unpaid balance on advances from the insurance fund, including any

recoverable cost charges paid for the borrower during the year.

(D) Any accrued interest on advances from the insurance fund.

(iii) The amounts owed on the insurance account must be paid by regular payments each year whether or not the note account is ahead of schedule.

(b) *Schedule status.* For direct and insured loans, a borrower will be on schedule when the sum of regular payments through the last preceding due date of the note equals the sum of installments on the note and other charges due through the same date. Such a borrower will be ahead of schedule or behind schedule when the sum of such regular payments is larger or smaller, respectively, than the sum of such installments on the note and other charges.

(c) *Real estate payments.* A borrower may make regular payments ahead of schedule at any time and use them later to forego payments or to supplement the amount available during any year for payment on the annual installment on the note and other charges. Refunds and extra payments will not be used in this way.

§§ 1951.17-1951.24 [Reserved]

§ 1951.25 Review of limited resource FO, OL, and SW loans.

(a) *Frequency of reviews.* OL, FO, and SW loans will be reviewed each year at the time the analysis is conducted in accordance with subpart B of part 1924 of this chapter and any time a servicing action such as consolidation, rescheduling, reamortization or deferral is taken. The interest rate may not be changed more often than quarterly.

(b) *Method of review.* (1) Each loan will be considered on its own merit.

(2) The County Supervisor should consider:

(i) The borrower's income and repayment record during the preceding years;

(ii) The projections shown on the most recent Farm and Home Plan or other similar plan or operation acceptable to RD, in light of the previous year's projected figures and actual figures; (See subpart B of part 1924 of this chapter)

(iii) Whether improved production practices have been or need to be implemented;

(iv) The borrower's progress as a farmer; and

(v) All other factors which the County Supervisor believes should be considered.

(3) The Farm and Home Plan projections for the coming year must show that the "balance available to pay debts" exceeds the amount needed to pay debts by at least 10 percent before an increase in interest rate is put into effect. Borrowers that continually purchase unplanned items without the County Supervisor's approval will have the interest rate on their loans increased to the current rate for that loan type. Borrowers that fail to provide the County Supervisor with the information needed to conduct the analysis required in subpart B of part 1924 of this chapter will have their interest rate on their loan increased to the current rate for the OL, FO, or SW loan as applicable. The rate may increase in increments of whole numbers to the current regular interest rate for borrowers. In the borrower's case file, the County Supervisor must document the unplanned purchases and the failure to provide information in a timely manner. The County Supervisor must write the borrower a letter which sets out the facts documented in the case file and advises the borrower that the interest rate will be increased unless the unplanned purchases cease or unless the borrower provides information in a timely manner. Whenever it appears that the borrower has a substantial increase in income and repayment ability or ceases farming, either the interest rate may be increased to the current rate for FO, OL or SW loans, as applicable, or the borrower will be graduated from the program as provided in subpart F of this part.

(4) The County Office will be responsible for scheduling and completing the reviews.

(5) Borrowers who have received a deferral under Subpart S of this part will not have the interest rate increased on their limited resource loans during the deferral period.

(c) *Processing.* (1) If, after the review, the interest rate is to remain the same, no further action needs to be taken.

(2) When the interest rate is increased to the current rate, the loan will be recorded as a regular loan and will no longer be considered a limited resource loan. The borrower must be notified in writing at least 30 days prior to the date of the change. Exhibit B of this subpart may be used as a guide. The effective date of the change in interest rate will be the effective date on Exhibit B. The borrower must be informed of the following for each loan:

(i) The authorization for the change,

(ii) Reason for change (repayment ability, etc.),

(iii) The effective date and rate of the increase in interest,

(iv) Amount of the new installments and dates due,

(v) Right to appeal.

(3) It is not necessary to obtain a new promissory note for this change in interest rate.

[50 FR 45764, Nov. 1, 1985, as amended at 53 FR 35717, Sept. 14, 1988; 56 FR 3395, Jan. 30, 1991; 58 FR 15074, Mar. 19, 1993]

§§ 1951.26–1951.49 [Reserved]

§ 1951.50 OMB control number.

The collection of information requirements in Subpart A of part 1951 have been approved by the Office of Management and Budget and assigned OMB control number 0575–0075.

[52 FR 26137, July 13, 1987]

EXHIBIT A TO SUBPART A OF PART 1951— NOTICE TO AGENCY BORROWERS

Agency borrowers with community program loan types made under the Consolidated Farm and Rural Development Act may request a loan summary statement which shows the calendar year account activity for each loan. Interested borrowers may request these statements through their local Rural Development office.

[80 FR 9891, Feb. 24, 2015]

EXHIBIT B TO SUBPART A OF PART 1951— NOTICE OF CHANGE IN INTEREST RATE

(insert date)

Notice of Change in Interest Rate

(insert borrower's address)

Re: □ □
Fund code
□ □
Loan number
□ □
Kind code

Dear (*insert borrower's name and case number*): Your promissory note dated _____, for the original amount of _____ dollars ($_____) provides for a change in interest rate for a limited resource loan in accordance with the Farmers Home Administration or its successor agency under Public Law 103–354 regulations.

Effective (insert date) the interest rate on this loan will be ____ percent (%) on the unpaid principal balance. Your installment due January 1, 19 , will be _____ dollars ($_____). This change in interest rate is for the reason indicated below.

□ Increase in repayment ability as per Farm and Home Plan dated _____.

□ (*insert reason if other than above for increase in interest rate*).

You may appeal this action by writing to (*hearing officer*), (*address*), within 30 calendar days of the date of this letter, giving the reason why you believe this matter should be decided differently. This time may be extended if you cannot notify the hearing officer within 30 days for reasons beyond your control.

[56 FR 3396, Jan. 30, 1991]

Subpart B [Reserved]

Subpart C—Offsets of Federal Payments to USDA Agency Borrowers

§ 1951.101 General.

Federal debt collection statutes provide for the use of administrative, salary, and Internal Revenue Service (IRS) offsets by government agencies, including the Farm Service Agency (FSA), Rural Housing Service (RHS) for its community facility program, and Rural Business-Cooperative Service (RBS), herein referred to collectively as "United States Department of Agriculture (USDA) Agency," to collect delinquent debts. Any money that is or may become payable from the United States to an individual or entity indebted to a USDA Agency may be subject to offset for the collection of a debt owed to a USDA Agency. In addition, money may be collected from the debtor's retirement payments for delin-

quent amounts owed to the USDA Agency if the debtor is an employee or retiree of a Federal agency, the U.S. Postal Service, the Postal Rate Commission, or a member of the U.S. Armed Forces or the Reserve. Amounts collected will be processed as regular payments and credited to the borrower's account. USDA Agencies will process requests by other Federal agencies for offset in accordance with § 1951.102 of this subpart. This subpart does not apply to direct single family housing loans, direct multi-family housing loans, and the Rural Utilities Service. Section 1951.136 of this subpart only applies to RHS for its community facility program and RBS for the offset of Federal payments. Nothing in this subpart affects the common law right of set off available to USDA Agencies.

[67 FR 69671, Nov. 19, 2002]

§ 1951.102 Administrative offset.

(a) *General.* Collections of delinquent debts through administrative offset will be taken in accordance with 7 CFR part 3, subpart B and § 1951.106.

(b) *Definitions.* In this subpart:

(1) *Agency* means Farm Service Agency, Farm Loan Programs; Rural Housing Service, except direct Single Family Housing loans and direct Multi-Family Housing loans; and Rural Business-Cooperative Service, or any successor agency.

(2) *Contracting officer* is any person who, by appointment in accordance with applicable regulations, has the authority to enter into and administer contracts and make determinations and findings with respect thereto. The term also includes the authorized representative of the contracting officer, acting within the limits of the representative's authority.

(3) *County Committee* means the local committee elected by farmers in the county, as authorized by the Soil Conservation and Domestic Allotment Act and the Department of Agriculture Reorganization Act of 1994, to administer FSA programs approved for the county as appropriate.

(4) *Creditor agency* means a Federal agency to whom a debtor owes a monetary debt. It need not be the same agency that effects the offset.

(5) *Debt management officer* means an agency employee responsible for collection by administrative offset of debts owed the United States.

(6) *Delinquent or past-due* means a payment that was not made by the due date.

(7) *Entity* means a corporation, joint stock company, association, general partnership, limited partnership, limited liability company, irrevocable trust, revocable trust, estate, charitable organization, or other similar organization participating in the farming operation.

(8) *FP* means Farm Programs.

(9) *FLP* means Farm Loan Programs.

(10) *FSA* means Farm Service Agency.

(11) *National Appeals Division* means the organization within the Department of Agriculture that conducts appeals of adverse decisions for program participants under the purview of 7 CFR part 11.

(12) *Offsetting agency* means an agency that withholds from its payment to a debtor an amount owed by the debtor to a creditor agency, and transfers the funds to the creditor agency for application to the debt.

(13) *Propriety* means the offset is feasible. It includes offsetting a debtor's payments due any entity in which the debtor participates either directly or indirectly equal to the debtor's interest in the entity. To be feasible the debt must exist and be 90 days past due or the borrower must be in default of other obligations to the Agency, which can be cured by the payment.

(14) *Reviewing officer* means an agency employee responsible for conducting a hearing or documentary review on the existence of debt and the propriety of administrative offset in accordance with 7 CFR 3.29. FSA District Directors or other State Executive Director designees are designated to conduct the hearings or reviews.

[65 FR 50602, Aug. 21, 2000, as amended at 67 FR 69671, Nov. 19, 2002; 69 FR 5267, Feb. 4, 2004]

§§ 1951.103–1951.105 [Reserved]

§ 1951.106 Offset of payments to entities related to debtors.

(a) *General.* Collections of delinquent debts through administrative offset will be in accordance with 7 CFR part 3, subpart B, and paragraphs (b) and (c) of this section.

(b) *Offsetting entities.* Collections of delinquent debts through administrative offset may be taken against a debtor's pro rata share of payments due any entity in which the debtor participates when:

(1) It is determined that FSA has a legally enforceable right under state law or Federal law, including program regulations at 7 CFR 792.7(1) and 1403.7(q), to pursue the entity payment;

(2) A debtor has created a shell corporation before receiving a loan, or after receiving a loan, established an entity, or has reorganized, transferred ownership of, or otherwise changed in some manner the debtor's operation or the operation of a related entity for the purpose of avoiding payment of the FSA, FLP debt or otherwise circumventing Agency regulations;

(3) Assets used in the entity's operation include assets pledged as security to the Agency which have been transferred to the entity without payment to the Agency of the value of the security or Agency consent to transfer of the assets;

(4) A corporation to which a payment is due is the alter ego of a debtor; or

(5) A debtor participates in, either directly or indirectly, the entity as determined by FSA.

(c) *Other remedies.* Nothing in this section shall be deemed to limit remedies otherwise available to the Agency under other applicable law.

[65 FR 50603, Aug. 21, 2000]

§§ 1951.107–1951.110 [Reserved]

§ 1951.111 Salary offset.

Salary offset may be used to collect debts arising from delinquent USDA Agency loans and other debts which arise through such activities as theft, embezzlement, fraud, salary overpayments, under withholding of amounts payable for life and health insurance, and any amount owed by former employees from loss of federal funds through negligence and other matters. Salary offset may also be used by other Federal agencies to collect delinquent debts owed to them by employees of the USDA Agency, excluding county committee members. Administrative offset, rather than salary offset, will be used to collect money from Federal employee retirement benefits. For delinquent Farm Loan Programs direct loans, salary offset will not begin until the borrower has been notified of servicing options in accordance with 7 CFR part 766. In addition, for Farm Loan Programs direct loans, salary offset will not be instituted if the Federal salary has been considered on the farm operating plan, and it was determined the funds were to be used for another purpose other than payment on the USDA Agency loan. For Farm Loan Programs guaranteed debtors, salary offset can not begin until a final loss claim has been paid. When salary offset is used, payment for the debt will be deducted from the employee's pay and sent directly to the creditor agency. Not more than 15 percent of the employee's disposable pay can be offset per pay period, unless the employee agrees to a larger amount. The debt does not have to be reduced to judgment or be undisputed, and the payment does not have to be covered by a security instrument. This section describes the procedures which must be followed before the USDA Agency can ask a Federal agency to offset any amount against an employee's salary.

(a) *Authorities.* The following authorities are granted to USDA Agency employees in order that they may initiate and implement salary offset:

(1) Certifying Officials are authorized to certify to the debtor's employing agency that the debt exists, the amount of the delinquency or debt, that the procedures in USDA Agency and United States Department of Agriculture's (USDA's) regulations regarding salary offsets have been followed, that the actions required by the Debt Collection Act have been taken; and to request that salary offset be initiated by the debtor's employing agency. This authority may not be redelegated.

(2) Certifying Officials are authorized to advise the Finance Office to establish employee defalcation accounts and non-cash credits to borrower accounts in cases involving other debts, such as those arising from theft, fraud, embezzlement, loss of funds through negligence, and similar actions involving USDA Agency employees.

(3) The Finance Office is authorized to establish defalcation accounts and non-cash credits to borrower accounts upon receipt of requests from the Certifying Officials.

(b) *Definitions*—(1) *Certifying Officials*—State Directors; State Executive Directors; the Assistant Administrator; Finance Office; Financial Management Director; Financial Management Division, and the Deputy Administrator for Management, National Office.

(2) *Debt or debts.* A term that refers to one or both of the following:

(i) *Delinquent debts.* A past due amount owed to the United States from sources which include, but are not limited to, insured or guaranteed loans, fees, leases, rents, royalties, services, sales of real or personal property, overpayments, penalties, damages, interest, fines and forfeitures (except those arising under the Uniform Code of Military Justice).

(ii) *Other debts.* An amount owed to the United States by an employee for pecuniary losses where the employee has been determined to be liable due to the employee's negligent, willful, unauthorized or illegal acts, including but not limited to:

(A) Theft, misuse, or loss of Government funds;

(B) False claims for services and travel;

(C) Illegal, unauthorized obligations and expenditures of Government appropriations;

(D) Using or authorizing the use of Government owned or leased equipment, facilities supplies, and services for other than official or approved purposes;

(E) Lost, stolen, damaged, or destroyed Government property;

(F) Erroneous entries on accounting record or reports; and,

(G) Deliberate failure to provide physical security and control procedures for accountable officers, if such failure is determined to be the proximate cause for a loss of Government funds.

(3) *Defalcation account.* An account established in the Finance Office for

other debts owed the Federal government in the amount missing due to the action of an employee or former employee.

(4) *Disposable pay.* Pay due an employee that remains after required deductions for Federal, State and local income taxes; Social Security taxes, including Medicare taxes; Federal retirement programs; premiums for life and health insurance benefits, and such other deductions required by law to be withheld.

(5) *Hearing Officer.* An Administrative Law Judge of the USDA or another individual not under the supervision or control of the USDA, designated by the Certifying Official to review the determination of the alleged debt.

(6) *Non-cash credit.* The accounting action taken by the Finance Office to credit and make a borrower's account whole for funds paid by the borrower but missing due to an employee's or former employee's actions.

(7) *Salary Offset.* The collection of a debt due to the U.S. by deducting a portion of the disposable pay of a Federal employee without the employee's consent.

(c) *Feasibility of salary offset.* The first step the Certifying Official must take to use this offset procedure is to decide, on a case by case basis, whether offset is feasible. If an offset is feasible, the directions in the following paragraphs of this section will be used to collect by salary offset. If the official making this determination decides that salary offset is not feasible, the reasons supporting this decision will be documented in the borrower's running case record in the case of delinquent debts, or the "For Official Use Only" file in cases of other debts. Ordinarily, and where possible, debts should be collected in one lump-sum; but payments may be made in installments. Installment deductions can be made over a period not greater than the anticipated period of employment. However, the amount deducted for a pay period will not exceed 15 percent of the disposable pay from which the deduction is made. If possible, the installment payment will be sufficient in size and frequency to liquidate the debt in approximately 3 years. Based on the Comptroller General's decisions, other debts by employ-

ees cannot be forgiven. If the employee retires or resigns, or if employment ends before collection of the debt is completed, final salary payment, lump-sum leave, etc. may be offset to the extent necessary to liquidate the debt. Salary offset is feasible if:

(1) The cost to the Government of collecting salary offset does not exceed the amount of the debt. County Committee members are exempt from salary offset because the amount collected by salary offset would be so small as to be impractical.

(2) There are not any legal restrictions to the debt, such as the debtor being under the jurisdiction of a bankruptcy court, or the statute of limitations having expired. The Debt Collection Act of 1982 permits offset of claims that have not been outstanding for more than 10 years.

(d) *Notice to debtor.* (1) After the Certifying Official determines that collection by salary offset is feasible, the debtor should be notified within 15 calendar days after the salary offset determination. This notice will notify the debtor of intended salary offset at least 30 days before the salary offset begins. For Farm Loan Programs direct loans, this notice will be sent after the borrower is over 90 days past due and immediately after sending notification of servicing rights in accordance with 7 CFR part 766. For Farm Loan Programs guaranteed debtors, this notice will be sent after a final loss claim has been paid. The salary offset determination notice will be delivered to the debtor by regular mail.

(2) The Debt Collection Act of 1982 requires that the hearing officer issue a written decision not later than 60 days after the filing of the petition requesting the hearing; thus, the evidence upon which the decision to notify the debtor is based, to the extent possible, should be sufficient for Rural Development to proceed at a hearing, should the debtor request a hearing under paragraph (f) of this section.

(e) *Notice requirement before salary offset.* Salary offset will not be made unless the employee receives 30 calendar days written notice. This Notice of Intent (RD Guide Letter 1951-C-4) will be addressed to the debtor or the debtor's representative. The Notice of Intent must be modified if it is addressed to the debtor's representative. In either case, the Notice of Intent will state:

(1) It has been determined that the debt is owed, the amount of the debt, and the facts giving rise to the debt;

(2) The cost to the Government of collecting salary offset does not exceed the amount of the debt;

(3) There are not any legal restrictions that would bar collecting the debt;

(4) The debt will be collected by means of deduction of not more than 15 percent from the employee's current disposable pay until the debt and all accumulated interest are paid in full;

(5) The amount, frequency, approximate beginning date, and duration of the intended deductions;

(6) An explanation of the requirements concerning interest, penalties and administrative costs, unless such payments are waived;

(7) The employee's right to inspect and request a copy of records relating to the debt;

(8) The employee's right to voluntarily enter into a written agreement for a repayment schedule with the agency different from that proposed by Rural Development, if the terms of the repayment proposed by the employee are agreeable with the agency;

(9) That the employee has a right to a hearing conducted by an Administrative Law Judge of USDA or a hearing official not under the supervision or control of the Secretary of Agriculture, concerning the agency's determination of the existence or amount of the debt and the percentage of disposable pay to be deducted each pay period, if a petition for a hearing is filed by the employee as prescribed by Rural Development;

(10) The timely filing of a petition for hearing will stay the collection proceedings;

(11) That a final decision will be issued at the earliest practical date, but not later than 60 calendar days after the filing of petition requesting the hearing;

(12) That any knowingly false or frivolous statements may subject the employee to disciplinary procedures, or penalties, under the applicable statutory authority;

(13) Any other rights and remedies available to the employee under statutes or regulations governing the program for which the collection is being made;

(14) That amounts paid on or deducted for the debt which are later waived or found not owed to the United States will be promptly refunded to the employee unless there are provisions to the contrary;

(15) The method and time period for requesting a hearing; and

(16) The name and address of an official of USDA to whom communications should be directed.

(f) *Debtor's request for records, offer to repay, request for a hearing or request for information concerning debt settlement—*

(1) If a debtor responds to RD Guide Letter 1951–C–4 by asking to review and copy Rural Development's records relating to the debt, the Certifying Official will promptly respond by sending a letter which tells the debtor the location of the debtor's Rural Development files and that the files may be reviewed and copied within the next 30 days. Copying costs (see subpart F of part 2018 of this chapter) will be set out in the letter, as well as the hours the files will be available each day. If a debtor asks to have Rural Development copy the records, a copy will be made within 30 days of the request.

(2) If a debtor responds to RD Guide Letter 1951–C–4 by offering to repay the debt, the offer may be accepted by the Certifying Official, if it would be in the best interest of the government. RD Form Letter 1951–8 will be used if a repayment offer for an Rural Development loan or grant is accepted. Upon receipt of an offer to repay, the Certifying Official will delay institution of a hearing until a decision is made on the repayment offer. Within 60 days after the initial offer to repay was made, the Certifying Official must decide whether to accept or reject the offer. This decision will be documented in the running case record or the "For Official Use Only" file, as appropriate, and the debtor will be sent a letter which sets out the decision to accept or reject the offer to repay. The decision to accept or reject a repayment offer should be based upon a realistic budget or farm and home plan and according to the

servicing regulations for the type of loan(s) involved.

(3) If a debtor responds to RD Guide Letter 1951–C–4 by asking for a hearing on Rural Development's determination that a debt exists and/or is due, or on the percentage of net pay to be deducted each pay period, the Certifying Official will notify the debtor in accordance with paragraph (g)(3) of this section and request the debtor's case file or the "For Official Use Only" file.

(4) If a debtor is willing to have more than 15 percent of the disposable pay sent to Rural Development, a letter prepared and signed by the debtor clearly stating this must be placed in the debtor's case file or the "For Official Use Only" file.

(5) If a debtor who is an Rural Development borrower requests debt settlement, the account must be in collection-only status or be an inactive account for which there is no security. The Certifying Official must inform the borrower of how to apply for debt settlement. Any application will be considered independently of the salary offset. A salary offset should not be delayed because the borrower applied for debt settlement.

(6) The time limits set in RD Guide Letter 1951–C–4 and in paragraphs (f) (1), (2), and (3) of this section run concurrently. In other words, if a debtor asks to review the Rural Development file and offers to repay the debt, the debtor cannot take 30 days to ask to review the file and then take another 30 days to offer to repay. The request to review the file and the offer to repay must both be made within 30 days of the date the debtor receives the notification letter.

(7) If an employee is included in a bargaining unit which has a negotiated grievance procedure that does not specifically exclude salary offset proceedings, the employee must grieve the matter in accordance with the negotiated procedure. Employees who are not covered by a negotiated procedure must utilize the salary offset proceedings as outlined in RD Guide Letter 1951–C–4. The employee must be informed, in writing, which procedure to follow and, as appropriate, reference should be made to the appropriate sections of the negotiated agreement.

(g) *Hearings.* (1) A hearing officer must be a USDA Administrative Law Judge or a person who is not a USDA employee. In order to ensure that a hearing officer will be available promptly when needed, Certifying Officials need to make appropriate arrangements with officials of nearby federal agencies for the use of each other's employees as hearing officers.

(2) Not later than 30 days from the date the debtor receives the Notice of Intent (RD Guide Letter 1951–C–4), the employee must file with the Certifying Official issuing the notice, a written petition establishing his/her desire for a hearing on the existence and amount of the debt or the proposed offset schedule. The employee's petition must fully identify and explain all the information and evidence that supports his/her position. In addition, the petition must bear the employee's original signature and be dated upon receipt by the Certifying Official.

(3) Certifying Officials are responsible for determining if the employee's petition for a hearing has been submitted in a timely fashion. Petitions received from employees after the 30-day time limitation expires will be accepted only if the employee can show the delay was because of circumstances beyond his/her control or because of failure to receive notice of the time limitation. Certifying Officials are required to provide written notification to the employee of the acceptance or non-acceptance of the employee's petition for hearing.

(4) For those petitions accepted, Rural Development will arrange for a hearing officer and notify the employee of the time and place of the hearing. The hearing location should be convenient to all parties involved. The employee will also be notified that the acceptance of the petition for hearing will stay the commencement of collection proceedings. Any payments collected in error due to untimely or delayed filing beyond the employee's control will be refunded unless there are applicable contractual or statutory provisions to the contrary.

(5) The hearing will be based on written submissions and documentation provided by the debtor and Rural Development unless:

(i) A statute authorizes or requires consideration of waiving the debt, the debtor requests waiver of the debt, and the waiver determination turns on an issue of credibility or truth.

(ii) The debtor requests reconsideration of the debt and the hearing officer determines that the question of the indebtedness cannot be resolved by a review of the documentary evidence; for example, when the validity of the debt turns on an issue of credibility or truth.

(iii) The hearing officer determines that an oral hearing is appropriate.

(6) Oral hearings may be conducted by conference call at the request of the debtor or at the discretion of the hearing officer. The hearing officer's determination that the offset hearing is on the written record is final and is not subject to review.

(7) The hearing officer will issue a written decision not later than 60 days after the filing of the petition requesting the hearing, unless the employee requests and the Certifying Official grants a delay in the proceedings. The written decision will state the facts supporting the nature and origin of the debt, the hearing officer's analysis, findings and conclusions as to the amount and validity of the debt, and repayment schedule. Both the employee and Rural Development will be provided with a copy of the hearing officer's written decision on the debt.

(h) *Processing delinquent debts.* (1) Form AD–343, "Payroll Action Request," and RD Form Letter 1951–6 will be prepared and submitted by the Certifying Official to the National Office, FMAS, for coordination and forwarding to the debtor's employing agency if:

(i) The borrower does not respond to RD Guide Letter 1951–C–4 within 30 days.

(ii) The borrower responds to RD Guide Letter 1951–C–4 within 30 days and

(A) Has had an opportunity to review the file, if requested,

(B) Has received a hearing, if requested, and

(C) A decision has been made by the hearing officer to uphold the offset.

(2) A copy of Form AD–343 and the Form letter 1951–6 will be sent to the Finance Office, St. Louis, MO 63103, Attn: Account Settlement Unit.

(3) If the debtor is an Rural Development employee, Form AD–343 will be sent to the National Office, FMAS, and a copy to the Finance Office, St. Louis, MO, Attn: Account Settlement Unit. This form can be signed for the Certifying Official by an employment officer, an Administrative Officer, or a personnel management specialist, or signed by the Certifying Official.

(4) If the debtor has agreed to have more or less than 15 percent of the disposable pay sent to Rural Development, a copy of the debtor's letter (RD Form Letter 1951–8) authorizing this must be attached to Form AD–343.

(5) Field offices will be notified of payments received from salary offset by receipt of a transaction record from the Finance Office.

(i) *Deduction percentage.* (1) Generally, installment deductions will be made over a period not greater than the anticipated period of employment. If possible, the installment payment will be sufficient in size and frequency to liquidate the debt in approximately 3 years. The size and frequency of installment deductions will bear a reasonable relation to the size of the debt and the employee's ability to pay. Certifying Officials are responsible for determining the size and frequency of the deductions. However, the amount deducted for any period will not exceed 15 percent of the disposable pay from which the deduction is made, unless the employee has agreed in writing to the deduction of a greater amount. Installment payments of less than $25 per pay period or $50 a month will be accepted only in the most unusual circumstances.

(2) Deductions will be made only from basic pay, incentive pay, retainer pay, or, in the case of an employee not entitled to basic pay, other authorized pay. If there is more than one salary offset, the maximum deduction for all salary offsets against an employee's disposable pay is 15 percent unless the employee has agreed in writing to a greater amount.

(j) *Agency/NFC responsibility for other debts.* (1) Rural Development will inform NFC about other indebtedness by transmitting to NFC an AD–343. NFC will process the documents through the Payroll/Personnel System, calculate the net amount of the adjustment and generate a salary offset notice. This notice will be sent to the employee's employing office along with a duplicate copy for the Rural Development's records. Rural Development is responsible for completing the necessary information and forwarding the employee's notice to the employee.

(2) Other indebtedness falls into two categories:

(i) An agency-initiated indebtedness (i.e. personal telephone calls, property damages, etc.).

(ii) An NFC-initiated indebtedness (i.e. duplicate salary payments, etc.). NFC will send the salary offset notice to the employing office.

(k) *Establishing employees or former employees defalcation accounts and non-cash credits to borrower accounts.* In cases where a borrower made a payment on an Rural Development account(s) and, due to theft, embezzlement, fraud, negligence, or some other action on the part of an Rural Development employee or employees, the payment is not transmitted to the Finance Office for application to the borrower's account(s), certain accounting actions must be taken by the Finance Office to establish non-cash credits to the borrower's account and an employee defalcation account.

(1) The Certifying Official will advise the Assistant Administrator, Finance Office by memorandum to establish a defalcation account. The memorandum must state the following information:

(i) Employee's name (or former),

(ii) Social Security Number,

(iii) Present or last known address,

(iv) Date of Payment, and

(v) Amount of the defalcation account.

(2) If a non-cash credit to a borrower's account(s) is required, the letter to the Finance Office will include:

(i) Borrower's name and case number,

(ii) Fund Code and Loan Code,

(iii) Date and amount of missing payment,

(iv) Copy of receipt issued for the missing payment, and

(v) Name of employee who last had custody of the missing funds.

(3) To assist and assure proper accounting for defalcation accounts and non-cash credits, the request should be made at the same time. Should requests be made separately, be sure to identify appropriately.

(4) The Certifying Official shall furnish a copy of the memorandum and supporting documentation for paragraphs (k) (1) and (2) of this section to the Deputy Administrator for Management for distribution to the Financial and Management Analysis Staff (FMAS) and Employee Relations Branch, Personnel Division.

(l) *Application of payments, refunds and overpayments.* (1) If a debtor is delinquent or indebted on more than one Rural Development loan or debt, amounts collected by offset will be applied as specified on Form AD–343, based on the advantage to agency or debtor. The check date will be used as the date of credit in applying payments to the borrower's accounts.

(2) If a court or agency orders Rural Development to refund the amount obtained by salary offset, a refund will be requested promptly by the Certifying Official in accordance with the order by sending RD Form Letter 1951–5 to the Finance Office. Processing RD Form Letter 1951–5 in the Finance Office will cause a refund to be sent to the debtor through the county office or other appropriate Rural Development office. The debtor is not entitled to any payment of interest, on the refunded amount.

(3) If a debtor does not request a hearing within the required time and it is later determined that the delay was due to circumstances beyond the debtor's control, any amount collected before the hearing decision is made will be refunded promptly by the Certifying Official in accordance with paragraphs (l) (1) and (2) of this section.

(4) If Rural Development receives money through an offset but the debtor is not delinquent or indebted at the time or the amount received is in excess of the delinquency or indebtedness, the entire amount or the amount in excess of the delinquency or indebtedness will be refunded promptly to the debtor by the Certifying Official in accordance with paragraphs (l) (1) and (2) of this section.

(m) *Cancellation of offset.* If a debtor's name has been submitted to another agency for offset and the debtor's account is brought current or otherwise satisfied, the Certifying Official will complete Form AD–343 and send it to the National Office, FMAS. FMAS will notify the paying agency with Form AD–343 that the debtor is no longer delinquent or indebted and to cancel the offset. A copy of the cancellation document will be sent to the debtor and the Finance Office, Attn: Account Settlement Unit.

(n) *Intra-departmental transfer.* When an Rural Development employee who is indebted to one agency in USDA transfers to another agency within USDA, a copy of the repayment schedule should be forwarded by the agency personnel office to the new employing agency. The NFC will continue to make deductions until full recovery is effected.

(o) *Liquidation from final checks.* Upon the determination that an employee owing a debt to Rural Development is to retire, resign, or employment otherwise ends, the Certifying Official should forward a telegram with the appropriate employee identification and amount of the debt to the NFC. The telegram should request that the debt be collected from final salary/lump sum leave or other funds due the employee, and, if necessary, to put a hold on the retirement funds. The telegram information should be confirmed by completion of Form AD–343. Collection from retirement funds will be in accordance with Departmental Administrative Offset procedures (7 CFR Part 3, Subpart B, § 3.32).

(p) *Coordination with other agencies.* (1) If Rural Development is the creditor agency but not the paying agency, the Certifying Official will submit Form AD–343 to the National Office, FMAS, to begin salary offset against an indebted employee. The request will include a certification as to the determination of indebtedness, and that Rural Development has complied with applicable regulations and instruction for submitting the funds to the Finance Office. (See RD Form Letter 1951–6).

(2) When an employee of Rural Development owes a debt to another Federal agency, salary offset may be used only when the Federal agency certifies that the person owes the debt and that the Federal agency has complied with its regulations. The request must include the creditor agency's certification as to the indebtedness, including the amount, and that the employee has been given the due process entitlements guaranteed by the Debt Collection Act of 1982. When a request for offset is received, Rural Development will notify the employee and NFC and arrange for offset. (See RD Form Letter 1951–7).

(q) *Deductions by the National Finance Center (NFC).* The NFC will automatically deduct the full amount of the delinquency or indebtedness if less than 15 percent of disposable pay or 15 percent of disposable pay if the delinquency or indebtedness exceeds 15 percent, unless the creditor agency advises otherwise. Deductions will begin the second pay period after the 30-day notification period has expired unless Rural Development issues the notice. If Rural Development issues the notice, the NFC will begin deductions on the first pay period after receipt of the Form AD–343.

(r) *Interest, penalties and administrative costs.* Interest and administrative costs will normally be assessed on outstanding claims being collected by salary offset. However, penalties should not be charged routinely on debts being collected in installments by salary offsets, since it is not to be construed as a failure to pay within a given time period. Additional interest, penalties, and administrative costs, will not be assessed on delinquent loans until Rural Development publishes regulations permitting such charges. ·

(s) *Adjustment in rate of repayment.* (1) When an employee who is indebted receives a reduction in basic pay that would cause the current deductions to exceed 15 percent of disposable pay, and the employee has not consented in writing to a greater amount, Rural Development must take action to reduce the amount of the deductions to 15 percent of the new amount of disposable pay. Upon an increase in basic pay which results in the current deductions to be less than the specified percentage, Rural Development may increase the amount of the deductions accordingly. In either case, when a change is made the employee will be notified in writing.

(2) When an employee has an existing reduced repayment schedule because of financial hardship, the creditor agency may arrange for a new repayment schedule.

[52 FR 18544, May 18, 1987, as amended at 53 FR 44178, Nov. 2, 1988; 54 FR 26945, June 27, 1989; 62 FR 41799, Aug. 1, 1997; 65 FR 50603, Aug. 21, 2000; 67 FR 69671, Nov. 19, 2002; 72 FR 64122, Nov. 15, 2007; 80 FR 9891, Feb. 24, 2015]

§§ 1951.112–1951.132 [Reserved]

§ 1951.133 Establishment of Federal Debt.

Any amounts paid by RBS on account of liabilities of a business and industry (B&I) program guaranteed loan borrower will constitute a Federal debt owing to RBS by the B&I guaranteed loan borrower. In such case, the RBS may use all remedies available to it, including offset under the Debt Collection Improvement Act of 1996 (DCIA), to collect the debt from the borrower. Interest charges will be established at the note rate of the guaranteed loan on the date a loss claim is paid. RBS may, at its option, refer such debt in all or part to the Department of the Treasury, before a final loss claim is determined.

[69 FR 3000, Jan. 22, 2004]

§§ 1951.134–1951.135 [Reserved]

§ 1951.136 Procedures for Department of Treasury offset and cross-servicing for the Rural Housing Service (Community Facility Program only) and the Rural Business-Cooperative Service.

(a) The National Offices of the Rural Housing Service (RHS), Community Facilities (CF) and the Rural Business-Cooperative Service (RBS) will refer past due, legally enforceable debts which are over 180 days delinquent to the Secretary of the Treasury for collection by centralized administrative offset (TOP), Internal Revenue Service offset administered through TOP and Treasury's Cross-Servicing (Cross-Servicing) Program, which centralizes all Government debt collection actions. A borrower with a workout

agreement in place, in bankruptcy or litigation, or meeting other exclusion criteria, may be excluded from TOP or Cross-Servicing.

(b) A 60 day due process notice will be sent to borrowers subject to TOP or Cross-Servicing. The borrower will be given 60 days to resolve any delinquency before the debt is reported to Treasury. The notice will include:

(1) The nature and amount of the debt, the intention of the Agency to collect the debt through TOP or Cross-Servicing, and an explanation of the debtor's rights;

(2) An opportunity to inspect and copy the records related to the debt from the Agency;

(3) An opportunity to review the matter within the Agency or the National Appeals Division, if there has not been a previous opportunity to appeal the offset; and

(4) An opportunity to enter into a written repayment agreement.

(c) In referring debt to the Department of Treasury the Agency will certify that:

(1) The debt is past due and legally enforceable in the amount submitted and the Agency will ensure that collections are properly credited to the debt;

(2) Except in the case of a judgment debt or as otherwise allowed by law, the debt is referred for offset within 10 years after the Agency's right of action accrues;

(3) The Agency has made reasonable efforts to obtain payment; and

(4) Payments that are prohibited by law from being offset are exempt from centralized administrative offset.

[67 FR 69672, Nov. 19, 2002]

§ 1951.137 Procedures for Treasury offset and cross-servicing for the Farm Service Agency (FSA) farm loan programs.

(a) The Farm Service Agency, Farm Loan Programs, will refer past due, legally enforceable debts which are over 180 days delinquent to the Secretary of the Treasury for collection by centralized administrative offset (TOP), Internal Revenue Service offset administered through TOP and Treasury's

Cross-Servicing (Cross-Servicing) Program, which centralizes all Government debt collection actions. A borrower with a workout agreement in place, in bankruptcy or litigation, or meeting other exclusion criteria, may be excluded from TOP or Cross-Servicing. Guaranteed debtors will only be referred to TOP upon confirmation of payment on a final loss claim.

(b) A 60 day due process notice will be sent to borrowers subject to TOP or Cross-Servicing by the Director of Kansas City Finance Office. The borrower will be given 60 days to resolve any delinquency before the debt is reported to Treasury. The notice will include:

(1) The nature and amount of the debt, the intention of the Agency to collect the debt through TOP or Cross-Servicing, and an explanation of the debtor's rights;

(2) An opportunity to inspect and copy the records related to the debt, from the Agency;

(3) An opportunity to review the matter within the Agency; and

(4) An opportunity to enter into a written repayment agreement.

(c) In referring debt to the Department of Treasury the Agency will certify that:

(1) The debt is past due and legally enforceable in the amount submitted and the Agency will ensure that collections are properly credited to the debt;

(2) Except in the case of a judgment debt or as otherwise allowed by law, the debt is referred for offset within 10 years after the Agency's right of action accrues;

(3) The Agency has made reasonable efforts to obtain payment; and

(4) Payments that are prohibited by law from being offset are exempt from centralized administrative offset.

[67 FR 69672, Nov. 19, 2002]

§§ 1951.138–1951.149 [Reserved]

§ 1951.150 OMB control number.

The collection of information requirements in this regulation have been approved by the Office of Management and Budget and assigned OMB control number 0575–0119.

[51 FR 42821, Nov. 26, 1986]

Subpart D—Final Payment on Loans

SOURCE: 57 FR 774, Jan. 9, 1992, unless otherwise noted.

§ 1951.151 Purpose.

This subpart prescribes authorizations, policies, and procedures of the Rural Housing Service (RHS), and Rural Business-Cooperative Service (RBS), herein referred to as "Agency," for processing final payment on all loans. This subpart does not apply to Direct Single Family Housing customers or to the Rural Rental Housing, Rural Cooperative Housing, or Farm Labor Housing Program of the RHS. This subpart does not apply to Water and Waste Programs of the Rural Utilities Service, Watershed loans, and Resource Conservation and Development loans, which are serviced under part 1782 of this title. In addition, this subpart is inapplicable to Farm Service Agency, Farm Loan Programs.

[72 FR 55018, Sept. 28, 2007, as amended at 72 FR 64123, Nov. 15, 2007]

§ 1951.152 Definition.

As used in this subpart:

Mortgage. Includes real estate mortgage, deed of trust or any other form of security instrument or lien on real property.

§ 1951.153 Chattel security or note-only cases.

(a) If a loan secured by both real estate and chattels is paid in full, the chattel security instrument will be satisfied or released in accordance with subpart A of part 1962 of this chapter.

(b) When a loan is evidenced by only a note and the note is paid in full, RD will deliver the note to the borrower in the manner prescribed in § 1951.155(c) of this subpart.

§ 1951.154 Satisfaction and release of documents.

(a) *Authorization.* RD is authorized to execute the necessary releases and satisfactions and return security instruments and related documents to borrowers. Satisfaction and release of security documents takes place:

(1) Upon receipt of payment in full of all amounts owed to the Government including any amounts owed to the loan insurance account, subsidy recapture amounts, all loan advances and/or other charges to the borrower's account;

(2) Upon verification that the amount of payment received is sufficient to pay the full amount owed by the borrower; or

(3) When a compromise or adjustment offer has been accepted and approved by the appropriate Government official in full settlement of the account and all required funds have been paid.

(b) [Reserved]

(c) *Lost note.* If the original note is lost RD will give the borrower an affidavit of lost note so that the release or satisfaction may be processed.

§1951.155 County and/or District Office actions.

(a) *Funds remaining in supervised bank accounts.* When a borrower is ready to pay an insured or direct loan in full, any funds remaining in a supervised bank account will be withdrawn and remitted for application to the borrower's account. If the entire principal of the loan is refunded after the loan is closed, the borrower will be required to pay interest from the date of the note to the date of receipt of the refund.

(b) *Determining amount to be collected.* RD will compute and verify the amount to be collected for payment of an account in full. Requests for payoff balances on all accounts will be furnished in writing in a format specified by RD (available in any Rural Development office).

(c) *Delivery of satisfaction, notes, and other documents.* When the remittance which paid an account in full has been processed by RD, the paid note and satisfied mortgage may be returned to the borrower. If other provisions exist, the mortgage will not be satisfied until the total indebtedness secured by the mortgage is paid. For instance, in a situation where a rural housing loan is paid-in-full and there is a subsidy recapture receivable balance that the borrower elects to delay repaying, the amount of recapture to be repaid will be determined when the principal and interest balance is paid. The mortgage

securing the RHS, RBS, RUS, and/or FSA or its successor agency under Public Law 103–354 debt will not be released of record until the total amount owed the Government is repaid. To permit graduation or refinancing by the borrower, the mortgage securing the recapture owed may be subordinated.

(1) If RD receives final payments in a form other than cash, U.S. Treasury check, cashier's check, certified check, money order, bank draft, or check issued by an institution determined by RD to be financially responsible, the mortgage and paid note will not be released until after a 30-day waiting period. If other indebtedness to RD is not secured by the mortgage, RD will execute the satisfaction or release. When the stamped note is delivered to the borrower, RD will also deliver the real estate mortgage and related title papers such as title opinions, title insurance binders, certificates of title, and abstracts which are the property of the borrower. Any water stock certificates or other securities that are the property of the borrower will be returned to the borrower. Also, any assignments of income will be terminated as provided in the assignment forms.

(2) Delivery of documents at the time of final payment will be made when payment is in the form of cash, U.S. Treasury check, cashier's check, certified check, money order, bank draft, or check issued by an institution determined by RD to be responsible. RD will not accept payment in the form of foreign currency, foreign checks or sight drafts. RD will execute the satisfaction or release (unless other indebtedness to RD is covered by the mortgage) and mark the original note with a paid-in-full legend based upon receipt of the full payment balance of the borrower's account(s), computed as of the date final payment is received. In unusual cases where an insured promissory note is held by a private holder, RD can release the mortgage and deliver the note when it is received.

(d)–(e) [Reserved]

(f) *Cost of recording or filing of satisfaction.* The satisfaction or release will be delivered to the borrower for recording and the recording costs will be paid by the borrower, except when State law requires the mortgagee to record or file satisfactions or release and pay the recording costs.

(g) *Property insurance.* When the borrower's loan has been paid-in-full and the satisfaction or release of the mortgage has been executed, FD may release the mortgage interest in the insurance policy as provided in subpart A of part 1806 of this chapter (RD Instruction 426.1).

(h) [Reserved]

(i) *Outstanding Loan Balance(s).* RD will attempt to collect any account balance(s) that may result from an error by RD in handling final payments according to paragraph 1951.155(b) of this section. If collection cannot be made, the debt will be settled according to subpart B of part 1956 of this chapter or reclassified to collection-only. A deficiency judgment may be considered if the balance is a significant amount ($1,000 or more) and the borrower has known assets.

[57 FR 774, Jan. 9, 1992, as amended at 60 FR 55145, Oct. 27, 1995]

§§ 1951.156–1951.200 [Reserved]

Subpart E—Servicing of Community and Direct Business Programs Loans and Grants

Source: 55 FR 4399, Feb. 8, 1990, unless otherwise noted.

§ 1951.201 Purposes.

This subpart prescribes the Rural Development mission area policies, authorizations, and procedures for servicing the following programs: Community Facility loans and grants, Rural Business Enterprise/Television Demonstration grants; Association Recreation loans; Direct Business loans; Economic Opportunity Cooperative loans; Rural Renewal loans; Energy Impacted Area Development Assistance Program grants; National Nonprofit Corporation grants; System for Delivery of Certain Rural Development Programs panel grants; in part 4284 of this title, Rural and Cooperative Development Grants, Value-Added Producer Grants, and Agriculture Innovation Center Grants. Rural Development State Offices act on behalf of the Rural Business-Cooperative Service and the Rural Housing Service as to loan and grant programs formerly administered by the Farmers Home Administration and the Rural Development Administration. Loans sold without insurance to the private sector will be serviced in the private sector and will not be serviced under this subpart. The provisions of this subpart are not applicable to such loans. Future changes to this subpart will not be made applicable to such loans. This subpart does not apply to Water and Waste Programs of the Rural Utilities Service, Watershed loans, and Resource Conservation and Development Loans, which are serviced under part 1782 of this title.

§ 1951.202 Objectives.

The purpose of loan and grant servicing functions is to assist recipients to meet the objectives of loans and grants, repay loans on schedule, comply with agreements, and protect Rural Development's financial interest. Supervision by Rural Development includes, but is not limited to, review of budgets, management reports, audits and financial statements; performing security inspections and providing, arranging for, or recommending technical assistance; evaluating environmental impacts of proposed actions by the borrower; and performing civil rights compliance reviews.

§ 1951.203 Definitions.

(a) *Approval official.* An official who has been delegated loan and/or grant approval authorities within applicable programs.

(b) *Assumption of debt.* The agreement by one party to legally bind itself to pay the debt incurred by another.

(c) *CONACT.* The Consolidated Farm and Rural Development Act, as amended.

(d) *Eligible applicant.* An entity that would be legally qualified for financial assistance under the loan or grant program involved in the servicing action.

(e) *Ineligible applicant.* An entity or individual that would not be considered eligible for financial assistance under the loan or grant program involved in the servicing action.

(f) *Nonprogram (NP) loan.* An NP loan exists when credit is extended to an ineligible applicant and/or transferee in connection with loan assumptions or sale of inventory property; any recipient in cases of unauthorized assistance; or a recipient whose legal organization has changed as set forth in §1951.220(e) of this subpart resulting in the borrower being ineligible for program benefits.

(g) *Servicing office.* The State, District, or County Office responsible for immediate servicing functions for the borrower or grantee.

(h) *Transfer fee.* A one-time nonrefundable application fee, charged to ineligible applicants for Rural Development services rendered in the processing of a transfer and assumption.

[55 FR 4399, Feb. 8, 1990, as amended at 69 FR 70884, Dec. 8, 2004]

§1951.204 Nondiscrimination.

Each instrument of conveyance required for a transfer, assumption, or other servicing action under this subpart will contain the following covenant.

The property described herein was obtained or improved with Federal financial assistance and is subject to the nondiscrimination provisions of title VI of the Civil Rights Act of 1964, title IX of the Education Amendments of 1972, section 504 of the Rehabilitation Act of 1973, and other similarly worded Federal statutes, and the regulations issued pursuant thereto that prohibit discrimination on the basis of race, color, national origin, handicap, religion, age, or sex in programs or activities receiving Federal financial assistance. Such provisions apply for as long as the property continues to be used for the same or similar purposes for which the Federal assistance was extended, for so long as the purchaser owns it, whichever is later.

§1951.205 Redelegation of authority.

Servicing functions under this subpart which are specifically assigned to the State Director may be redelegated in writing to an appropriate sufficiently trained designee.

§1951.206 Forms.

Forms utilized for actions under this subpart are to be modified appropriately where necessary to adapt the forms for use by corporate recipients rather than individuals.

§1951.207 State supplements.

State supplements developed to carry out the provisions of this subpart will be prepared in accordance with subpart B of part 2006 of this chapter (available in any Rural Development office) and applicable State laws and regulations. State supplements are to be used only when required by National Instructions or necessary to clarify the impact of State laws or regulations, and not to restate the provisions of National Instructions. Advice and guidance will be obtained as needed from the Office of the General Counsel (OGC).

§§1951.208–1951.209 [Reserved]

§1951.210 Environmental requirements.

Servicing actions as defined in §1970.6 of this chapter are part of the financial assistance already provided and do not require additional NEPA review. Actions such as lien subordinations, sale or lease of Agency-owned real property, or approval of a substantial change in the scope of a project, as defined in §1970.8, must comply with the environmental review requirements in accordance with 7 CFR part 1970.

[81 FR 11032, Mar. 2, 2016]

§1951.211 Refinancing requirements.

In accordance with the CONACT, Rural Development requires for most loans covered by this subpart that if at any time it shall appear to the Government that the borrower is able to refinance the amount of the indebtedness then outstanding, in whole or in part, by obtaining a loan for such purposes from responsible cooperative or private credit sources, at reasonable rates and terms for loans for similar purposes and periods of time, the borrower will, upon request of the Government, apply for and accept such loan in sufficient amount to repay the Government and will take all such actions as may be required in connection with such loan. Applicable requirements are set forth in subpart F of part 1951 of this chapter. A civil rights impact analysis is required.

[55 FR 4399, Feb. 8, 1990, as amended at 63 FR 16089, Apr. 2, 1998]

§ 1951.212 Unauthorized financial assistance.

Subpart O of part 1951 of this chapter prescribes policies for servicing the loans and grants covered under this subpart when it is determined that a borrower or grantee was not eligible for all or part of the financial assistance received in the form of a loan, grant, subsidy, or any other direct financial assistance.

§ 1951.213 Debt settlement.

Subpart C of part 1956 of this chapter prescribes policies and procedures for debt settlement actions for loans covered under this subpart when it is determined that a debt is eligible for settlement except as provided in §§ 1951.216 and 1951.231.

§ 1951.214 Care, management, and disposal of acquired property.

Property acquired by Government or its successor agency under Public Law 103–354 will be handled according to subparts B and C of part 1955 of this chapter.

[55 FR 4399, Feb. 8, 1990, as amended at 63 FR 16089, Apr. 2, 1998]

§ 1951.215 Grants.

No monitoring action by Rural Development is required after grant closeout. Grant closeout is when all required work is completed, administrative actions relating to the completion of work and expenditure of funds have been accomplished, and Rural Development accepts final expenditure information. However, grantees remain responsible in accordance with the terms of the grant for property acquired with grant funds.

(a) *Applicability of requirements.* Servicing actions relating to Rural Development or its successor agency under Public Law 103–354 grants are governed by the provisions of this subpart, the terms of the Grant Agreement and, if applicable, the provisions of 2 CFR parts 200, 400, 415, 417, 418, and 421.

(1) Servicing actions will be carried out in accordance with the terms of the "Association Water or Sewer System Grant Agreement," and RUS Bulletin 1780–12, "Water and Waste Grant Agreement" (available from any USDA/Rural Development office or the Rural Utilities Service, United States Department of Agriculture, Washington, DC 20250–1500). Grant agreements with a revision date on or after January 29, 1979, require that the grantee request disposition instructions from the Agency before disposing of property which is no longer needed for original grant purposes.

(2) When facilities financed in part by Rural Development grants are transferred or sold, repayment of all or a portion of the grant is not required if the facility will be used for the same purposes and the new owner provides a written agreement to abide by the terms of the grant agreement.

(b) *Authorities.* Subject to the requirements of § 1951.215(a), authority to approve servicing actions is as follows:

(1) For water and waste disposal grants, the State Director is authorized to approve any servicing actions needed, except that prior approval of the Administrator is required when property acquired with grant funds is disposed of in accordance with §§ 1951.226, 1951.230, or 1951.232 of this subpart and the buyer or transferee refuses to assume all terms of the grant agreement.

(2) All other grants will be serviced in accordance with the Grant Agreement and this subpart. Prior approval of the Administrator is required except for actions covered in the preceding paragraph.

[55 FR 4399, Feb. 8, 1990, as amended at 63 FR 16089, Apr. 2, 1998; 79 FR 76012, Dec. 19, 2014]

§ 1951.216 Nonprogram (NP) loans.

Borrowers with NP loans are not eligible for any program benefits, including appeal rights. However, Rural Development may use any servicing tool under this subpart necessary to protect the Government's security interest, including reamortization or rescheduling. The refinancing requirements of subpart F of part 1951 of this chapter do not apply to NP loans. Debt settlement actions relating to NP loans must be handled under the Federal Claims Collection Act; proposals will be submitted to the National Office for review and approval. Any exception to the servicing requirements of NP loans under this subpart must have prior concurrence of the National Office.

§ 1951.217 Public bodies.

Servicing actions involving public bodies will be carried out to the extent feasible according to the provisions of this subpart. With prior National Office approval, the State Director is authorized to vary from such provisions if necessary and approved by OGC, provided such variation will not violate other regulatory or statutory provisions. To request approval, the case file, including copies of applicable documents, recommendations, and OGC comments, will be forwarded to the Administrator, Attention: (appropriate program division).

§ 1951.218 Use of Rural Development loans and grants for other purposes.

(a) If, after making a loan or a grant, the Administrator determines that the circumstances under which the loan or grant was made have sufficiently changed to make the project or activity for which the loan or grant was made available no longer appropriate, the Administrator may allow the loan borrower or grant recipient to use property (real and personal) purchased or improved with the loan or grant funds, or proceeds from the sale of property (real and personal) purchased with such funds, for another project or activity that:

(1) Will be carried out in the same area as the original project or activity;

(2) Meets the criteria for a loan or grant described in section 381E(d) of the Consolidated Farm and Rural Development Act, as amended; and

(3) Satisfies such additional requirements as are established by the Administrator.

(b) For the purpose of this section, Administrator means the Administrator of the Rural Housing Service or Rural Business-Cooperative Service that has the delegated authority to administer the loan or grant program that covers the property or the proceeds from the sale of property proposed to be used in another way.

(c) If the new use of the property is under the authority of another Administrator, the other Administrator will be consulted on whether the new use will meet the criteria of the other program. Since the new project or activity must be carried out in the same area as the original project or activity, a new rural area determination will not be necessary.

(d) Borrowers and grantees that wish to take advantage of this option may make their request through the appropriate Rural Development State Office. Permission to use this option will be exercised on a case-by-case-basis on applications submitted through the State Office to the Administrator for consideration. If the proposal is approved, the Administrator will issue a memorandum to the State Director outlining the conditions necessary to complete the transaction.

[72 FR 55018, Sept. 28, 2007]

§ 1951.219 [Reserved]

§ 1951.220 General servicing actions.

(a) *Payment in full.* Payment in full of a loan is handled according to subpart D of part 1951 of this chapter. When a loan is paid in full, the servicing official will:

(1) Notify the company providing fidelity bond coverage in writing that the government no longer has an interest in the bond if the government is named co-obligee on the bond.

(2) Release Rural Development's interest in insurance policies according to applicable provisions of subpart A of part 1806 (RD Instruction 426.1).

(3) Release Rural Development's interest in any other security as appropriate, consulting with OGC if necessary.

(b) *Loan summary statements.* Upon request of a borrower, Rural Development will issue a loan summary statement showing account activity for each loan made or insured under the CONACT. Field offices will post a notice on the bulletin board informing borrowers of the availability of loan summary statements. See exhibit A of subpart A of this part for a sample of the required notice.

(1) The loan summary statement period is from January 1 through December 31. The Finance Office forwards to field offices a copy of Form RD 1951-9, "Annual Statement of Loan Account,"

to be retained in borrower files as a permanent record of account activity for the year.

(2) Quarterly Form RD 1951-9 are retained in the Finance Office on microfiche. These statements reflect cumulative data from the beginning of the current year through the end of the most recent quarter. Servicing offices may request copies of these quarterly or annual statements by sending Form RD 1951-57, "Request for Loan Summary Statement," to the Finance Office.

(3) The servicing office will provide a copy of the applicable loan summary statement to the borrower on request. A copy of Form RD 1951-9 and, for loans with unamortized installments, a printout of future installments owed obtained using the borrower status screen option in the Automated Discrepancy Processing System (ADPS), will constitute the loan summary statement to be provided to the borrower.

(c) *Insurance.* Rural Development borrowers shall maintain insurance coverage as follows:

(1) Community and Insured Business Programs borrowers shall continuously maintain adequate insurance coverage as required by the loan agreement and § 1942.17(j)(3) of subpart A of part 1942 of this chapter. Insurance coverage must be monitored in accordance with the above-referenced section to determine that adequate policies and bonds are in force.

(2) For all other types of loans covered by this subpart, property insurance will be serviced according to subpart A of part 1806 of this chapter (RD Instruction 426.1) in real estate mortgage cases, and according to the loan agreement in other cases.

(d) *Property taxes.* Real property taxes are serviced according to Subpart A of part 1925 of this chapter. If State statutes permit a personal property tax lien to have priority over Rural Development's lien, such taxes are serviced according to §§ 1925.3 and 1925.4 of subpart A of part 1925 of this chapter.

(e) *Changes in borrower's legal organization.* (1) The State Director may approve, with OGC's concurrence, changes in a recipient's legal organization, including revisions of articles of incorporation or charter and bylaws, when:

(i) The change does not provide for a sole member type of organization;

(ii) The borrower retains control over its assets and the operation, management, and maintenance of the facility, and continues to carry out its responsibilities as set forth in § 1942.17(b)(4) of subpart A of part 1942 of this chapter; and

(iii) The borrower retains significant local ties with the rural community.

(2) The State Director may approve, with prior concurrence of the Administrator, changes in a recipient's legal organization which result in a sole member type of organization, or any other change which results in a recipient's loss of control over its assets and/or the operation, management and maintenance of the facility, provided all of the following have been or will be met:

(i) The change is in the best interest of the Government;

(ii) The State Director determines and documents that other servicing options under this subpart, such as sale or transfer and assumption, have been explored and are not feasible;

(iii) The loan is classified as a nonprogram loan;

(iv) The borrower is notified that it is no longer eligible for any program benefits, but will remain responsible under the loan agreement; and

(v) Prior concurrence of the Administrator is obtained. Requests will be forwarded to the Administrator: Attention (appropriate program division), and will include the case file; Exhibit A of this subpart (available in any Rural Development office), appropriately completed; the proposed changes; OGC comments; and any other necessary supporting information.

(f) *Membership liability.* As a loan approval requirement, some borrowers may have special agreements with members of the purchase of shares of stock or for payment of a pro rata share of the loan in the event of default, or they may have authority in their corporate instruments to make special assessments in that event. Such agreements may be referred to as individual liability agreements and may be assigned to and held by Rural Development as additional security. In other cases the borrower's note may be endorsed by individuals. The liability instruments will be serviced in a manner indicated by their contents and the advice of OGC to adequately protect Rural Development's interest. Servicing actions necessary due to such provisions will be tracked in the Multi-Family Housing Information System (MFIS).

(g) *Other security*. Other security such as collateral assignments, water stock certificates, notices of lienholder interest (Bureau of Land Management grazing permits) and waivers of grazing privileges (Forest Service grazing permits) will be serviced to protect the interest of Rural Development and in compliance with any special servicing actions developed by the State Director with OGC assistance. Evidence of the security will be filed in the servicing office case file. Necessary servicing actions will be noted in MFIS.

(h) *Correcting errors in security instruments*. Land, buildings, or chattels included in a mortgage through mutual mistake may be released from the mortgage by the State Director when substantiated by the factual situation. The release is contingent on the State Director determining, with OGC advice, that the property was included due to mutual error.

(i) *Present market value determination*. For purposes of this subpart, the value of security is determined by the approval official as follows:

(1) *Security representing a relatively small portion of the total value of the security property*. The approval official will determine that the real estate and chattels are disposed of at a reasonable price. A current appraisal report may be required.

(2) *Security representing a relatively large portion of the total value of the security property*. The approval official will require a current appraisal report, and the sale prices of the real estate and chattels disposed of will at least equal the present market value as determined by this appraisal.

(3) *Appraisal report*. If required, a current appraisal report will be completed in accordance with §1942.3 of subpart A of part 1942 of this chapter. The ap-

praisal will be completed by a qualified Rural Development employee or an independent appraiser as determined appropriate by the approval official.

[55 FR 4399, Feb. 8, 1990, as amended at 57 FR 775, Jan. 9, 1992; 57 FR 21199, May 19, 1992; 57 FR 36591, Aug. 14, 1992; 69 FR 69105, Nov. 26, 2004]

§1951.221 **Collections, payments and refunds.**

Payments and refunds are handled in accordance with the following:

(a) *Community and Insured Business Programs*. (1) Field offices can obtain data on principal installments due for Community and Insured Business Programs loans with unamortized installments using the borrower status screen option in the ADPS.

(2) Regular payments for Community and Insured Business Programs borrowers are all payments other than extra payments and refunds. Such payments are usually derived from facility revenues, and do not include proceeds from the sale of security. They also include payments derived from sources which do not decrease the value of Rural Development's security.

(i) Distribution of such payments is made as follows:

(A) First, to the Rural Development loan(s) in proportion to the delinquency existing on each. Any excess will be distributed in accordance with paragraphs (a)(2)(i) (B) and (C) of this section.

(B) Second, to the Rural Development loan or loans in proportion to the approximate amounts due on each. Any excess will be distributed according to paragraph (a)(2)(i)(C) of this section.

(C) Third, as advance payments on Rural Development loans. In making such distributions, consider the principal balance outstanding on each loan, the security position of the liens securing each loan, the borrower's request, and related circumstances.

(ii) Unless otherwise established by the debt instrument, regular payments will be applied as follows:

(A) For amortized loans, first to interest accrued (as of the date of receipt of the payment), and then to principal.

(B) For principal-plus-interest loans, first to the interest due through the date of the next scheduled installment of principal and interest and then to principal due, with any balance applied to the next scheduled principal installment.

(3) Extra payments are derived from sale of basic chattel or real estate security; refund of unused loan funds; cash proceeds of property insurance as provided in § 1806.5(b) of subpart A of part 1806 (paragraph V B of RD Instruction 426.1); and similar actions which reduce the value of basic security. At the option of the borrower, regular facility revenue may also be used as extra payments when regular payments are current. Unless otherwise established in the note or bond, extra payments will be distributed and applied as follows:

(i) First to the account secured by the lowest priority of lien on the property from which the extra payment was obtained. Any balance will be applied to other Rural Development loans in ascending order of priority.

(ii) For amortized loans, first to interest accrued to the date payment is received, and then to principal. For debt instruments with installments of principal plus interest, such payments will be applied to the final unpaid principal installment.

(b) *Soil and Water Conservation Loans.* (1) Regular payments for such loans are defined in § 1951.8(a) of subpart A of part 1951 of this chapter, and are distributed according to § 1951.9(a) of that subpart unless otherwise established by the note or bond.

(2) Extra payments are defined in § 1951.8(b) of subpart A of part 1951 of this chapter, and are distributed according to § 1951.9(b) of that subpart.

[55 FR 4399, Feb. 8, 1990, as amended at 66 FR 1569, Jan. 9, 2001; 68 FR 61331, Oct. 28, 2003; 68 FR 69952, Dec. 16, 2003]

§ 1951.222　Subordination of security.

When a borrower requests Rural Development to subordinate a security instrument so that another creditor or lender can refinance, extend, reamortize, or increase the amount of a prior lien; be on parity with; or place a lien ahead of the Rural Development lien, it will submit a written request to the servicing office as provided below. For purposes of this subpart, subordination is defined to include cases where a parity security position is being considered.

(a) *General.* The following requirements must normally be met:

(1) The request must be for subordination of a specific amount of the Rural Development indebtedness.

(2) It must be determined that the borrower cannot refinance its Rural Development debt in accordance with subpart F of part 1951 of this chapter.

(3) The transaction will further the purposes for which the Rural Development loan was made, not adversely affect the borrower's debt-paying ability, and result in the Rural Development debt being adequately secured.

(4) The terms and conditions of the prior lien will be such that the borrower can reasonably be expected to meet them as well as the requirements of all other debts.

(5) Any proposed development work will be planned and performed according to § 1942.18 of subpart A of part 1942 of this chapter or in a manner directed by the creditor which reasonably attains the objectives of that section.

(6) All contracts, pay estimates, and change orders will be reviewed and concurred in by the State Director.

(7) In cases involving land purchase, the Rural Development will obtain a mortgage on the purchased land.

(8) When the transaction involves more than $10,000 or the approval official considers it necessary, a present market value appraisal report will be obtained. However, a new report need not be obtained if there is an appraisal report not over one year old which permits a proper determination of the present market value of the total property after the transaction.

(9) The proposed action must not change the nature of the borrower's activities so as to make it ineligible for Rural Development loan assistance.

(10) Necessary consent and subordination of all other outstanding security interests must be obtained.

(b) *Authorities.* Proposals not meeting one or more of the above requirements will be submitted to the Administrator, Attention (appropriate program division) for prior concurrence. All other proposals may be approved by the official with loan approval authority under subpart A of part 1901 of this chapter.

(c) *Processing.* The case file is to include:

(1) The borrower's written request on Form RD 465–1, "Application for Partial Release, Subordination, or Consent," if appropriate, or in other acceptable format. The request must contain the purpose of the subordination; exact amount of money or property involved; description of security property involved; type of security instrument; name, address, line of business and other general information pertaining to the party in favor of which the request is made; and other pertinent information to evaluate the need for the request;

(2) Current balance sheet;

(3) If development work is involved, an operating budget on Form RD 442–7, "Operating Budget," or similar form which projects income and expenses through the first full year of operation following completion of planned improvements; or if no development work is involved, an income statement and budget on Form RD 442–2, "Statement of Budget, Income, and Equity," schedules 1 and 2, or similar form;

(4) Copy of proposed security instrument;

(5) Appraisal report, when applicable;

(6) OGC opinion on the request;

(7) Exhibit A of this subpart (available in any Rural Development office), appropriately completed;

(8) Appropriate environmental review; and

(9) Any other necessary supporting information.

(d) *Closing.* All requests for subordination will be closed according to instructions from OGC except those which affect only chattel liens other than pledges of revenue. Rural Development's consent on Form RD 465–1 will be signed concurrently with Form RD 460–2, "Subordination by the Government," when applicable.

[55 FR 4399, Feb. 8, 1990, as amended at 66 FR 1569, Jan. 9, 2001; 69 FR 70884, Dec. 8, 2004]

§ 1951.223 Reamortization.

(a) *State Director authorization.* The State Director is authorized to approve reamortization of loans under the following conditions:

(1) The account is delinquent and cannot be brought current within one year while maintaining a reasonable reserve;

(2) The borrower has demonstrated for at least one year by actual performance or has presented a budget which clearly indicates that it is able to meet the proposed payment schedule;

(3) The amount being reamortized is within the State Director's loan approval authorization; and

(4) There is no extension of the final maturity date.

(b) *Requests requiring National Office approval.* Reamortizations not meeting the above conditions require prior National Office approval. Requests will be forwarded to the National Office with the case file, including:

(1) Current budget and cash flow prepared on RD 442–2, schedules 1 and 2, or similar form;

(2) Current balance sheet and income statement;

(3) Exhibit A of this subpart, appropriately completed;

(4) Form RD 1951–33, "Reamortization Request," completed in accordance with § 1951.223(c)(3) of this subpart, when applicable; and

(5) Any other necessary supporting information.

(c) *Processing.* When legally permissible and administratively acceptable, the total outstanding principal and interest balances will be reamortized rather than only the delinquent amount. Accrued interest will be at the rate currently reflected in Finance Office records.

(1) Reamortizations will be perfected in accordance with OGC closing instructions.

(2) When debt instruments are being modified or new debt instruments executed, bond counsel or local counsel, as appropriate, must provide an opinion indicating any effect on Rural Development's security position. The Rural Development's approval official must determine that the government's interest will remain adequately protected if the security position will be affected.

(3) *Notes.* Except as provided in § 1951.223(c)(4), loans evidenced by notes

will be reamortized through a new evidence of debt unless OGC recommends that the terms of the existing document be modified. Form RD 1951-33 may be used to effect such modifications, if legally adequate, or other forms may be used if acceptable to Rural Development. The original of a new note or any endorsement required by OGC is to be attached to the existing note, filed in the servicing office, and retained until the account is paid in full or otherwise satisfied. A copy will be forwarded to the Finance Office.

(4) *Bonds and notes with other than real or chattel security pledged to Rural Development.* Loans evidenced by bonds, or by notes with other than real or chattel security pledged to Rural Development, may be reamortized using procedures acceptable to the State Director and legally permissible under State statutes in the opinion of the borrower's counsel and the OGC.

(i) The procedure may consist of a new debt instrument or agreement for the total Rural Development indebtedness, including the delinquency, or a new instrument or agreement whereby the borrower agrees to repay the delinquency plus interest. If a new instrument or agreement for only the delinquent amount is used, a new loan number will be assigned to the delinquent amount, and the borrower will be required to pay the amounts due under both the original and the new instruments.

(ii) When a delinquent or problem loan cannot be reamortized by issuing a new debt instrument due to State statutes, or the cost of preparation and closing is prohibitive, the rescheduling agreement provided as Exhibit H of this subpart (available in any Rural Development office), may be used.

(iii) Section 1942.19 of subpart A of part 1942 of this chapter applies to any new bonds issued unless precluded by State statutes or an exception is approved by the National Office.

(iv) If State statutes do not require the release of existing bonds, they will be retained with the new bond instrument or agreement in the Rural Development office authorized to store such documents. If State statutes require release of existing bonds, the exchange will be accomplished by the District

Director, and the new bond and/or agreement will be retained in the appropriate office.

(5) *New debt instruments or agreements.* (i) A copy will be sent to the Finance Office after execution, except that if serial bonds are used, the original bond(s) will be submitted to the Finance Office.

(ii) Any agreement used will contain:

(A) The amount delinquent, which must equal the total delinquency on the account and net advances (the unpaid principal on any advance and the accrued interest on any advance through the date of reamortization, less interest payments credited on the advance account);

(B) The effective date of the reamortization;

(C) The number of years over which the delinquency will be amortized;

(D) The repayment schedule; and

(E) The interest rate.

(iii) A payment will be due on the next scheduled due date. Deferment of interest and/or principal payments is not authorized.

(iv) A separate new instrument will be required for each loan being reamortized.

(v) If amortized payments are not used, the schedule of principal installments developed will be such that combined payments of principal and interest closely ⌐pproximate an amortized payment.

(d) *Reamortization with interest rate adjustment—Water and waste borrowers only.* A borrower that is seriously delinquent in loan payments may be eligible for loan reamortization with interest rate adjustment. The purpose of loan reamortization with interest rate adjustment is to provide relief for a borrower that is unable to service the outstanding loan in accordance with its existing terms and to enhance recovery on the loan. A borrower must meet the conditions of this subpart to be considered eligible for this provision.

(1) *Eligibility determination.* The State Director, Rural Development, may submit to the Administrator for approval an adjustment in the rate of interest charged on outstanding loans only for those borrowers who meet the following requirements:

(i) The borrower has exhausted all other servicing provisions contained in this subpart;

(ii) The borrower is experiencing severe financial problems;

(iii) Any management deficiencies must have been corrected or the borrower must submit a plan acceptable to the State Office to correct any deficiencies before an interest rate adjustment may be considered;

(iv) Borrower user rates must be comparable to similar systems. In addition, the operating expenses reported by the borrower must appear reasonable in relation to similar system expenses;

(v) The borrower has cooperated with Rural Development in exploring alternative servicing options and has acted in good faith with regard to eliminating the delinquency and complying with its loan agreements and agency regulations; and

(vi) The borrower's account must be delinquent at least one annual debt payment for 180 days.

(2) *Conditions of approval.* All borrowers approved for an adjustment in the rate of interest by the Administrator shall agree to the following conditions:

(i) The borrower shall agree not to maintain cash or cash reserves beyond what is reasonable at the time of interest rate adjustment to meet debt service, operating, and reserve requirements.

(ii) A review of the borrower's management and business operations may be required at the discretion of the State Director. This review shall be performed by an independent expert who has been recommended by the State Director and approved by the National Office. The borrower must agree to implement all recommendations made by the State Director as a result of the review.

(iii) If requested, a copy of the latest audited financial statements or management report must be submitted to the Administrator.

(3) *Reamortization.* At the discretion of the Administrator, the interest rate charged on outstanding loans of eligible borrowers may be adjusted to no less than the poverty interest rate and the term of the loans may be extended up to a new 40 year term or the remaining useful life of the facility, whichever is less.

[55 FR 4399, Feb. 8, 1990, as amended at 56 FR 25351, June 4, 1991; 63 FR 41714, Aug. 5, 1998; 69 FR 69105, Nov. 26, 2004; 73 FR 8008, Feb. 12, 2008]

§1951.224 Third party agreements.

The State Director may authorize all or part of a facility to be operated, maintained or managed by a third party under a contract, management agreement, written lease, or other third party agreement as follows:

(a) *Leases*—(1) *Lease of all or part of a facility (except when liquidation action is pending).* The State Director may consent to the leasing of all or a portion of security property when:

(i) Leasing is the only feasible way to provide the service and is the customary practice as required under §1942.17(b)(4) of subpart A of part 1942 of this chapter;

(ii) The borrower retains ultimate responsibility for operating, maintaining, and managing the facility and for its continued availability and use at reasonable rates and terms as required under §1942.17(b)(4) of subpart A of part 1942 of this chapter. The lease agreement must clearly reflect sufficient control by the borrower over the operation, maintenance, and management of the facility to assure that the borrower maintains this responsibility;

(iii) The lease agreement contains provisions prohibiting any amendments to the lease or any subleasing arrangements without prior written approval from Rural Development;

(iv) The lease document contains nondiscrimination requirements as set forth in §1951.204 of this subpart;

(v) The lease contains a provision which recognizes that Rural Development is a lienholder on the subject facility and, as such, the lease is subordinate to the rights and claims of Rural Development as lienholder; and

(vi) The lease does not constitute a lease/purchase arrangement, unless permitted under §1951.232 of this subpart.

(2) *Lease of all or part of a facility (pending liquidation action).* The State Director may consent to the leasing of

all or a portion of security property when:

(i) The lease will not adversely affect the repayment of the loan or the Government's rights under the security or other instruments;

(ii) The State Director has determined that liquidation will likely be necessary and the lease is necessary until liquidation can be accomplished;

(iii) Leasing is not an alternative to, or means of delaying, liquidation action;

(iv) The lease and use of any proceeds from the lease will further the objective of the loan;

(v) Rental income is assigned to Rural Development in an amount sufficient to make regular payments on the loan and operate and maintain the facility unless such payments are otherwise adequately secured;

(vi) The lease is advantageous to the borrower and is not disadvantageous to the Government;

(vii) If foreclosure action has been approved and the case has been submitted to OGC, consent to lease and use of proceeds will be granted only with OGC's concurrence; and

(viii) The lease does not exceed a one-year period. The property may not be under lease more than two consecutive years without authorization from the National Office. Long-term leases may be approved, with prior authorization from the National Office, if necessary to ensure the continuation of services for which the loan was made and if other servicing options contained in this subpart have been determined inappropriate for servicing the loan.

(b) *Mineral leases.* Unless liquidation is pending, the State Director is authorized to approve mineral leases when:

(1) The lessee agrees, or is liable without any agreement, to pay adequate compensation for any damage to the real estate surface and improvements. Damage compensation will be assigned to Rural Development or the prior lienholder by the use of Form FD 443-16, "Assignment of Income from Real Estate Security," or other appropriate instrument;

(2) Royalty payments are adequate and are assigned to Rural Development on Form RD 443-16 in an amount deter-

mined by the State Director to be adequate to protect the Government's interest;

(3) All or a portion of delay rentals and bonus payments may be assigned on Form RD 443-16 if needed for protection of the Government's interest;

(4) The lease, subordination, or consent form is acceptable to OGC;

(5) The lease will not interfere with the purpose for which the loan or grant was made; and

(6) When Rural Development consent is required, the borrower submits a completed Form RD 465-1. The form will include the terms of the proposed agreement and specify the use of all proceeds, including any to be released to the borrower.

(c) *Management agreements.* Management agreements should contain the minimum suggested contents contained in Guide 24 of part 1942, subpart A of this chapter (available in any Rural Development office).

(d) *Affiliation agreements.* An affiliation agreement between the borrower and a third party may be approved by the State Director, with OGC concurrence, if it provides for shared services between the parties and does not result in changes to the borrower's legal organizational structure which would result in its loss of control over its assets and/or over the operation, management, and maintenance of the facility to the extent that it cannot carry out its responsibilities as set forth in § 1942.17(b)(4) of subpart A of part 1942 of this chapter. However, affiliation agreements which result in a loss of borrower control may be approved with prior concurrence of the Administrator if the loan is reclassified as a nonprogram loan and the borrower is notified that it is no longer eligible for any program benefit. Requests forwarded to the Administrator will contain the case file, the proposed affiliation agreement, and necessary supporting information.

(e) *Processing.* The consent of other lienholders will be obtained when required. When National Office approval is required, or if the State Director wishes to have a transaction reviewed prior to approval, the case file will be forwarded to the National Office and will include:

(1) A copy of the proposed agreement;

(2) Exhibit A of this subpart (available in any Rural Development office), appropriately completed;

(3) Any other necessary supporting information.

[55 FR 4399, Feb. 8, 1990, as amended at 57 FR 21199, May 19, 1992]

§1951.225 Liquidation of security.

When the District Director believes that continued servicing will not accomplish the objectives of the loan, he or she will complete Exhibit A of this subpart (available in any Rural Development office), and submit it with the District Office file to the State Office. If the State Director determines the account should be liquidated, he or she will encourage the borrower to dispose of the Rural Development security voluntarily through a sale or transfer and assumption, and establish a specified period, not to exceed 180 days, to accomplish the action. If a transfer or voluntary sale is not carried out, the loan will be liquidated according to subpart A of part 1955 of this chapter.

§1951.226 Sale or exchange of security property.

A cash sale of all or a portion of a borrower's assets or an exchange of security property may be approved subject to the conditions set forth below.

(a) *Authorities.* (1) The District Director is authorized to approve actions under this section involving only chattels.

(2) The State Director is authorized to approve real estate transactions except as noted in the following paragraph.

(3) Approval of the Administrator must be obtained when a substantial loss to the Government will result from a sale; one or more members of the borrower's organization proposes to purchase the property; it is proposed to sell the property for less than the appraised value; or the buyer refuses to assume all the terms of the Grant Agreement. It is not Rural Development policy to sell security property to one or more members of the borrower's organization at a price which will result in a loss to the Government.

(b) *General.* Approval may be given when the approval official determines and documents that:

(1) The consideration is adequate;

(2) The release will not prevent carrying out the purpose of the loan;

(3) The remaining property is adequate security for the loan or the transaction will not adversely affect Rural Development's security position;

(4) If the property to be sold or exchanged is to be used for the same or similar purposes for which the loan or grant was made, the purchaser will:

(i) Execute Form RD 400–4, "Assurance Agreement." The covenants involved will remain in effect as long as the property continues to be used for the same or similar purposes for which the loan or grant was made. The instrument of conveyance will contain the covenant referenced in §1951.204 of this subpart; and

(ii) Provide to Rural Development a written agreement assuming all rights and obligations of the original grantee if grant funds were provided. See §1951.215 of this subpart for additional guidance on grant agreements.

(5) The proceeds remaining after paying any reasonable and necessary selling expenses are used for one or more of the following purposes:

(i) To pay on Rural Development debts according to §1951.221 of this subpart; on debts secured by a prior lien; and on debts secured by a subsequent lien if it is to Rural Development's advantage.

(ii) To purchase or acquire through exchange property more suited to the borrower's needs, if the Rural Development debt will be as well secured after the transaction as before.

(iii) To develop or enlarge the facility if necessary to improve the borrower's debt-paying ability; place the operation on a sounder basis; or otherwise further the loan objectives and purposes.

(6) Disposition of property acquired in whole or part with Rural Development grant funds will be handled in accordance with the grant agreement.

(c) *Processing.* (1) The case file will contain the following:

(i) Except for actions approved by the District Director, Exhibit A of this subpart (available in any Rural Development office), appropriately completed;

(ii) The appraisal report, if appropriate;

(iii) Name of purchaser, anticipated sales price, and proposed terms and conditions;

(iv) Form RD 1965–8, "Release from Personal Liability," including the County Committee memorandum and the State Director's recommendations;

(v) An executed Form RD 400–4, if applicable;

(vi) An executed Form RD 465–1, if applicable;

(vii) Form RD 460–4, "Satisfaction," if a debt has been paid in full or satisfied by debt settlement action. For cases involving real estate, a similar form may be used if approved by OGC; and

(viii) Written approval of the Administrator when required under § 1951.226(a)(3) of this subpart;

(2) *Releasing security.* (i) The District Director is authorized to satisfy or terminate chattel security instruments when § 1951.226(b) of this subpart and § 1962.17 and § 1962.27 of subpart A of part 1962 of this chapter have been complied with. Partial release may be made by using Form RD 460–1, "Partial Release," or Form RD 462–12, "Statements of Continuation, Partial Release, Assignment, Etc."

(ii) Subject to § 1951.226(b) of this subpart, the State Director is authorized to release part or all of an interest in real estate security by approving Form RD 465–1. Partial release of real estate security may be made by use of Form RD 460–1 or other form approved by OGC.

(3) Rural Development liens will not be released until the sale proceeds are received for application on the Government's claim. In states where it is necessary to obtain the insured note from the lender to present to the recorder before releasing a portion of the land from the mortgage, the borrower must pay any cost for postage and insurance of the note while in transit. The District Director will advise the borrower when it requests a partial release that it must pay these costs. If the borrower is unable to pay the costs from its own funds, the amounts shown on the state-ment of actual costs furnished by the insured lender may be deducted from the sale proceeds.

(d) *Release from liability.* (1) When an Rural Development debt is paid in full from the proceeds of a sale, the borrower will be released from liability by use of Form RD 1965–8.

(2) When sale proceeds are not sufficient to pay the Rural Development debt in full, any balance remaining will be handled in accordance with procedures for debt settlement actions set forth in subpart C of part 1956 of this chapter.

(i) In determining whether a borrower should be released from liability, the State Director will consider the borrower's debt-paying ability based on its assets and income at the time of the sale.

(ii) Release from liability will be accomplished by using Form RD 1965–8 and obtaining from the County Committee a memorandum recommending the release which contains the following statement:

_____ in our opinion does not have reasonable debt-paying ability to pay the balance of the debt after considering its assets and income at the time of the sale. The borrower has cooperated in good faith, used due diligence to maintain the security against loss, and otherwise fulfilled the covenants incident to the loan to the best of its ability. Therefore, we recommend that the borrower be released from liability upon the completion of the sale.

[55 FR 4399, Feb. 8, 1990, as amended at 69 FR 70884, Dec. 8, 2004]

§ 1951.227 Protective advances.

The State Director is authorized to approve, without regard to any loan or total indebtedness limitation, vouchers to pay costs, including insurance and real estate taxes, to preserve and protect the security, the lien, or the priority of the lien securing the debt owed to or insured by Rural Development if the debt instrument provides that Rural Development may voucher the account to protect its lien or security. The State Director must determine that authorizing a protective advance is in the best interest of the government. For insurance, factors such as the amount of advance, occupancy of the structure, vulnerability to damage and present value of the structure and contents will be considered.

(a) Protective advances are considered due and payable when advanced. Advances bear interest at the rate specified in the most recent debt instrument authorizing such an advance.

(b) Protective advances are not to be used as a substitute for a loan.

(c) Vouchers are prepared in accordance with applicable procedures set forth in RD Instruction 2024–A (available in any Rural Development office).

[55 FR 4399, Feb. 8, 1990, as amended at 57 FR 36591, Aug. 14, 1992]

§§ 1951.228–1951.229 [Reserved]

§ 1951.230 Transfer of security and assumption of loans.

(a) *General.* It is Rural Development policy to approve transfers and assumptions to transferees which will continue the original purpose of the loan in accordance with the following and specific requirements relating to eligible and ineligible borrowers set forth below:

(1) The present borrower is unable or unwilling to accomplish the objectives of the loan.

(2) The transfer will not be disadvantageous to the Government or adversely affect either Rural Development's security position or the Rural Development program in the area.

(3) Transfers to eligible applicants will receive preference over transfers to ineligible applicants if recovery to Rural Development is not less than it would be if the transfer were to an ineligible applicant.

(4) If the Rural Development debt(s) exceed the present market value of the security as determined by the State Director, the transferee will assume an amount at least equal to the present value.

(5) If the transfer and assumption is to one or more members of the borrower's organization, there must not be a loss to the government.

(6) Rural Development concurs in plans for disposition of funds in the transferor's debt service, reserve, operation and maintenance, and any other project account, including supervised bank accounts.

(7) When the property to be transferred is to be used for the same or similar purposes for which the loan was made, the transferee will execute Form RD 400–4 to continue nondiscrimination covenants and provide to Rural Development a written certification assuming all terms of the Grant Agreement executed by the transferor. All instruments of conveyance will contain the covenant referenced in § 1951.204 of this subpart.

(8) This subpart does not preclude the transferor from receiving equity payments when the full account of the Rural Development debt is assumed. However, equity payments will not be made on more favorable terms than those on which the balance of the Rural Development debt will be paid.

(9) Transferees must have the ability to pay the Rural Development debt as provided in the assumption agreement and the legal capacity to enter into the contract. The applicant will submit a current balanced sheet using Form RD 442–3, "Balance Sheet," and budget and cash flow information using Form RD 442–2, or similar forms. For ineligible applicants, such information may be supplemented by a credit report from an independent source or verified by an independent certified public accountant.

(10) For purposes of this subpart, transfers to eligible applicants will include mergers and consolidations. Mergers occur when two or more corporations combine in such a manner that only one remains in existence. In a consolidation, two or more corporations combine to form a new, consolidated corporation, with all of the original corporations ceasing to exist. In both mergers and consolidations, the surviving or emerging corporation takes the assets and assumes the liabilities of the corporation(s) which ceased to exist. Such transactions must be distinguished from transfers and assumptions, in which a transferor will not necessarily go out of existence and the transferee will not always take all assets or assume all liabilities of the transferor.

(11) A current appraisal report to establish the present market value of the security will be completed in accordance with § 1951.220(i) of this subpart

when the full debt is not being assumed.

(12) There must be no lien, judgment, or similar claims of other parties against the Rural Development security being transferred unless the transferee is willing to accept such claims and the Rural Development approval official determines that they will not prevent the transferee from repaying the Rural Development debt, meeting all operating and maintenance costs, and maintaining required reserves. The written consent of any other lienholder will be obtained where required.

(b) *Authorities.* The State Director is authorized to approve transfers and assumptions of Rural Development loans in accordance with the provisions of paragraphs (c) and (d) of this section, except for the following, which require prior approval of the Administrator:

(1) Proposals which will involve a loss to the Government;

(2) Proposals involving a transfer to one or more members of the present borrower's organization;

(3) Proposals involving rates and terms which are more liberal than those set forth in § 1951.230(c) of this subpart;

(4) Proposals involving a cash payment to the present borrower which exceeds the actual sales expenses;

(5) The transferee refuses to assume all terms of the Grant Agreement for a project financed in part with Rural Development grant funds; and

(6) Proposed transfers to ineligible applicants when there is no significant downpayment and/or the repayment period is to exceed 25 years.

(c) *Eligible applicants.* Except as noted in § 1951.230(b) of this subpart, the State Director is authorized to approve transfers of security property to and assumptions of Rural Development debts by transferees who would be eligible for financial assistance under the loan program involved for the type of loan being transferred. The State Director must determine and document that eligibility requirements have been satisfied.

(1) If a loan is evidenced and secured by a note and lien on real or chattel property, Form RD 1951-15, "Community Programs Assumption Agreement," will be executed by the transferee. When the terms of the loan are changed, the new repayment period may not exceed the lesser of the repayment period for a new loan of the type involved or the expected life of the facility. Interest will accrue at the rate currently reflected in Finance Office records.

(2) If the loan is evidenced and secured by a bond, procedures will be followed which are acceptable to the State Director and legally permissible under State law in the opinion of the borrower's counsel and OGC. The interest rate will be the rate currently reflected in Finance Office records. Any new repayment period provided may not exceed the lesser of the repayment period for a new loan of the type involved or the expected life of the facility.

(3) Loans being transferred and assumed may be combined when the security is the same, new terms are being provided, a new debt instrument will be issued, and the loans have the same interest rate and are for the same purpose. If applicable, § 1942.19(h)(11) will govern the preparation of any new debt instruments required.

(4) A loan may be made in connection with a transfer if the transferee meets all eligibility and other requirements for the kind of loan being made. Such a loan will be considered as a separate loan, and must be evidenced by a separate debt instrument. However, it is permissible to have one authorizing loan resolution or ordinance if permitted by State statutes.

(5) Any development funds remaining in a supervised bank account which are not to be refunded to Rural Development will be transferred to a supervised bank account for the transferee simultaneously with the closing of the transfer for use in completing planned development.

(d) *Ineligible applicants.* Except as noted in § 1951.230(b) of this subpart, the State Director is authorized to approve transfer and assumptions to transferees who would not be eligible for financial assistance under the loan program involved for the type of loan being transferred. However, the State Director is authorized to approve all transfers of incorporated Economic Opportunity Cooperative loans to ineligible applicants without regard to the requirements set forth in § 1951.230(b). Such transfers are considered only when an eligible transferee is not available or when the recovery to Rural Development from a transfer to an available eligible transferee would be less. Transfers are not to be considered as a means by which members of the transferor's governing body can obtain an equity or as a method of providing a source of easy credit for purchasers.

(1) Ineligible applicants must pay a one-time nonrefundable transfer fee when they submit an application or proposal.

(i) The National Office will issue a directive annually advising the field of the amount of the fee. Any cost for appraisals performed by non-Rural Development personnel will be handled in accordance with RD Instruction 2024-A (available in any Rural Development office), and will be added to the basic fee.

(ii) Transfer fees will be deposited in accordance with current instructions governing the handling of collections. The fees will be identified as transfer fees on Form RD 451-2, "Schedule of Remittances," and will be included on the Daily Activity Report. The amount will be credited to the Rural Development Insurance Fund.

(iii) If the State Director determines waiver of the transfer fee is in the best interest of the government, he or she will request prior approval by submitting the transfer case file established in accordance with processing requirements set forth below to the National Office, Attention (appropriate program division).

(2) Any funds remaining in a supervised bank account will be refunded to Rural Development and applied to the debt as a condition of transfer.

(3) The interest rate will be the greater of the rate specified for the note in current Finance Office records or the market rate for Community Programs as of the transfer closing date.

(4) The transferred loan will be identified as an NP loan and serviced in accordance with § 1951.216 of this subpart.

(5) Form RD 465-5, "Transfer of Real Estate Security," will be used, and will be modified as appropriate before execution.

(6) Consideration will be given to obtaining individual liability agreements from members of the transferee organization.

(e) *Release from liability.* Except when nonprogram loans or Economic Opportunity Cooperative loans are involved, transferors may be released from liability in accordance with the following:

(1) If the full amount of the debt is assumed, the State Director may approve the release from liability by use of Form RD 1965-8.

(2) If less than the full amount of the debt is assumed, any balance remaining will be handled in accordance with procedures for debt settlement actions set forth in subpart C of part 1956 of this chapter.

(i) In determining whether a borrower should be released from liability, the State Director will consider the borrower's debt-paying ability based on its assets and income at the time of the sale.

(ii) Release from liability will be accomplished by using Form RD 1965-8 and obtaining from the County Committee a memorandum recommending the release which contains the statement set forth in § 1951.226(d)(2)(ii) of this subpart.

(f) *Processing.* Transfers and assumptions will be processed in accordance with the following:

(1) A transfer case file organized in accordance with RD Instruction 2033-A (available in any Rural Development office) will be established, and will contain all documents and correspondence relating to the transfer. The forms utilized for transfers and assumptions are listed in Exhibit D (available in any Rural Development office). All forms listed must be completed and included in the case file unless inappropriate for the particular situation.

(2) A letter of conditions establishing requirements to be met in connection with the transfer and assumption will be issued, and the transferee will be required to execute an Agency approved

form, "Letter of Intent to Meet Conditions," prior to the closing of the transfer.

(3) Both the transferee and transferor are responsible for obtaining the legal services necessary to accomplish the transfer.

(4) Transfers will be closed in accordance with instructions provided by OGC.

(5) When the transferee is a public body and Form RD 1951-15 is not suitable, the transferee's attorney will prepare the documents necessary to effect the transfer and assumption and submit them for approval by Rural Development and OGC.

(6) Accrued interest to be entered in either Table 1 of Form RD 1951-15 or other appropriate assumption agreement is to be obtained using the status screen option in ADPS.

(7) The following forms, if utilized, will be sent immediately to the Finance Office:

(i) Form RD 1951-15 or other appropriate assumption agreement;

(ii) A conformed copy of Form RD 1965-8.

(8) If an Rural Development grant was made in conjunction with the loan being transferred, the transferee must agree in writing to assume all rights and obligations of the original grantee. See § 1951.215 for additional guidance on grant agreements.

(9) The transferee will obtain insurance according to requirements for the loan(s) being transferred unless the approval official requires additional insurance. When the entire Rural Development debt is being assumed and an amount has been advanced for insurance premiums or any other purposes, the transfer will not be completed until the Finance Office has charged the advance to the transferor's account.

(10) *Rates and terms.* (i) If the transfer will be closed at the same rates and terms, the transferee will be informed of the amount needed to be on schedule by the next installment due date.

(ii) If the transfer will be closed at new rates and terms, the transferee will be informed of the amount of principal and interest owed based on information obtained using the ADPS status screen option.

(11) The effective date of a transfer is the actual date the transfer is closed, which is the same date Form RD 1951-15 or other appropriate assumption agreement is signed.

(12) Title to all assets will be conveyed from the transferor to the transferee unless other arrangements are agreed upon by all parties concerned, including Rural Development. All instruments of conveyance will contain the covenant referenced in § 1951.204 of this subpart.

(13) If an insured loan being held by an investor is involved, the Finance Office will have to repurchase the note prior to processing the assumption agreement.

(14) When National Office approval is required, the transfer case file will be submitted to the Administrator, Attention: (appropriate program division), with Exhibit A of this subpart (available in any Rural Development office), appropriately completed, and a cover memorandum which denotes any unusual circumstances.

(15) The District Director must review Form RD 1910-11, "Applicant Certification, Federal Collection Policies for Consumer or Commercial Debts," with the applicant, and the form must be signed by the applicant and included in the file.

[55 FR 4399, Feb. 8, 1990, as amended at 57 FR 36590, Aug. 14, 1992; 66 FR 1569, Jan. 9, 2001; 69 FR 70884, Dec. 8, 2004]

§ 1951.231 Special provisions applicable to Economic Opportunity (EO) Cooperative Loans.

(a) *Withdrawal of member and transfer to and assumption by new members of Unincorporated Cooperatives.* (1) Withdrawal of a member who is no longer utilizing the services of an association and transfer of withdrawing member interest in the association to a new member who will assume the entire unpaid balance of the indebtedness of the withdrawing member may be permitted, if the remaining members agree to accept the new member and the transfer will not adversely affect collection of the loan. The servicing office will submit to the State Office the borrow case file and the following:

(i) Form RD 1951-15 executed by the proposed new member;

(ii) Statement of the current amount of the indebtedness involved;

(iii) A description and statement of the value of the security property;

(iv) A memorandum to justify the transaction;

(v) Form RD 440-2, "County Committee Certification or Recommendation;"

(vi) Exhibit B of this subpart, "Agreement for New Member (With or Without Withdrawing Member)," (available in any Rural Development office), executed by the remaining members of the association, the proposed new member, and the withdrawing member; and

(vii) Form RD 450-12, "Bill of Sale (Transfer by Withdrawing Member)," executed by the withdrawing member.

(2) If the State Director determines after review of the above information that the proposed new member is eligible and the transfer is justified, the State Director may approve the transfer and assumption by executing Form RD 1951-15.

(3) Upon completion of the above actions, the State Director may release the outgoing member from personal liability using Form RD 1965-8.

(4) If Finance Office records must be changed due to changes in borrower name, address and/or case number, necessary documents, including Form RD 1951-15 and, if applicable, Form RD 1965-8, will be forwarded to the Finance Office immediately with a memorandum indicating that the purpose of the submission is only to establish liability for a new member and release an old member from liability.

(b) *Withdrawal of members from Unincorporated Cooperatives when new member not available.* Withdrawal of a member who no longer utilizes the services of an association may be permitted even though a new member is not available, provided:

(1) The State Director determines that the remaining members have sufficient need for the property, and that the withdrawal of the member will not adversely affect collection of the loan; and

(2) The remaining members obtain from the outgoing member an agreement conveying his or her interest in the cooperative property to them. They

may also wish to agree to protect the outgoing member against liability on the debt owed to Rural Development as well as any other debts. Exhibit C of this subpart, "Agreement for Withdrawal of Member (Without New Member)," (available in any Rural Development office), may be used by the cooperative. Rural Development will not be a party to the agreement.

(c) *Addition of new members (no withdrawing member or transfer involved) for both Incorporated and Unincorporated Cooperatives.* (1) A new member may be admitted to the association even though there is no withdrawing member, if:

(i) The members of the association agree to accept the proposed new member, and

(ii) The State Director determines that the association owns adequate facilities to provide service to the new member.

(2) The servicing office will submit to the State Office the case file and items (i) through (vi) of §1951.231(a)(1).

(3) If the State Director determines after the review of the above information that the proposed new member is eligible and the transaction is justified, the State Director may approve the transaction by executing Form RD 1951-15.

(4) Form RD 1951-15 will be forwarded immediately to the Finance Office with a memorandum indicating that the form is intended only to establish liability for a new member.

(d) *Deceased members of Unincorporated Cooperatives.* Form RD 442-24, "Operating Agreement," (now obsolete) was executed by recipients of these loans. Paragraph 10 of that form provides that in case of the death of any member, the heirs or personal representative of the deceased member shall take the deceased member's place in the association. This provision also covers sale of the decedent's interest in the association if the sale is necessary to pay debts of the estate.

(1) If the heirs or personal representative do not wish to continue membership in the association, the remaining members may be permitted to continue to operate the property if Rural Development's financial interest will not be jeopardized. The remaining members

should obtain from the deceased member's estate an agreement conveying the estate's interest in the cooperative property to them. The remaining members may wish to agree to protect the estate against liability on the debt to Rural Development as well as any other debts of the cooperative.

(2) The requirement of § 1962.46(h) of subpart A of part 1962 will also be followed.

(e) *Action which affects individual members of Unincorporated EO Cooperative security.* The borrower will be expected to protect its own interest in condemnation, trespass, quiet title, and other cases affecting the security. The servicing office will immediately furnish the complete facts concerning any action taken against individual members of Unincorporated Cooperatives to the State Director together with the case file.

(f) *Debt Settlement.* Debt settlement actions for Economic Opportunity Cooperative loans must be handled under the Federal Claims Collection Act; proposals will be submitted to the National Office for review and approval.

§ 1951.232 Water and waste disposal systems which have become part of an urban area.

A water and/or waste disposal system serving an area which was formerly a rural area as defined in § 1942.17(b)(2)(iii) and (iv) of subpart A of part 1942 of this chapter, but which has become in its entirety part of an urban area, will be serviced in accordance with this section.

(a) *Curtailment or limitation of service.* Service may not be curtailed or limited by the inclusion of a system within an urban area.

(b) *Sale or transfer and assumption.* (1) The urban community or another entity may purchase the facility involved and immediately pay the Rural Development debt in full; or

(2) The urban community or another entity may accept a transfer of the Rural Development debt on an ineligible applicant basis.

(3) When a grant is involved, the entity will agree in writing to assume all rights and obligations of the original grantee. See § 1951.215 for additional guidance on grant agreements.

(c) *Lease-purchase arrangement.* If § 1951.232(b) (1) and (2) of this section are not practicable, the urban community may, with prior approval of the National Office, operate and maintain the system under a lease-purchase arrangement which provides that:

(1) The urban community will:

(i) Assume responsibility for operation and maintenance of the facility, subject to nondiscrimination and all other requirements which are applicable to the borrower, which are to be specified in the agreement between the parties; and

(ii) Pay the association annually an amount sufficient to enable it to meet all its obligations, including reserve account requirements.

(2) The Rural Development borrower will:

(i) Meet its debt service and reserve account requirements to Rural Development;

(ii) Retain its corporate existence until Rural Development has been paid in full; and

(iii) If agreed upon by both parties, convey title to the facility to the urban community when the Rural Development debt has been paid in full.

(d) *Processing.* (1) Sale of a borrower's assets will be handled in accordance with § 1951.226 of this subpart.

(2) Transfer and assumption of a borrower's assets and indebtedness will be handled in accordance with § 1951.230 of this subpart.

(3) Lease-operation-to-purchase arrangements are not permitted.

(4) When a lease-purchase arrangement is proposed, the State Director will obtain a proposed agreement drafted by either the borrower or the urban community. The following will be forwarded to the Administrator, Attention: Water and Waste Disposal Division, for review and approval authorization:

(i) A copy of the proposed agreement;

(ii) Exhibit A of this subpart (available in any Rural Development office), appropriately completed;

(iii) OGC comments;

(iv) The case file, including all documentation appropriate for the type of servicing action involved.

[55 FR 4399, Feb. 8, 1992, as amended at 57 FR 21199, May 19, 1992]

§§ 1951.233–1951.239 [Reserved]

§ 1951.240 State Director's additional authorizations and guidance.

(a) *Promote financing purposes and improve or maintain collectibility.* The State Director is authorized to perform the following functions when the action is determined likely to promote the loan or grant purposes without jeopardizing collectibility of the loan or imparing the adequacy of the security; will strengthen the security; or will facilitate, improve, or maintain the orderly collection of the loan:

(1) Approve requests for permission to modify bylaws, articles of incorporation, or other rules and regulations of recipients, including changes in rate or fee schedules. Changes affecting the recipient's legal organizational structure must be approved by OGC.

(2) Consent to requests by the recipient to incur additional indebtedness, subject to applicable Rural Development instructions and covenants in the loan or grant agreement.

(3) Renew existing security instruments.

(4) Approve the extension or expansion of facilities and services.

(5) Require additional security when:

(i) Existing security is inadequate and the loan or security instruments obligate the borrower to give additional security; or

(ii) The loan is in default and additional security is acceptable in lieu of other servicing actions.

(6) Release properties being sold by the borrower from mortgages securing Rural Renewal loans if the amount of the notes and mortgages given by the purchaser to the borrower equal the present market value and are assigned and pledged to Rural Development, and any money payable to the borrower is applied as an extra payment on the Rural Renewal loan.

(7) Approve requests for rights-of-way and easements and any subordination necessary in connection with such requests.

(b) *Referrals to National Office.* All proposed servicing actions which the State Director is not authorized by this subpart to approve will be referred to the National Office.

(c) *Defeasance of Rural Development indebtedness.* Defeasance is the use of invested proceeds from a new bond issue to repay outstanding bonds in accordance with the repayment schedule of the outstanding bonds. The new issue supersedes the contractual agreements the borrower agreed to in the prior issue. Defeasance, or amending outstanding loan instruments and agreements to permit defeasance, of Rural Development debt instruments is not authorized, since defeasance limits, or eliminates entirely, the borrower's ability to comply with statutory refinancing requirements implemented by subpart F of part 1951 of this chapter.

§ 1951.241 Special provision for interest rate change.

(a) *General.* Effective October 1, 1981, and thereafter, upon request of the borrower, the interest rate charged by Rural Development to water and waste disposal and community facility borrowers shall be the lower of the rates in effect at either the time of loan approval or loan closing. Pub. L. 99–88 provides that any Rural Development grant funds associated with such loans shall be set in the amount based on the interest rate in effect at the time of loan approval. Loans closed October 1, 1981, through October 25, 1985, were closed at the interest rate in effect at the time of loan approval and that interest rate is reflected in the borrower's debt instrument. For community facility and water and waste disposal loans closed on or after October 1, 1981, and for which the interest rate in effect at the time of loan closing is lower than the interest rate in effect at the time of loan approval, the borrower may request to be charged the lower interest rate. The loan closing interest rate will be determined by Rural Development based upon requirements in effect at the date of loan closing. Exhibit E of this subpart (available in any Rural Development office) contains a summary of interest rate requirements for specific time periods. Exhibit C of Subpart O of this part (available in any Rural Development office) will be used to determine the interest rate and effective dates by category of poverty, intermediate, and market rates. Exhibit F of this subpart (available in any

Rural Development office) contains the instructions on how to process a change of interest rate. Loans meeting the criteria of this section that have been paid in full are eligible for the borrower to request the lower interest rate. For loan(s) that involved multiple advances of Rural Development funds using temporary debt instruments, wherein the borrower requests the interest rate in effect at loan closing, the interest rate charged shall be the rate in effect on the date when the first temporary debt instrument was issued.

(b) *Notification to borrower and borrower selection of interest rate.* (1) Rural Development servicing officials will notify each borrower meeting the provisions of this section of the availability of a choice of interest rate. The notification will be made in writing at the earliest possible date, utilizing Exhibit G of this subpart (available in any Rural Development office), and sent by certified mail, return receipt requested. Borrowers will be advised at the time of notification that if a change of interest rate is requested, the change will be accomplished administratively by Rural Development. The effect of the change on the loan account will also be fully explained to the borrower.

(2) Borrowers must notify Rural Development within 90 calendar days of the date of Rural Development notification indicating their election to retain the rate in effect at loan approval or to change the rate to the rate in effect at the time of loan closing. If the borrower does not respond within the 90-day period, Rural Development will not consider a future request for a lower interest rate under the provisions of this subpart.

(3) The borrower is responsible for assuring that the official executing the letter requesting the change of interest rate is duly authorized and any action(s) necessary for this authorization have been taken as required. Any costs associated with a change of interest rate will be the responsibility of the borrower.

(c) *Processing loan interest rate change.* The State Director is authorized to approve loan interest rate changes which meet the requirements of this section.

Loan interest rate changes will be accomplished as follows:

(1) All loan payments already applied to the account(s) will be reversed and reapplied by Rural Development utilizing the changed interest rate. The balance remaining after the completion of the reversal and reapplication procedures will be applied first to any delinquency on the account and then to principal.

(2) For paid-in-full accounts which meet the criteria of § 1951.241(a) of this subpart, the balance of loan payments after completion of the reversal and reapplication procedures will be returned to the borrower unless the borrower is delinquent on another Rural Development loan of the same type. In those cases the amount will be applied to the delinquent amount owed, with any balance refunded to the borrower.

(3) The Finance Office will administratively change the interest rate on a borrower's account in accordance with notification from the servicing official. The installment schedule set forth in each borrower's debt instrument will not change. The original principal schedule for principal-plus-interest accounts where principal *only* is stipulated will continue to be used for payment calculation by the Finance Office. Amortized accounts will adhere to the original payment schedule and amount. The last scheduled principal installment will be reduced by the amount of the balance previously generated by the reversal and reapplication of payments.

(4) When Rural Development has processed a change of interest rate for an amortized loan and a reduction in installment amounts is needed to provide for a sound operation, the borrower may request reamortization in accordance with § 1951.223 of this subpart.

(5) The borrower will be notified in writing of the new interest rate as changed.

§ 1951.242 Servicing delinquent Community Facility loans.

(a) For the purpose of this section, a loan is delinquent when a borrower fails to make all or part of a payment by the due date.

(b) The delinquent loan borrower and the Agency, at its discretion, may enter into a written workout agreement.

(c) For loans that are delinquent, the borrower must provide, monthly comparative financial statements in a format that is acceptable to the Agency by the 15th day of the following month. The Agency may waive this requirement if it would cause a hardship for the borrower or the borrower is actively marketing the security property.

[69 FR 70884, Dec. 8, 2004]

§§ 1951.243–1951.249 [Reserved]

§ 1951.250 OMB control number.

The reporting and recordkeeping requirements contained in this regulation have been approved by the Office of Management and Budget and have been assigned OMB Control Number 0575–0066. Public reporting burden for this collection of information is estimated to vary from fifteen minutes to three hours per response including time for reviewing instructions, searching existing data sources, gathering and maintaining the data needed, and completing and reviewing the collection of information.

[55 FR 4399, Feb. 8, 1990, as amended at 69 FR 70884, Dec. 8, 2004]

EXHIBITS TO SUBPART E OF PART 1951

EDITORIAL NOTE: Exhibits A through H are not published in the Code of Federal Regulations.

EXHIBIT A—REPORT ON SERVICING ACTION

EXHIBIT B—AGREEMENT FOR NEW MEMBER (WITH OR WITHOUT WITHDRAWING MEMBER)

EXHIBIT C—AGREEMENT FOR WITHDRAWAL OF MEMBER (WITHOUT NEW MEMBER)

EXHIBIT D—ITEMS TO BE INCLUDED IN TRANSFER AND ASSUMPTION DOCKETS (IF APPLICABLE)

EXHIBIT E—INTEREST RATE REQUIREMENTS AND EFFECTIVE DATES

EXHIBIT F—INSTRUCTION TO FMHA OR ITS SUCCESSOR AGENCY UNDER PUBLIC LAW 103–354 PERSONNEL TO IMPLEMENT PUBLIC LAW 100–233

EXHIBIT G—LETTER TO BORROWER NOTIFYING OF CHOICE OF INTEREST RATE

EXHIBIT H—RESCHEDULING AGREEMENT—PUBLIC BODIES

Subpart F—Analyzing Credit Needs and Graduation of Borrowers

SOURCE: 61 FR 35927, July 9, 1996, unless otherwise noted.

§ 1951.251 Purpose.

This subpart prescribes the policies to be followed when analyzing a direct borrower's need for continued Agency supervision, further credit, and graduation. All loan accounts will be reviewed for graduation in accordance with this subpart, with the exception of Guaranteed, Rural Development Loan Funds, and Rural Rental Housing loans made to build or acquire new units pursuant to contracts entered into on or after December 15, 1989, and Intermediary Relending Program loans. The term "Agency" used in this subpart refers to theRural Housing Service (RHS), or Rural Business-Cooperative Service (RBS), depending upon the loan program discussed herein. This subpart does not apply to Farm Service Agency, Farm Loan Programs and to RHS direct single family housing (SFH) customers. In addition, this subpart does not apply to Water and Waste Programs of the Rural Utilities Service, Watershed loans, Resource Conservation and Development loans, which are serviced under part 1782 of this title.

[72 FR 55018, Sept. 28, 2007, as amended at 72 FR 64123, Nov. 15, 2007]

§ 1951.252 Definitions.

Commercial classified. The Agency's highest quality Farm Credit Programs (FCP) accounts. The financial condition of the borrowers is strong enough to enable them to absorb the normal adversities of agricultural production and marketing. There is ample security for all loans, there is sufficient cash flow to meet the expenses of the agricultural enterprise and the financial needs of the family, and to service debts. The account is of such quality that commercial lenders would likely view the loans as a profitable investment.

Farm Credit Programs (FCP) loans. FSA Farm Ownership (FO), Operating (OL), Soil and Water (SW), Recreation (RL), Emergency (EM), Economic

Emergency (EE), Economic Opportunity (EO), Special Livestock (SL), Softwood Timber (ST) loans, and Rural Housing loans for farm service buildings (RHF).

Graduation, FCP. The payment in full of all FCP loans or all FCP loans of one type (i.e., all loans made for chattel purposes or all loans made for real estate purposes) by refinancing with other credit sources either with or without an Agency loan guarantee. A loan made for both chattel and real estate purposes, for example an EM loan, will be classified according to how the majority of the loan's funds were expended. Borrowers must continue with their farming operations to be considered as graduated.

Graduation, other programs. The payment in full of any direct loan for Community and Business Programs, and all direct loans for housing programs, before maturity by refinancing with other credit sources. Graduated housing borrowers must continue to hold title to the property. Graduation, for other than FCP, does not include credit which is guaranteed by the United States.

Prospectus, FCP. Consists of a transmittal letter with a current balance sheet and projected year's budget attached. The applicant's or borrower's name and address need not be withheld from the lender. The prospectus is used to determine lender interest in financing or refinancing specific Agency direct loan applicants and borrowers. The prospectus will provide information regarding the availability of an Agency loan guarantee and interest assistance.

Reasonable rates and terms. Those commercial rates and terms which borrowers are expected to meet when borrowing for similar purposes and similar periods of time. The "similar periods of time" of available commercial loans will be measured against, but need not be the same as, the remaining or original term of the loan. In the case of Multi-Family Housing (MFH) loans, "reasonable rates and terms" would be considered to mean financing that would allow the units to be offered to eligible tenants at rates consistent with other multi-family housing.

Servicing official. The district or county office official responsible for the immediate servicing functions of the borrower.

Standard classified. These loan accounts are fully acceptable by Agency standards. Loan risk and potential loan servicing costs are higher than would be acceptable to other lenders, but all loans are adequately secured. Repayment ability is adequate, and there is a high probability that all loans will be repaid as scheduled and in full.

§ 1951.253 Objectives.

(a) [Reserved]

(b) Borrowers must graduate to other credit at reasonable rates and terms when they are able to do so.

(c) If a borrower refuses to graduate, the account will be liquidated under the following conditions:

(1) The borrower has the legal capacity and financial ability to obtain other credit.

(2) Other credit is available from a commercial lender at reasonable rates and terms. In the case of Labor Housing (LH), Rural Rental Housing (RRH), and Rural Cooperative Housing (RCH) Programs, reasonable rates and terms must also permit the borrowers to continue providing housing for low and moderate income persons at rental rates tenants can afford considering the loss of any subsidy which will be canceled when the loan is paid in full.

(d) The Agency will enforce borrower graduation.

§ 1951.254 [Reserved]

§ 1951.255 Nondiscrimination.

All loan servicing actions described in this subpart will be conducted without regard to race, color, religion, sex, familial status, national origin, age, or physical or mental handicap.

§§ 1951.256-1951.261 [Reserved]

§ 1951.262 Farm Credit Programs—graduation of borrowers.

(a)–(d) [Reserved]

(e) *Graduation candidates.* Borrowers who are classified "commercial" or "standard" are graduation candidates. At least every 2 years, all borrowers who have a current classification of commercial or standard must submit a year-end balance sheet, actual financial performance information for the most recent year, and a projected budget for the current year to enable the Agency to reclassify their status and determine their ability to graduate.

(f) *Sending prospectus information to lenders.* (1) The Agency will distribute a borrower's prospectus to local lenders for possible refinancing. The borrower's permission is not required, however, the borrower must be notified of this action.

(2) The borrower is responsible for any application fees. The borrower has 30 days from the date the borrower is notified of lender interest in refinancing to make application, if required by the lender, and refinance the FLP loan. For good cause, the borrower may be granted a reasonable amount of additional time by the Agency.

[61 FR 35927, July 9, 1996, as amended at 62 FR 10120, Mar. 5, 1997]

§ 1951.263 Graduation of non-Farm Credit programs borrowers.

(a)–(b) [Reserved]

(c) *The thorough review.* Borrowers are required to supply such financial information as the Agency deems necessary to determine whether they are able to graduate to other credit. At a minimum, the financial statements requested from the borrower must include a balance sheet and a statement of income and expenses. Ordinarily, the financial statements will be those normally required at the end of the particular borrower's fiscal year. For borrowers who are not requested to furnish audited financial statements, the balance sheet and statement of income and expenses may be of the borrower's own format if the borrower's financial situation is accurately reflected. The borrower has 60 days for group type loans and 30 days for individual type loans to supply the financial information requested.

(d) [Reserved]

(e) *Requesting the borrower to graduate.* (1) The Agency will send written notice to borrowers found able to graduate requesting them to graduate. The borrower must seek a loan only in the amount necessary to repay the unpaid balance.

(2) Borrowers must provide evidence of their ability or inability to graduate within 30 days for RH borrowers, and 90 days for group type borrowers, after the date of the request. The Agency may allow additional time for good cause, for example when a borrower expects to receive income in the near future for the payment of accounts which would substantially reduce the amount required for refinancing, or when a borrower is a public body and must issue bonds to accomplish graduation.

(3) If a borrower is unable to graduate the full amount of the loan, the borrower must furnish evidence to the Agency, showing:

(i) The names of other lenders contacted;

(ii) The amount of loan requested by the borrower and the amount, if any, offered by the lenders;

(iii) The rates and terms offered by the lenders or the specific reasons why other credit is not available; and

(iv) The purpose of the loan request.

(4) The difference in interest rates between the Agency and other lenders will not be sufficient reason for failure to graduate if the other credit is available at rates and terms which the borrower can reasonably be expected to pay. An exception is made where there is an interest rate ceiling imposed by Federal law or contained in the note or mortgage.

(5) The Agency will notify the borrower in writing if it determines that the borrower can graduate. The borrower must take positive steps to graduate within 15 days for individual loans and 60 days for group loans from such notice to avoid legal action. The servicing official may grant a longer period where warranted.

§ 1951.264 Action when borrower fails to cooperate, respond or graduate.

(a) When borrowers with other than FCP loans fail to:

(1) Provide information following receipt of both FmHA Guide Letters 1951–1 and 1951–2 (available in any Agency office), or letters of similar format, they are in default of the terms of their

security instruments. The approval official may, when appropriate, accelerate the account based on the borrower's failure to perform as required by this subpart and the loan and security instruments.

(2) Apply for or accept other credit following receipt of both FmHA Guide Letters 1951– 5 and 1951–6 (available in any Agency office), or letters of similar format, they are in default under the graduation requirement of their security instruments. If the Agency determines the borrower is able to graduate, foreclosure action will be initiated in accordance with § 1955.15(d)(2)(ii). If the borrower's account is accelerated, the borrower may appeal the decision.

(b) If an FCP borrower fails to cooperate after a lender expresses a willingness to consider refinancing the Agency loan, the account will be referred for legal action.

§ 1951.265 Application for subsequent loan, subordination, or consent to additional indebtedness from a borrower who has been requested to graduate.

(a) Any borrower who appears to meet the local commercial lending standards, taking into consideration the Agency's loan guarantee program, will not be considered for a subsequent loan, subordination, or consent to additional indebtedness until the borrower's ability or inability to graduate has been confirmed. An exception may be made where the proposed action is needed to alleviate an emergency situation, such as meeting applicable health or sanitary standards which require immediate attention.

(b) If the borrower has been requested to graduate and has also been denied a request for a subsequent loan, subordination, or consent to additional indebtedness, the borrower may appeal both issues.

§ 1951.266 Special requirements for MFH borrowers.

All requirements of 7 CFR part 3560, subpart K must be met prior to graduation and acceptance of the full payment from an MFH borrower.

[69 FR 69105, Nov. 26, 2004]

§§ 1951.267–1951.299 [Reserved]

§ 1951.300 OMB control number.

The reporting requirements contained in this regulation have been approved by the Office of Management and Budget (OMB) and have been assigned OMB control number 0575–0093.

EXHIBIT A TO SUBPART F OF PART 1951
[RESERVED]

EXHIBIT B TO SUBPART F OF PART 1951—
SUGGESTED OUTLINE FOR SEEKING
INFORMATION FROM LENDERS ON
CREDIT CRITERIA FOR GRADUATION
OF SINGLE FAMILY HOUSING LOANS

Date: _____
Name of Lender: _____
Title: _____
Address: _____
Name of County Supervisor: _____
Service Area: _____

1. Is the lender interested in making loans to refinance rural housing borrowers? Yes:___; No:___.
If later, when? _____

How much credit does the lender expect to have available in the next three to four months for making such loans? $_____
In the next twelve (12) months? $_____

2. What are the loan terms? _____

3. What is the current interest rate? _____ □ Variable rate. □ Fixed rate.
If variable, how is it determined? _____

4. Is a risk differential used in establishing interest rates charged for new customers? Yes: ___; No: ___.
If yes, explain: _____

5. What can a typical loan applicant be expected to pay for:

	Dollars	Or percent
a. Filing an application.		
b. Real estate appraisal.		
c. Credit report.		
d. Loan orgination fee.		
e. Loan closing costs	....	....

6. Is mortgage guarantee insurance required? Yes: ___; No: ___. If yes, how many years? ___. Cost? _____.

7. Is there a minimum or maximum loan size policy? Yes: ___; No: ___.
If yes, explain: _____

8. Is there a minimum and maximum home value the lender will loan on? Yes: ___; No: ___. If yes, minimum: $_____; maximum: $_____.

9. Does the lender use a loan to market value ratio? _____

10. Is there a minimum net and gross income criteria? Yes: ___; No: ___. If yes, net: $_____; gross: $_____.

11. Does the lender use a minimum loan or home value to income ratio? Yes: ___; No: ___. If yes, loan to income ratio: _____ Value to income ratio: _____

12. Is there a percentage of gross income a typical applicant should have available to pay housing costs? _____

a. To pay for principal, interest, taxes and insurance (PITI)? ___%.

b. To pay for the total housing costs and other credit obligations? ___%.

13. Are there any age of home, housing type, site size, and/or geographic restriction policies? Yes: ___; No: ___. If yes, List: _____

14. Other Comments: _____

15. For the purpose of reducing the number of inappropriate referrals, would the lender like the opportunity to review specific borrower financial information prior to the borrower being asked to file a formal application? Yes: ___; No: ___. If the answer is yes, *only* those borrowers who are listed on Form RD 1951–24 will be referred to the bank. The lenders should be advised, however, the information supplied to them will not include the borrower's name, social security number, exact address, or place of employment that could be used to link a specific borrower to the information being provided by Rural Development.

[48 FR 40203, Sept. 6, 1983; 48 FR 41142, Sept. 14, 1983]

Subparts G–N [Reserved]

Subpart O—Servicing Cases Where Unauthorized Loan(s) or Other Financial Assistance Was Received—Community and Insured Business Programs

Source: 71 FR 75852, Dec. 19, 2006, unless otherwise noted.

§ 1951.701 Purpose.

This subpart prescribes the policies and procedures for servicing Community and Business Program loans and/or grants made by Rural Development when it is determined that the borrower or grantee was not eligible for all or part of the financial assistance received in the form of a loan, grant, or subsidy granted, or any other direct financial assistance. It does not apply to guaranteed loans. Loans sold without

insurance by Rural Development to the private sector will be serviced in the private sector and will not be serviced under this subpart. The provisions of this subpart are not applicable to such loans. Future changes to this subpart will not be made applicable to such loans. This subpart does not apply to Water and Waste Programs of the Rural Utilities Service, Watershed loans, and Resource Conservation and Development Loans, which are serviced under part 1782 of this title.

[72 FR 55018, Sept. 28, 2007]

§ 1951.702 Definitions.

As used in this subpart, the following definitions apply:

Active borrower. A borrower who has an outstanding account in the records of the Office of the Deputy Chief Financial Officer (ODCFO), including collection-only or an unsatisfied account balance where a voluntary conveyance was accepted without release from liability of foreclosure did not satisfy the indebtedness.

Assistance. Finance assistance in the form of a loan, grant, or subsidy received.

Debt instrument. Used as a collective term to include promissory note, assumption agreement, grant agreement, or bond.

False information. Information, known to be incorrect, provided with the intent to obtain benefits which would not have been obtainable based on correct information.

Inaccurate information. Incorrect information provided inadvertently without intent to obtain benefits fraudulently.

Inactive borrower. A former borrower whose loan(s) has been paid in full or assumed by another party(ies) and who does not have an outstanding account in the records of the ODCFO.

Recipient. "Recipient" refers to an individual or entity that received a loan, or portion of a loan, an interest subsidy, a grant, or a portion of a grant which was unauthorized.

Rural Development. A mission area within the U.S. Department of Agriculture consisting of the Office of the Under Secretary for Rural Development, Office of Community Development, Rural Business-Cooperative

Service, Rural Housing Service, and Rural Utilities Service and their successors.

Unauthorized assistance. Any loan, interest subsidy, grant, or portion thereof received by a recipient for which there was no regulatory authorization or for which the recipient was not eligible. Interest subsidy includes subsidy benefits received because a loan was closed at a lower interest rate than that to which the recipient was entitled, whether the incorrect interest rate was selected erroneously by the approval official or the documents were prepared in error.

§ 1951.703   Policy.

When unauthorized assistance has been received, an expeditious effort must be made to collect from the recipient the sum which is determined to be unauthorized, regardless of amount.

§§ 1951.704–1951.705   [Reserved]

§ 1951.706   Initial determination that unauthorized assistance was received.

Unauthorized assistance may be identified through audits conducted by the USDA Office of Inspector General (OIG), through reviews made by Rural Development personnel, or through other means such as information provided by a private citizen who documents that unauthorized assistance has been received by a recipient of Rural Development assistance.

§ 1951.707   Determination of the amount of unauthorized assistance.

(a) *Unauthorized loan amount.* The unauthorized loan amount will be the unauthorized principal plus any interest accruing on the unauthorized principal at the note interest rate until the date paid unless otherwise agreed in writing by Rural Development.

(b) *Unauthorized grant amount.* The unauthorized amount will be the unauthorized grant amount actually expended under the grant agreement plus interest accrued beginning on the date of the demand letter at the interest rate stipulated in the applicable grant agreement, or, if none is stated, the default rate established by the U.S. Department of the Treasury, until the

date paid unless otherwise agreed in writing by Rural Development.

§ 1951.708   Notification to recipient.

(a) Upon determination that unauthorized assistance was received, Rural Development will send a demand letter to the recipient that:

(1) Specifies the amount of unauthorized assistance, including any accrued interest to be repaid, and the standards for imposing accrued interest;

(2) States the amount of penalties and administrative costs to be paid, the standards for imposing them, and the date on which they will begin to accrue;

(3) Provides detailed reason(s) why the assistance was determined to be unauthorized;

(4) States the amount is immediately due and payable to Rural Development;

(5) Describes the rights the recipient has for seeking review of Rural Development's determination pursuant to 7 CFR part 11;

(6) Describes the Agency's available remedies regarding enforced collection, including referral of debt delinquent more than 180 days for Federal salary, benefit, and tax offset under the Department of Treasury Offset Program (TOP); and

(7) Provides an opportunity for the recipient to meet with Rural Development to provide facts, figures, written records, or other information which might refute Rural Development's determination.

(b) If the recipient meets with Rural Development, Rural Development will outline to the recipient why the assistance was determined to be unauthorized. The recipient will be given an opportunity to provide information to refute Rural Development's findings. When requested by the recipient, Rural Development may grant additional time for the recipient to assemble documentation. Such extension of time for payment will be valid only if Rural Development documents the extension in writing and specifies the period in days during which period the payment obligation created by the demand letter (but not the ongoing accrual of interest) will be suspended. Interest and other charges will continue to accrue pursuant to the demand letter during any extension period unless the terms of the demand letter are modified in writing by Rural Development.

(c) Unless Rural Development modifies the original demand, it will remain in full force and effect.

§1951.709 Decision on servicing actions.

(a) *Payment in full.* If the recipient agrees with Rural Development's determination or will pay the amount in question, Rural Development may allow a reasonable period of time (usually not to exceed 90 days) for the recipient to arrange for repayment. The amount due will be determined according to §1951.707.

(b) *Continuation with recipient.* If the recipient agrees with Rural Development's determination or is willing to pay the amount in question but cannot repay the unauthorized assistance within a reasonable period of time, continuation is authorized and servicing actions outlined in §1951.711 may be taken provided all of the following conditions are met:

(1) The recipient did not provide false information as defined in §1951.702.

(2) It would be highly inequitable to require prompt repayment of the unauthorized assistance.

(3) Failure to collect the unauthorized assistance in full will not adversely affect Rural Development's financial interest.

(c) *Appeals.* Appeals resulting from the letter prescribed in §1951.708 will be handled according to 7 CFR Part 11. All appeal provisions will be concluded before proceeding with further actions.

(d) *Liquidation of loan(s) or legal action to enforce collection.* When a case cannot be handled according to the provisions of paragraph (a) or (b) of this section, or if the recipient refuses to execute the documents necessary to establish an obligation to repay the unauthorized assistance as provided in §1951.711, one or more of the following actions will be taken:

(1) *Active borrower with a secured loan.* (i) Rural Development will attempt to have the recipient liquidate voluntarily. If the recipient does not agree to voluntary liquidation, or agrees but it cannot be accomplished within a reasonable period of time (usually not

more than 90 days), forced liquidation action will be initiated in accordance with applicable provisions of subpart A of part 1955 of this chapter unless:

(A) The amount of unauthorized assistance outstanding, including principal, accrued interest, and any recoverable costs charged to the account, is less than $1,000; or

(B) It would not be in the best financial interest of the Government to force liquidation.

(ii) When all of the conditions of paragraph (a) or (b) of this section are met, but the recipient does not repay or refuses to execute documents to effect necessary account adjustments according of the provisions of §1951.711, forced liquidation action will be initiated as provided in paragraph (d)(1)(i) of this section.

(iii) When forced liquidation would be initiated, except that the loan is being handled in accordance with paragraph (d)(1)(i)(A) or (d)(1)(i)(B) of this section, continuation with the loan on existing terms may be provided.

(iv) If the debt is not otherwise resolved, Rural Development will take appropriate debt collection actions in accordance with 7 CFR Part 3, subparts B and C, and the Federal Claims Collection Standards at 31 CFR Chapter IX, Parts 900–904.

(2) *Grantee, inactive borrower, or active borrower with unsecured loan (such as collection-only, or unsatisfied balance after liquidation).* Rural Development may pursue all reasonable legal remedies.

§1951.710 [Reserved]

§1951.711 Servicing options in lieu of liquidation or legal action to collect.

When the conditions outlined in §1951.709(b) are met, the servicing options outlined in this section will be considered.

(a) *Continuation on modified terms.* When the recipient has the legal and financial capabilities, the case will be serviced according to one of the following, as appropriate.

(1) *Unauthorized loan.* A loan for the unauthorized amount determined according to §1951.707(a) will remain accelerated per the demand letter sent in

accordance with § 1951.708 unless modified terms are timely reached with the recipient and accrued at the interest rate specified in the outstanding debt instrument or at the present market interest rate, whichever is greater, for the respective Community and Business program area. The loan will be amortized per a repayment schedule satisfactory to Rural Development, but in no event may the revised repayment schedule exceed a period of fifteen (15) years, the remaining term of the original loan, or the remaining useful life of the facility, whichever is shorter.

(2) *Unauthorized grant.* The unauthorized grant amount determined according to § 1951.707(b) will be converted to an account receivable, with interest payable at the market interest rate for the respective Community Facilities or Business and Industry Program area in effect on the date the financial assistance was provided. In all cases, the receivable will be amortized per a repayment schedule satisfactory to Rural Development, but in no event may the amortization period exceed fifteen (15) years. The recipient will be required to execute a debt instrument to evidence this receivable, and the best security position available to adequately protect Rural Development's interest during the repayment period will be taken as security.

(3) *Unauthorized subsidy benefits received.* When the recipient was eligible for the loan but should have been charged a higher interest rate than that in the debt instrument, which resulted in the receipt of unauthorized subsidy benefits, the case will be handled as follows:

(i) The recipient will be given the option to submit a written request that the interest rate be corrected to the lower of the rate for which they were eligible that was in effect at the date of loan approval or loan closing.

(ii) Any accrued unauthorized subsidy will be handled in accordance with § 1951.709.

(b) *Continuation on existing terms.* When the recipient does not have the legal and/or financial capabilities for the options outlined in paragraph (a)(1), (a)(2), or (a)(3) of this section, the recipient may be allowed to continue to meet the loan obligations out-

lined in the existing loan instruments. Rural Development will not continue with unauthorized grants on existing terms.

§§ 1951.712–1951.716 [Reserved]

§ 1951.717 Exception authority.

The Administrator may, in individual cases, make an exception to any requirement or provision of this subpart, provided that any such exception is not inconsistent with any applicable law or opinion of the Comptroller General, and provided further, the Administrator determines that the application of the requirement or provision would adversely affect the Government's interest.

§§ 1951.718–1951.750 [Reserved]

Subparts P–Q [Reserved]

Subpart R—Rural Development Loan Servicing

SOURCE: 53 FR 30656, Aug. 15, 1988, unless otherwise noted.

§ 1951.851 Introduction.

(a) This subpart contains regulations for servicing or liquidating loans or other assistance made by the Rural Business-Cooperative Service or its successor agency under the IRP and the RMAP. All debt settlement cases under this subpart will be settled in accordance with the debt settlement provisions set forth in 7 CFR part 1956, subpart C. The provisions of this subpart supersede conflicting provisions of any other subpart.

(b) This subpart also contains regulations for servicing the existing Rural Development Loan Fund (RDLF) loans previously approved and administered by the U.S. Department of Health and Human Services (HHS) under 45 CFR part 1076. This action is needed to implement the provisions of Section 1323 of the Food Security Act of 1985, Pub. L. 99–198, which provides for the transfer of the loan servicing authority for those loans from the HHS to the U.S. Department of Agriculture (USDA).

(c) These regulations do not negate contractual arrangements that were previously made by the HHS, Office of

Community Services (OCS), or the intermediaries operating relending programs that have already been entered into with ultimate recipients under previous regulations.

(d) The loan program is administered by the Rural Development National Office. The Director, Business and Industry Division, is the point of contact for servicing activities unless otherwise delegated by the Administrator.

[53 FR 30656, Aug. 15, 1988, as amended at 79 FR 31847, June 3, 2014; 80 FR 13201, Mar. 13, 2015]

§1951.852 Definitions and abbreviations.

(a) *General definitions.* The following definitions are applicable to the terms used in this subpart.

(1) *Intermediary* (Borrower). The entity receiving Rural Development loan funds for relending to ultimate recipients. Rural Development becomes an intermediary in the event it takes over loan servicing and/or liquidation.

(2) *Loan Agreement.* The signed agreement between Rural Development and the intermediary setting forth the terms and conditions of the loan.

(3) *Low-income.* The level of income of a person or family which is at or below the Poverty Guidelines as defined in section 673(2) of the Community Services Block Grant Act (42 U.S.C. 9902(2)).

(4) *Market value.* The most probable price which property should bring, as of a specific date in a competitive and open market, assuming the buyer and seller are prudent and knowledgeable, and the price is not affected by undue stimulus such as forced sale or loan interest subsidy.

(5) *Principals of intermediary.* Includes members, officers, directors, and other entities directly involved in the operation and management of an intermediary organization.

(6) *Ultimate recipient.* The entity receiving financial assistance from the intermediary. This may be interchangeable with the term "subrecipient" in some documents previously issued by HHS.

(7) *Rural area.* Includes all territory of a State that is not within the outer boundary of any city having a population of twenty-five thousand or more.

(8) *State.* Any of the fifty States, the Commonwealth of Puerto Rico, the Virgin Islands of the United States, Guam, American Samoa, and the Commonwealth of the Northern Mariana Islands.

(9) *Technical assistance or service.* Technical assistance or service is any function unreimbursed by Rural Development performed by the intermediary for the benefit of the ultimate recipient.

(10) *Working capital.* The excess of current assets over current liabilities. It identifies the liquid portion of total enterprise capital which constitutes a margin or buffer for meeting obligations within the ordinary operating cycle of the business.

(b) *Abbreviations.* The following abbreviations are applicable:

B&I—Business and Industry
CSA—Community Services Administration
EIS—Environmental Impact Statement
HHS—U.S. Department of Health and Human Services
IRP—Intermediary Relending Program
OCS—Office of Community Services
OIG—Office of Inspector General
OGC—Office of the General Counsel
RDLF—Rural Development Loan Fund
USDA—United States Department of Agriculture

[53 FR 30656, Aug. 15, 1988, as amended at 63 FR 6052, Feb. 6, 1998]

§§1951.853–1951.858 [Reserved]

§1951.859 Term of loans.

(a) No loans shall be extended for a period exceeding 30 years. Principal payments on loans will be made at least annually. The initial principal payment may be deferred not more than 3 years.

(b) The terms of loan repayment will be those stipulated in the loan agreement and/or promissory note.

§§1951.860–1951.865 [Reserved]

§1951.866 Security.

(a) *Loans from RDLF intermediaries to ultimate recipients.* Security requirements for loans from intermediaries to ultimate recipients will be negotiated between the intermediaries and ultimate recipients. Rural Development concurrence in the intermediary's security proposal is required only when security for the loan from the intermediary to the ultimate recipient will also serve as security for the Rural Development.

(b) *Additional security.* The Rural Development may require additional security at any time during the term of a loan to an intermediary if, after review and monitoring, an assessment indicates the need for such security.

(c) *Appraisals.* Real property serving as security for all loans to intermediaries and for loans to ultimate recipients serving as security for loans to intermediaries will be appraised by a qualified appraiser. For all other types of property, a valuation shall be made using any recognized, standard technique for the type of property involved (including standard reference manuals), and this valuation shall be described in the loan file.

§§ 1951.867–1951.871 [Reserved]

§ 1951.872 Other regulatory requirements.

Intergovernmental consultation. The RDLF program is subject to the provisions of Executive Order 12372 which requires intergovernmental consultation with State and local officials. For each ultimate recipient to be assisted with a loan under this subpart and for which the State in which the ultimate recipient is to be located has elected to review the program under their intergovernmental review process, the State Point of Contact must be notified. Notification, in the form of a project description, can be initiated by the intermediary or the ultimate recipient. Any comments from the State must be included with the intermediary's request to use the loan funds for the ultimate recipient. Prior to Rural Development's decision on the request, compliance with the requirements of intergovernmental consultation must be demonstrated for each ultimate recipient. These requirements should be carried out in accordance with the requirements set forth in U.S. Department of Agriculture regulations 2 CFR part 415, subpart C, and RD Instruction 1970–I, 'Intergovernmental Review,' available in any Agency office or on the Agency's Web site.

[79 FR 76012, Dec. 19, 2014]

§§ 1951.873–1951.880 [Reserved]

§ 1951.881 Loan servicing.

(a) These regulations do not negate contractual arrangements that were previously made by the HHS, Office of Community Services (OCS), or the intermediaries operating relending programs that have already been entered into with ultimate recipients under previous regulations. Pre-existing documents control when in conflict with these regulations. The loan is governed by terms of existing legal documents of each intermediary. The RDLF/IRP intermediary is responsible for compliance with the terms and conditions of the loan agreement. Other than 7 CFR 1951.709(d)(1)(B)(iv), intermediaries receiving an unauthorized loan or using their revolving fund for unauthorized purposes will be serviced in accordance with 7 CFR part 1951, subpart O.

(b) Each intermediary will be monitored by Rural Development based on progress reports submitted by the intermediary, audit findings, disbursement transactions, visitations, and other contract with the intermediary as necessary.

(c) Loan servicing is intended to be preventive rather than a curative action. Prompt followup on delinquent accounts and early recognition of potential problems and pursuing a solution to them are keys to resolving many problem loan cases.

(d) Written notices on payments coming due will be prepared and sent to the intermediary by the Rural Development Finance Office approximately 15 days in advance of the due date of the payments. A copy of the notice will be sent to the Rural Development Under Secretary or designee.

(e) If the scheduled payment is not made by the intermediary within 30 days after the due date of the payment, the Finance Office will send a past due notice to the intermediary. The notice will show the late charge amount, if applicable, and the interest amount past due. The late charge amount, if applicable, and the interest past due amount will be capitalized as principal due 30 days after the due date of the monthly payment unless existing loan documents prior to this regulation state otherwise. If the loan documents state when late charge amounts or interest accruals are to be capitalized, the loan documents will prevail.

(1) A per diem amount will be shown on the late notice sent to the intermediary. The Finance Office will send this notice to the Administrator or designee 30 days after the past due notice has been sent to the intermediary and the account remains delinquent. Thereafter, further notices by Rural Development designee will be sent to the intermediary on the late payments or any further payments until the account is in a current status.

(2) The Finance Office will notify the Administrator or designee on any payments due from the delinquent intermediary. It will be the responsibility of the Administrator or designee to follow up on delinquent payments to bring the account to a current status.

(3) A copy of any correspondence or notice generated by the Administrator or designee on any delinquent loan will be sent to the Finance Office.

(4) Interest will be computed on a 365-day basis unless legal documents state otherwise.

(f) It is the responsibility of the Finance Office to maintain complete accounting records for each intermediary. The Finance Office will:

(1) Coordinate with the Administrator or designee to assure that interest and principal payments received are in accordance with the promissory notes and its companion documents, and the effective amortization schedule. If the payments received appear to be incorrect, the Finance Office will advise the Administrator or designee. The Administrator or designee will take the necessary action to clear the issue and promptly advise the Finance Office of the proper accounting procedure.

(2) Send monthly statements to the National Office reflecting all payments received to date on each borrower.

(3) Send to the Administrator or designee a monthly summary of all intermediary loans as follows:

(i) Number and amount of all loans.

(ii) Total advanced on all loans.

(iii) Total interest and principal received on the loans.

(iv) Total outstanding balance on all loans.

(4) Prepare reamortization schedules needed as a result of restructuring any loans and send to the Administrator or designee.

(5) Furnish in writing to the Administrator or designee a per diem amount on the actual interest amount due when requested by the Administrator.

(g) It is the responsibility of the Administrator or designee to:

(1) Review and analyze the semiannual report of the intermediaries and reconcile same to the annual audits.

(2) Review the annual audits of intermediaries.

(3) Review the semiannual reports of the intermediaries and take appropriate action when necessary.

(4) Follow up on delinquent intermediaries to bring the account current.

(5) Notify the Finance Office in writing when a loan is determined to be uncollectible in order for the Finance Office to make provisions for an appropriate timely entry to the loss account.

(6) Furnish to the Finance Office the necessary information to produce reamortization schedules.

(7) Provide the Finance Office a copy of any correspondence in regard to the restructuring of the loans.

(8) Review reamortization schedules, the schedule will then be forwarded to the intermediary.

(9) Confirm account balances. Payment history of loans and any other related matter will be furnished to the requesting party, (i.e. third party auditing firms) if warranted and proper. If there are discrepancies in any loan balances being confirmed, the Finance Office should be consulted before the Administrator or designee writes the requested parties.

(10) Furnish upon request by the Finance Office, the information necessary to help reconcile account balances, obtain evidence of payments made by the borrower, and any other related data necessary to keep the financial records correct and in balance.

(11) Answer Congressional and other correspondence.

(12) Review intermediary's plans, cash flow projections, balance sheets, and operating statements.

[53 FR 30656, Aug. 15, 1988, as amended at 79 FR 31847, June 3, 2014]

§ 1951.882 [Reserved]

§ 1951.883 Reporting requirements.

(a) Intermediaries are to provide Rural Development with reports as required in their respective loan agreements, applicable statutes and as required by Rural Development. The report shall include the following:

(1) An annual audit; dates of audit report period need not necessarily coincide with other reports on the RDLF/IRP. Audits shall be due 90 days following the audit period. Audits must cover all of the intermediary's activities. Audits will be performed by an independent certified public accountant or by an independent public accountant licensed and certified on or before December 31, 1970, by a regulatory authority of a State or other political subdivision of the United States. An acceptable audit will be performed in accordance with generally accepted auditing standards and include such tests of the accounting records as the auditor considers necessary in order to express an opinion on the financial condition of the intermediary. Rural Development does not require an unqualified audit opinion as a result of the audit. Compilations or reviews do not satisfy the audit requirement.

(2) Quarterly or semiannual reports (due 30 days after the end of the period).

(i) Reports will be required quarterly during the first year after loan closing and, if all loan funds are not utilized during the first year, quarterly reports will be continued until at least 90 percent of the Agency IRP loan funds have been advanced to ultimate recipients. Thereafter, reports will be required semiannually. Also, the Agency may require quarterly reports if the intermediary becomes delinquent in repayment of its loan or otherwise fails to fully comply with the provisions of its work plan or Loan Agreement, or the Agency determines that the intermediary's IRP revolving fund is not adequately protected by the current sound worth and paying capacity of the ultimate recipients.

(ii) These reports shall contain only information on the IRP revolving loan fund, or if other funds are included, the IRP loan program portion shall be segregated from the others; and in the case where the intermediary has more than one IRP revolving fund from the Agency a separate report shall be made for each of the IRP revolving funds.

(iii) The reports will include, on a form provided by the Agency, information on the intermediary's lending activity, income and expenses, financial condition, and a summary of names and characteristics of the ultimate recipients the intermediary has financed.

(3) An annual report on the extent to which increased employment income and ownership opportunities are provided to low-income persons, farm families, and displaced farm families for each loan made by such intermediary.

(4) Proposed budget for the following year.

(5) Other reports as Rural Development may require from time to time.

(b) Intermediaries shall report to Rural Development whenever an ultimate recipient is more than 90 days in arrears in the repayment of principal or interest.

[53 FR 30656, Aug. 15, 1988, as amended at 63 FR 6053, Feb. 6, 1998]

§ 1951.884 Revolved funds.

For ultimate recipients assisted by the intermediary with Rural Development, revolved funds derived from IRP funds shall be required to comply with the provisions of these regulations and/or loan agreement.

§ 1951.885 Loan classifications.

All loans to intermediaries in the Rural Development portfolio will be classified by Rural Development at loan closing and again whenever there is a change in the loan which would impact on the original classification. No one classification should be viewed as more important than others. The uncollectibility aspect of Doubtful and Loss classifications is of obvious importance. However, the function of the Substandard classification is to indicate those loans that are unduly risky which may result in future losses. Substandard, Doubtful and Loss are adverse classifications. The special mention classification is for loans which are not adversely classified but which require the attention and followup of Rural Development. The loans will be classified as follows:

(a) *Seasoned loan classification.* To be classified as a seasoned loan, a loan must:

(1) Have a remaining principal loan balance of two-thirds or less of the original aggregate of all existing loans made to that intermediary.

(2) Be in compliance with all loan conditions and Rural Development regulations.

(3) Have been current on the loan(s) payments for 24 consecutive months.

(4) Be secured by collateral which is determined to be adequate to ensure there will be no loss on the loan.

(b) *Current non-problem classification.* This classification includes those loans which have been current for less than 24 consecutive months and are in compliance with the loan conditions and Rural Development regulations, and are not considered to pose a credit risk to Rural Development. These loans would be classified as seasoned but for the "24 months" and "two-thirds" requirements for seasoned loans.

(c) *Special mention classification.* This classification includes loans which do not presently expose Rural Development to a sufficient degree of risk to warrant a Substandard classification but do possess credit deficiencies deserving Rural Development's close attention because the failure to correct these deficiencies could result in greater risk in the future. This classification would include loans that may be high quality, but which Rural Development is unable to supervise properly because of an inadequate loan agreement, the condition or lack of control over the collateral, failure to obtain proper documentation or any other deviations from prudent lending practices. Adverse trends in the intermediary's operation or an imbalanced position in the balance sheet which has not reached a point that jeopardizes the repayment of the loan should be assigned to this classification. Loans in which actual, not poten-

tial, weaknesses are evident and significant should be considered for a Substandard classification.

(d) *Substandard classification.* This classification includes loans which are inadequately protected by the current sound worth and paying capacity of the obligor or of the collateral pledged, if any. Loans in this classification must have a well defined weakness or weaknesses that jeopardize the payment in full of the debt. If the deficiencies are not corrected, there is a distinct possibility that Rural Development will sustain some loss.

(e) *Doubtful classification.* This classification includes those loans which have all the weaknesses inherent in those classified Substandard with the added characteristic that the weaknesses make collection or liquidation in full, based on currently known facts, conditions and values, highly questionable and improbable.

(f) *Loss classification.* This classification includes those loans which are considered uncollectible and of such little value that their continuance as loans is not warranted. Even though partial recovery may be effected in the future, it is not practical or desirable to defer writing off these basically worthless loans.

§§ 1951.886–1951.888 [Reserved]

§ 1951.889 Transfer and assumption.

(a) All transfers and assumptions must be approved in advance in writing by Rural Development. Such transfers and assumptions must be to an eligible intermediary.

(b) Available transfer and assumption options to eligible intermediaries include the following:

(1) The total indebtedness may be transferred to another eligible intermediary on the same terms.

(2) The total indebtedness may be transferred to another eligible intermediary on different terms not to exceed those terms for which an initial loan can be made to an organization that would have been eligible originally.

(3) Less than total indebtedness may be transferred to another eligible intermediary on the same terms.

(4) Less than total indebtedness may be transferred to another eligible intermediary on different terms.

(c) The transferor will prepare the transfer document for Rural Development's review prior to the transfer and assumption.

(d) The transferee will provide Rural Development with a copy of its latest financial statement and a copy of its annual financial statement for the past 3 years if available; its Federal Tax Identification number; organizational charter; minutes from the Board of Directors authorizing the transaction; certification of good standing from the Secretary of State or whatever regulatory agency oversees nonprofit corporations for that State or Commonwealth where the entity is headquartered; and any other information that Rural Development deems necessary for its review.

(e) The assumption agreement will contain the Rural Development case nunber of the transferor and transferee.

(f) When the transferee makes a cash downpayment in connection with the transfer and assumption, any proceeds received by the transferor will be credited on the transferor's loan debt in inverse order of maturity.

(g) The Administrator or designee will approve or decline all transfers and assumptions.

§ 1951.890 Office of Inspector General and Office of General Counsel referrals.

When facts or circumstances indicate that criminal violations, civil fraud, misrepresentations, or regulatory violations may have been committed by an applicant or an intermediary, Rural Development will refer the case to the appropriate Regional Inspector General for Investigations, OIG, USDA, in accordance with RD Instruction 2012–B (available in any Rural Development office) for criminal investigation. Any questions as to whether a matter should be referred will be resolved through consultation with OIG and Rural Development and confirmed in writing. In order to assure protection of the financial and other interests of the Government, a duplicate of the notification will be sent to the OGC. OGC

will be consulted on legal questions. After OIG has accepted any matter for investigation, Rural Development staff must coordinate with OIG in advance regarding routine servicing actions on existing loans.

§ 1951.891 Liquidation; default.

(a) In the event that Rural Development takes over the servicing of the ultimate recipient of an intermediary, those loans will be serviced by this regulation and in accordance with the contractual arrangement between the intermediary and the ultimate recipient. Should Rural Development determine that it is necessary or desirable to take action to protect or further the interests of Rural Development in connection with any default or breach of conditions under any loan made hereunder, the Rural Development may:

(1) Declare that the loan is immediately due and payable.

(2) Assign or sell at public or private sale, or otherwise dispose of for cash or credit at its discretion and upon such terms and conditions as Rural Development shall determine to be reasonable, any evidence of debt, contract, claim, personal or real property or security assigned to or held by the Rural Development in connection with financial assistance extended hereunder.

(3) Adjust interest rates, use fixed or variable rates, grant moratoriums on repayment of principal and interest, collect or compromise any obligations held by Rural Development and take such actions in respect to such loans as are necessary or appropriate, consistent with the purpose of the program and this subpart. The Administrator will notify the Rural Development Finance Office of any change in payment terms, such as reamortizations or interest rate adjustments, and effective dates of any changes resulting from servicing actions.

(b) Failure by an ultimate recipient to comply with the provisions of these regulations and/or loan agreement shall constitute grounds for a declaration of default and the demand for immediate and full repayment of its loan.

(c) Failure by an intermediary to comply with the provisions of these regulations or to relend funds in accordance with an approved work plan or loan agreement shall constitute grounds for a declaration of default and the demand for immediate and full repayment of the loan.

(d) In the event of default, the intermediary will promptly be informed in writing of the consequences of failing to comply with loan covenant(s).

(e) Protective advances to the intermediary will not be made in lieu of additional loans, in particular working capital loans. Protective advances are advances made by Rural Development for the purpose of preserving and protecting the collateral where the intermediary has failed to and will not or cannot meet its obligations. The Administrator or designee must approve in writing all protective advances.

(f) In the event of bankruptcy by the intermediary and/or ultimate recipient, Rural Development is responsible for protecting the interests of the Government. All bankruptcy cases should be reported immediately to the Regional Attorney. The Administrator must approve in advance and in writing the estimated liquidation expenses on loans in liquidation bankruptcy. These expenses must be considered by Rural Development to be reasonable and customary.

(g) Liquidation, management, and disposal of inventory property will be handled in accordance with subparts A, B, and C of part 1955 of this chapter.

§§ 1951.892–1951.893   [Reserved]

§ 1951.894   Debt settlement.

Debt settlement of all claims will be handled in accordance with subpart C of part 1956 of this chapter.

[80 FR 13201, Mar. 13, 2015]

§ 1951.895   [Reserved]

§ 1951.896   Appeals.

Any appealable adverse decision made by FmHA or its successor agency under Public Law 103–354 which affects the borrower may be appealed upon written request of the aggrieved party in accordance with subpart B of part 1900 of this chapter.

§ 1951.897   Exception authority.

The Administrator may, in individual cases, grant an exception to any requirement or provision of this subpart which is not inconsistent with an applicable law or opinion of the Comptroller General, provided the Administrator determines that application of the requirement or provision would adversely affect the Government's interest. The basis for this exception will be fully documented. The documentation will: demonstrate the adverse impact; identify the particular requirement involved; and show how the adverse impact will be eliminated.

§§ 1951.898–1951.899   [Reserved]

§ 1951.900   OMB control number.

The information collection requirement obtained for this part is pending OMB approval at the time of this rule's publication in the FEDERAL REGISTER.

[81 FR 11032, Mar. 2, 2016]

PART 1955—PROPERTY MANAGEMENT

Subpart A—Liquidation of Loans Secured by Real Estate and Acquisition of Real and Chattel Property

AUTHORITY: 5 U.S.C. 301; 7 U.S.C. 1989; 42 U.S.C. 1480.

SOURCE: 50 FR 23904, June 7, 1985, unless otherwise noted.

EDITORIAL NOTE: Nomenclature changes to part 1955 appear at 80 FR 9895, Feb. 24, 2015.

Subpart A—Liquidation of Loans Secured by Real Estate and Acquisition of Real and Chattel Property

§1955.1 Purpose.

This subpart delegates authority and prescribes procedures for the liquidation of loans to individuals and to organizations as identified in §1955.3 of this subpart. It pertains to the Multi-Family Housing (MFH) and Community Facility (CF) programs of the Rural Housing Service (RHS), and direct programs of the Rural Business-Cooperative Service (RBS). Guaranteed RBS loans are liquidated upon direction from the Deputy Administrator, Business Programs, RBS. This subpart does not apply to Farm Service Agency, Farm Loan Programs, to RHS single family housing loans, or to CF loans sold without insurance in the private sector. These CF loans will be serviced in the private sector, and future revisions to this subpart no longer apply to such loans. This subpart does not apply to the Rural Rental Housing, Rural Cooperative Housing, or Farm Labor Housing Programs of RHS. In addition, this subpart does not apply to Water and Waste Programs of the Rural Utilities Service, Watershed loans, and Resource Conservation and Development loans, which are serviced under part 1782 of this title.

[72 FR 55019, Sept. 28, 2007, as amended at 72 FR 64123, Nov. 15, 2007]

§1955.2 Policy.

When it has been determined in accordance with applicable loan servicing regulations that further servicing will not achieve loan objectives and that voluntary sale of the property by the borrower (except for Multiple Family Housing (MFH) loans subject to prepayment restrictions) cannot be accomplished, the loan(s) will be liquidated through voluntary conveyance of the property to Rural Development or by foreclosure as outlined in this subpart. For MFH loans subject to the prepayment restrictions, voluntary liquidation may be accomplished only through voluntary conveyance to Rural Development in accordance with applicable portions of §1955.10 of this subpart. Nonprogram (NP) loans, except for Community and Business Programs, will be liquidated as provided in subpart J of part 1951 of this chapter, unless specifically referenced in this subpart.

[51 FR 4138, Feb. 3, 1986, as amended at 53 FR 27826, July 25, 1988; 58 FR 52652, Oct. 12, 1993]

§1955.3 Definitions.

As used in this subpart, the following definitions apply:

Closing agent. An attorney or title insurance company which is approved as a loan closing agent in accordance with subpart B of part 1927 of this chapter.

CONACT or CONACT property. Property acquired or sold pursuant to the Consolidated Farm and Rural Development Act. Within this subpart, it shall also be construed to cover property which secured loans made pursuant to the Agriculture Credit Act of 1978; the Emergency Agricultural Credit Adjustment Act of 1978; the Emergency Agricultural Credit Act of 1984; the Food Security Act of 1985; and other statutes giving agricultural lending authority to the government.

Farmer Programs loans. The term "Farmer Program loans" (FP) refers to the following types of loans: Farm Ownership (FO), Soil and Water (SW), Recreation (RL), Economic Opportunity (EO), Operating (OL), Emergency (EM), Economic Emergency (EE), Softwood Timber (ST), and Rural Housing Loans for farm service buildings (RHF).

Government. The United States of America acting through the RBS, RHS, and RUS of the U.S. Department of Agriculture;

Homestead protection. The Farmer Programs borrower-owner's right to lease with an option to purchase the principal residence located on or off the farm and up to 10 acres of adjoining land possessed and occupied by the borrower-owner, including a reasonable number of farm outbuildings located on the adjoining land that are useful to the occupants of the homestead.

Interest credit. The terms "interest credit" and "interest credit assistance," as they relate to Single Family Housing (SFH) loans, are interchangeable with the term "payment assistance." Payment assistance is the generic term for the subsidy provided to eligible SFH borrowers to reduce mortgage payments.

Loans to individuals. Farm Ownership (FO), Soil and Water (SW), Recreation (RL), Special Livestock (SL), Economic Opportunity (EO), Operating (OL), Emergency (EM), Economic Emergency (EE), Softwood Timber (ST), and Rural Housing loans for farm service buildings (RHF), whether to individuals or entities, referred to in this subpart as Farmer Programs (FP) loans; and Land Conservation and Development (LCD); and Single-Family Housing (SFH), including both Section 502 and 504 loans.

Loans to Native Americans. Farmer Program loans secured by real estate located within the boundaries of a federally recognized Indian reservation. The Native American borrower-owner is defined as the party who pledged real estate as collateral for an FP loan and is the tribe or a member of the tribe with control over the reservation.

Loans to organizations. Community Facility (CF); Water and Waste Disposal (WWD); Association Recreation; Watershed (WS); Resource Conservation and Development (RC&D); insured Business and Industrial (B&I) both to individuals and groups; Rural Development Loan Fund (RDLF); Intermediary Relending Program (IRP); Nonprofit National Corporations (NNC); loans to associations for Irrigation and Drainage (I&D) and other Soil and Water conservation measures; loans to Indian Tribes and Tribal Corporations; Shift-In-Land Use (Grazing Association); Economic Opportunity Cooperative (EOC); Rural Housing Site (RHS); Rural Cooperative Housing (RCH); Rural Rental Housing (RRH) and Labor Housing (LH) to both individuals and groups. The housing-type organization loans identified here are referred to in this subpart collectively as Multiple-family Housing (MFH) loans.

Market value. The most probable price which property should bring, as of a specific date, in a competitive and open market, assuming the buyer and seller are prudent and knowledgeable, and the price is not affected by undue stimulus such as forced sale or loan interest subsidy.

Nonrecoverable cost is a contractual or noncontractual program loan cost expense not chargeable to a borrower, property account, or part of the loan subsidy.

OGC. The Office of the General Counsel, U.S. Department of Agriculture; refers to the Regional Attorney or Attorney-in-Charge in an OGC field office unless otherwise indicated.

Prior lien. A security instrument (such as a mortgage or deed of trust) or a judgment which was of public record before the Rural Development security instrument(s) as well as real estate taxes or assessments which are or will become a lien against the property which is superior to Rural Development's security instrument(s).

Recoverable cost is a contractual or noncontractual program loan cost expense chargeable to a borrower, property account, or part of the loan subsidy.

Servicing official. For loans to individuals as defined in paragraph (d) of this section, the servicing official is the County Supervisor. For insured B&I loans, the servicing official is the State Director. For RDLF and IRP, the servicing official is the Director, Business and Industry Division. For NNC, the servicing official is the Director, Community Facility Division. For all other types of loans, the servicing official is the District Director.

[50 FR 23904, June 7, 1985, as amended at 50 FR 45782, Nov. 1, 1985; 52 FR 26138, July 13, 1987; 53 FR 27826, July 25, 1988; 53 FR 30664, Aug. 15, 1988; 53 FR 35762, Sept. 14, 1988; 56 FR 15821, Apr. 18, 1991; 56 FR 29402, June 27, 1991; 56 FR 67484, Dec. 31, 1991; 58 FR 68723, Dec. 29, 1993; 60 FR 55147, Oct. 27, 1995; 62 FR 44395, Aug. 21, 1997; 63 FR 41716, Aug. 5, 1998]

§ 1955.4 Redelegation of authority.

Authorities will be redelegated to the extent possible, consistent with program requirements and available resources.

(a) Except as provided in § 1900.6(c) of this chapter, any authority in this subpart which is specifically delegated to the Administrator or to an Deputy Administrator may only be delegated to a State Director. The State Director cannot redelegate such authority.

(b) Except as provided in paragraph (a) of this section, the State Director is authorized to redelegate, in writing, any authority delegated to the State Director in this subpart to a Program Chief, Program Specialist or Property Management Specialist on the State Office staff; except the authority to approve or disapprove foreclosure as outlined in § 1955.115(a)(2) of this subpart may not be redelegated. However, a duly-designated Acting State Director may approve or disapprove foreclosure.

(c) The District Director is authorized to redelegate, in writing, any authority delegated to the District Director in this subpart to an Assistant District Director or District Loan Specialist determined by the District Director to be qualified; except the authority to approve or disapprove foreclosure as outlined in § 1955.15(a)(1) of this subpart may not be redelegated. However, a duly designated Acting District Director may approve or disapprove foreclosure. Authority of District Directors in this subpart applies to Area Loan Specialists in Alaska and the Director for the Western Pacific Territories.

(d) The County Supervisor is authorized to redelegate, in writing, any authority delegated to the County Supervisor in this subpart to an Assistant County Supervisor, GS-7, or above, determined by the County Supervisor to be qualified. Authority of County Supervisors in this subpart applies to Area Loan Specialists in Alaska and Area Supervisors in the Western Pacific Territories and American Samoa.

(e) The monetary limitations on acceptance of voluntary conveyance as provided in § 1955.10(a) of this subpart may not be redelegated from a higher-level official to a lower level official.

[53 FR 27826, July 25, 1988, as amended at 54 FR 6875, Feb. 15, 1989; 59 FR 43441, Aug. 24, 1994; 62 FR 44395, Aug. 21, 1997]

§ 1955.5 General actions.

(a) *Assignment of notes to Rural Development.* When liquidation action is approved and the insured note is not held in the County or District Office, the approval official will request the Finance Office to purchase the note and forward it to the appropriate office. Voluntary conveyance may be closed pending receipt of the note(s), and foreclosure may also be processed pending receipt of the note(s), unless the original note is required in connection with the foreclosure action.

(b) *Execution of documents.* (1) After liquidation of loans to individuals has been approved by the appropriate official, the County Supervisor is authorized to execute all necessary forms and documents except notices of acceleration required to complete transactions covered by this subpart.

(2) After liquidation of loans to organizations has been approved by the appropriate official, the District Director is authorized to execute all forms and documents for completion of the liquidation except:

(i) Notice of acceleration; or

(ii) Other form or document which specifically required State or National Office approval because of monetary limits or policy statement established elsewhere in this subpart.

(c) *Unused loan funds.* (1) Funds remaining in a supervised bank account will be handed in accordance with § 1902.15 of subpart A of part 1902 of this chapter before a voluntary conveyance or foreclosure is processed.

(2) Funds remaining in a construction or other account will be applied to the borrower's Rural Development accounts.

(d) *Payment of costs.* Costs related to liquidation of a loan or acquisition of property will be paid according to RD Instruction 2024–A as either a recoverable or nonrecoverable cost as defined in § 1955.3 of this subpart.

(e) *Escrow funds.* Any funds remaining in the borrower's escrow account at the time of liquidation by voluntary conveyance or foreclosure are nonrefundable and will be credited to the borrower's loan account.

[50 FR 23904, June 7, 1985, as amended at 56 FR 6953, Feb. 21, 1991, 57 FR 36590, Aug. 14, 1992]

§§ 1955.6–1955.8 [Reserved]

§ 1955.9 Requirements for voluntary conveyance of real property located within a federally recognized Indian reservation owned by a Native American borrower-owner.

(a) The borrower-owner is a member of the tribe that has jurisdiction over the reservation in which the real property is located. An Indian tribe may also meet the borrower-owner criterion if it is indebted for Farm Credit Programs loans.

(b) A voluntary conveyance will be accepted only after all preacquisition primary and preservation servicing actions have been considered in accordance with subpart S of part 1951 of this chapter.

(c) When all servicing actions have been considered under subpart S of part 1951 of this chapter and a positive outcome cannot be achieved, the following additional actions are to be taken:

(1) The county official will notify the Native American borrower-owner and the tribe by certified mail, return receipt requested, and by regular mail if the certified mail is not received, that:

(i) The borrower-owner may convey the real estate security to FSA and FSA will consider acceptance of the property into inventory in accordance with paragraph (d) of this section.

(ii) The borrower-owner must inform FSA within 60 days from receipt of this notice of the borrower and owner's decision to deed the property to FSA;

(iii) The borrower-owner has the opportunity to consult with the Indian tribe that has jurisdiction over the reservation in which the real property is located, or counsel, to determine if

State or tribal law provides rights and protections that are more beneficial than those provided the borrower-owner under Agency regulations;

(2) If the borrower-owner does not voluntarily deed the property to FSA, not later than 30 days before the foreclosure sale, FSA will provide the Native American borrower-owner with the following options:

(i) The Native American borrower-owner may require FSA to assign the loan and security instruments to the Secretary of the Interior. If the Secretary of the Interior agrees to such an assignment, FSA will be released from all further responsibility for collection of any amounts with regard to the loans secured by the real property.

(ii) The Native American borrower-owner may require FSA to complete a transfer and assumption of the loan to the tribe having jurisdiction over the reservation in which the real property is located if the tribe agrees to the assumption. If the tribe assumes the loans, the following actions shall occur:

(A) FSA shall not foreclose the loan because of any default that occurred before the date of the assumption.

(B) The assumed loan shall be for the lesser of the outstanding principal and interest of the loan or the fair market value of the property as determined by an appraisal.

(C) The assumed loan shall be treated as though it is a regular Indian Land Acquisition Loan made in accordance with subpart N of part 1823 of this chapter.

(3) If a Native American borrower-owner does not voluntarily convey the real property to FSA, not less than 30 days before a foreclosure sale of the property, FSA will provide written notice to the Indian tribe that has jurisdiction over the reservation in which the real property is located of the following:

(i) The sale;

(ii) The fair market value of the property; and

(iii) The ability of the Native American borrower-owner to require the assignment of the loan and security instruments either to the Secretary of the Interior or the tribe (and the consequences of either action) as provided in § 1955.9(c)(2).

(4) FSA will accept the offer of voluntary conveyance of the property unless a hazardous substance, as defined in the Comprehensive Environmental Response, Compensation, and Liability Act of 1980, is located on the property which will require FSA to take remedial action to protect human health or the environment if the property is taken into inventory. In this case, a voluntary conveyance will be accepted only if FSA determines that it is in the best interests of the Government to acquire title to the property.

(d) When determining whether to accept a voluntary conveyance of a Native American borrower-owner's real property, the county official must consider:

(1) The cost of cleaning or mitigating the effects if a hazardous substance is found on the property. A deduction equal to the amount of the cost of a hazardous waste clean-up will be made to the fair market value of the property to determine if it is in the best interest of the Government to accept title to the property. FSA will accept the property if clear title can be obtained and if the value of the property after removal of hazardous substances exceeds the cost of hazardous waste clean-up.

(2) If the property is located within the boundaries of a federally recognized Indian reservation, and is owned by a member of the tribe with jurisdiction over the reservation, FSA will credit the Native American borrower-owner's account based on the fair market value of the property or the FSA debt against the property, whichever is greater.

[62 FR 44395, Aug. 21, 1997]

§ 1955.10 Voluntary conveyance of real property by the borrower to the Government.

Voluntary conveyance is a method of liquidation by which title to security is transferred to the Government. Rural Development will not make a demand on a borrower to voluntarily convey. If there is equity in the property, Rural Development should advise the borrower, in writing, that there is equity in the property before accepting an offer to voluntarily convey. If Rural Development receives an offer of voluntary conveyance, acceptance should only be considered when the Government will likely receive a recovery on its investment. In cases where there are outstanding liens, a full assessment should be made of the debts against the property compared to the current market value. Rural Development should refuse the voluntary conveyance, if the Rural Development lien has neither present nor prospective value or recovery of the value would be unlikely or uneconomical. Instead, for loans to individuals, Rural Development should release its lien as valueless in accordance with § 1965.25(d) of subpart A of part 1965 of this chapter or § 1965.118(c) of subpart C of this chapter, as appropriate. For non-FP borrowers, a voluntary conveyance should only be considered after all available servicing actions outlined in the respective servicing regulations have been used or considered and it is determined that the borrower will not be successful. For FP borrowers, if the borrower has not received exhibit A with attachments 1 and 2 of subpart S of part 1951 of this chapter, a voluntary conveyance should be accepted only after the borrower has been sent exhibit A with attachments 1 and 2 of subpart S of 1951 of this chapter; all available servicing actions outlined in the respective program servicing regulations have been used or considered; and it will be in the Government's best financial interest to accept the FP voluntary conveyance. Exhibit G of this subpart will be used to determine whether or not to accept an FP voluntary conveyance. In determining if the acceptance of the FP voluntary conveyance is in the best financial interest of the Government, the County Supervisor will determine if the borrower has exhausted all possibilities of restructuring the loan to where a feasible plan of operation may be developed, the borrower has acted in good faith in trying to service the debt and Rural Development may recover its investment in return for the acceptance of the voluntary conveyance. In addition, prior to acceptance of a voluntary conveyance of farm real property that collateralizes an FP loan, the

County Supervisor will remind the borrower-owner of possible deed restrictions and easement that may be placed on the property in the event the property contains wetlands, floodplains, historical sites and/or other federally protected environmental resources as set forth in part 1970 of this chapter and § 1955.137 of this part. When it is determined that all conditions of § 1951.558(b) of subpart L of part 1951 of this chapter have been met, loans for unauthorized assistance will be treated as authorized loans and exhibit A with attachments 1 and 2 of subpart S of part 1951 of this chapter will be sent prior to accepting a voluntary conveyance. Those borrowers who are indebted for nonprogram (NP) loans who wish to voluntarily convey property will not be sent exhibit A with attachments 1 and 2 of subpart S of part 1951 of this chapter. For Farmer Program borrowers who have received exhibit A with attachments 1 and 2 of subpart S of part 1951 of this chapter, a voluntary conveyance should only be accepted when it is determined to be in the Government's best financial interest. Rejection of an offer of voluntary conveyance made before or after acceleration from an FP borrower is appealable. For borrowers having both FP and non-FP loans secured by a farm tract, a voluntary conveyance should be handled as outlined above for non-FP loans secured by farm tracts, except that the applicable servicing option for the FP and non-FP loans should be considered separately. This separation of servicing options may permit a borrower to retain the nonfarm tract. For newly constructed SFH properties with major construction defects, see subpart F of part 1924 of this chapter.

(a) *Authority*—(1) *Loans to individuals*—(i) *SFH loans.* The County Supervisor is authorized to accept voluntary conveyances regardless of amount of indebtedness.

(ii) [Reserved]

(2) *Loans to organizations.* (i) The State Director is authorized to approve voluntary conveyance of property securing Farmer Programs and EOC loans regardless of amount of indebtedness.

(ii) The State Director is authorized to approve voluntary conveyance of property securing MFH loans if the total indebtedness against the property, including prior and junior liens, does not exceed his/her approval authority for the type loan involved. Loan approval authorities are outlined in exhibits A through E of RD Instruction 1901–A (available in any Rural Development office).

(iii) Offers to convey property securing loans other than those outlined in paragraphs (a)(2)(i) and (ii) of this section will be submitted to the Administrator for approval prior to acceptance of the conveyance offer. Submissions will include the case file; OGC's opinion on settling any other liens involved; a statement of essential facts; and recommendations of the State Director and Program Chief. Submissions are to be addressed to the Administrator, ATTN: (appropriate program division.)

(b) *Forms and documents.* All forms and documents in connection with voluntary conveyance will be prepared and distributed in accordance with the respective FMI or applicable OGC instructions. For loans to individuals when the County Supervisor has approval authority, the facts will be documented in the running record of the borrower's case file. For all other loans, the servicing official will submit the voluntary conveyance offer, the case file and a narrative report to the appropriate approval official.

(c) *Liens against the property other than Rural Development liens*—(1) *Prior liens.* (i) The approval official will determine whether or not prior liens will be paid. Normally, the Government will pay prior liens in full prior to acquisition if:

(A) A substantial recovery on the Government's investment plus the amount of the prior lien(s) can be obtained; and

(B) The holder of the prior lien(s) objects to the Government accepting voluntary conveyance subject to the prior lien(s), if consent of the prior lienholder(s) is required.

(ii) If property is acquired subject to prior lien(s), payment of installments on the lien(s) may be made while title to the property is held by the Government in accordance with § 1955.67 of subpart B of part 1955 of this chapter.

(2) *Junior liens.* The borrower must satisfy junior liens on the property (except Rural Development liens) and pay real estate taxes or assessments which are or will become a lien on the property. However, if the borrower is unable or unwilling to do so, settlement of the liens may be made by Rural Development if settlement would be in the best interest of the Government, considering all factors such as length of time required to foreclose, vandalism or other deterioration of the property which might occur, and effect on management of a MFH project and its tenants. An Rural Development official will contact junior lienholders, negotiate the most favorable settlement possible, and determine whether it is in the Government's best interest to settle the junior liens and accept the voluntary coveyance.

(i) For loans to individuals, the approval official is authorized to settle junior liens in the smallest amount possible, but not to exceed an aggregate amount of $1,000 in each SFH case or $5,000 for other type loans. For junior liens in greater amounts when the approval official is the County Supervisor or District Director, prior authorization must be obtained from the State Director.

(ii) For loans to organizations, the State Director will determine whether or not junior liens will be settled and voluntary conveyance accepted.

(3) *Payment of liens.* A lien to be settled in accordance with paragraph (c)(1)(i) or (c)(2) of this section will be paid as outlined in §1955.5(d) of this subpart and charged to the borrower's account as a recoverable cost.

(d) *Offer of voluntary conveyance.* An offer of voluntary conveyance will consist of the following:

(1) Form RD 1955-1, "Offer to Convey Security."

(2) Warranty deed, or other deed approved by OGC to comply with State Laws. The deed will not be recorded until it is determined the voluntary conveyance will be accepted. At the time of the offer, the borrowers will be informed that the conveyance will not be accepted until the property has been appraised and a lien search has been obtained. If the voluntary conveyance is not accepted, the deed and Form RD 1955-1, properly executed, will be returned to the borrower along with a memorandum stating the reason(s) for nonacceptance.

(3) A current financial statement containing information similar to that required to complete Forms RD 410-1, "Application for Rural Development Services" or RD 442-3, "Balance Sheet," and information on present income and potential earning ability. Exception for SFH loans: Rural Development requires a budget and/or financial statement and, if necessary to discover suspected undisclosed assets, a search of public records, only when the value of the security property may be less than the debt.

(4) For organization borrowers, a duly-adopted Resolution by the governing body authorizing the conveyance and certified by the attesting official with the corporate seal affixed. The Resolution will indicate which officials are authorized to execute the offer to convey and the deed on behalf of the borrower. If shareholder approval is necessary, the Resolution will specifically recite that shareholder approval has been obtained.

(5) If water rights, mineral rights, development rights, or other use rights are not fully covered in the deed, the advice of OGC will be obtained and appropriate documents to transfer rights to the Government will be obtained before the voluntary conveyance is accepted. The documents will be recorded, if necessary, in connection with closing the conveyance.

(6) If property is under lease, an assignment of the lease to the Government will be obtained with the effective date being the date the voluntary conveyance is closed. If an oral lease is in force, it will be reduced to writing and assigned to the Government.

(7) The borrower may be required to provide a title insurance policy or a final title opinion from a designated attorney when the State Director determines it is necessary to protect the Government's interest. Such title insurance policy or final title opinion will show title vested to the Government subject only to exceptions and liens approved by the County Supervisor.

(8) Farmer program loan borrowers who voluntarily convey after receiving the appropriate loan servicing notice(s) contained in the attachments of exhibit A of subpart S of part 1951 of this chapter, must properly complete and return the acknowledgement form sent with the notice.

(9) For MFH loans, assignment of Housing Assistance Payments (HAP) Contracts will be obtained. Rental Assistance will be retained until the State Director is advised by OGC that the Agency has title to the property. After a voluntary conveyance, the Agency may transfer Rental Assistance in accordance with 7 CFR part 3560, subpart F.

(e) *Appraisal of property.* After an offer of voluntary conveyance, but before acceptance by Rural Development, an appraisal of the property will be made to establish the current market value of the property. If a qualified Rural Development appraiser is not available to appraise property securing a loan other than MFH, the State Director may obtain an appraisal from a qualified appraiser outside Rural Development in accordance with RD Instruction 2024-A (available in any Rural Development office). For property securing MFH, prior authorization must be obtained by the Assistant Administrator, Housing, to secure an appraisal from a source outside Rural Development. For property securing FP loan(s), the contract appraiser must complete the appraisal in accordance with § 761.7 of this title for FP property, or subpart C of part 1922 for Single Family Housing property. Also, the appraiser must meet at least one of the following qualifications:

(1) Certification by a National or State Appraisal Society.

(2) If a certified appraiser is not available, the appraiser may be one who meets the criteria for certification in a National or State Appraisal Society.

(3) The appraiser has recent, relevant documented appraisal experience or training, or other factors clearly establishing the appraiser's qualifications.

(f) *Processing offer to convey security and acceptance by Rural Development.* If a borrower has both SFH and other type loans, the portion of this paragraph dealing with the loan(s) other than SFH will be followed.

(1) *SFH loans.* Rural Development does not solicit or encourage conveyance of SFH security property to the Government and will consider a borrower's offer to convey by deed in lieu of foreclosure only after the debt is accelerated and when it is in the Government's interest. Upon receipt of an offer to convey, the servicing official will remind the borrower of provisions for voluntary liquidation under 7 CFR part 3550, and the consequences of a conveyance by deed in lieu of foreclosure as follows: All costs related to the conveyance which Rural Development pays will be added to the debt; a credit equal to the market value of the property, as determined by Rural Development, less prior liens, will be applied to the debt; and if the credit does not satisfy the debt, the borrower will not automatically be released of liability. The unsatisfied debt, after acceleration under § 1955.10(h)(5) of this part, may be settled according to subpart B of part 1956 of this chapter; however, a deficiency judgment will not be pursued when the borrower was granted a moratorium if the borrower faithfully tried to meet loan obligations. The conveyance is processed as follows:

(i) Before accepting the offer, the County Supervisor will transmit the deed to a closing agent requesting a title search covering the period of time since the latest title opinion in the case file. The same agent who closed the loan should be used, if possible; otherwise one will be selected from the approved list of closing agents, taking care that cases are distributed fairly among approved agents. The closing agent may be instructed that the County Supervisor considers the voluntary conveyance offer conditionally approved, and the closing agent may record the deed after the title search if there are no liens against the property other than:

(A) The Rural Development lien(s);

(B) Prior liens when Rural Development has advised the closing agent that title will be taken subject to the prior lien(s) or has told the closing agent that the prior lien(s) will be handled in accordance with § 1955.10(c)(1) of this subpart; and/or

(C) Real estate taxes and/or assessments which must be paid when title to the property is transferred.

(ii) If junior liens are discovered, the closing agent will be requested to provide Rural Development with the lienholder's name, amount of lien, date recorded, and the recording information (recording office, book and page), return the unrecorded deed to Rural Development, and await further instructions from Rural Development. In such cases, the County Supervisor will proceed in accordance with §1955.10(c)(2) of this subpart. If agreement has been reached with the lienholder(s) for settling the junior lien(s) in order to accept the conveyance, the deed will be returned to the closing agent for a title update and recording.

(iii) The closing agent will be requested to provide a certification of title to Rural Development after recordation of the deed. A certification of title in a statement that fee title is vested in the Government subject only to the Rural Development lien(s) and prior liens previously approved by Rural Development. After receipt of the certification of title, the County Supervisor will notify the borrower that the conveyance has been accepted in accordance with §1955.10(g) of this subpart.

(2) *Consolidated Farm and Rural Development Act (CONACT) loans to individuals.* If the Agency indebtedness plus any prior liens exceeds the market value of the property, the indebtedness cannot be satisfied but a credit can be given equal to the market value less prior liens. Debt settlement will be considered in accordance with subpart B of part 1956 of this chapter.

(i) *Crediting accounts.* The Agency will credit an account by an amount equal to the market value less prior liens, unless the borrower is Native American. Native American borrower-owners will be credited with the fair market value or the Agency debt against the property, whichever is greater, provided:

(A) The borrower-owner is a member of a tribe or the tribe, and

(B) The property is located within the confines of a federally recognized Indian reservation.

(ii) *Agency approval.* The same procedure outlined in paragraphs (f)(1)(i) through (f)(1)(iii) of this section will be followed for approving the voluntary conveyance. The conveyance will be accepted in full satisfaction of the indebtedness unless the market value of the property to be conveyed is less than the total of Government indebtedness and prior liens, and the borrower has agreed to accept a credit in the amount of the market value of the security property less prior liens, if any.

(3) *Loans to organizations.* When an offer of voluntary conveyance is received from an organization borrower, and the market value of the property being conveyed (less prior liens, if any) is less than the Government debt, full consideration must be given to the borrower's present situation and future prospects for paying all or a part of the debt.

(g) *Closing of conveyance.* (1) The conveyance to the Government will be considered closed when the recorded deed has been returned to Rural Development, a certification of title is received from the closing agent that title is vested in the Government with no outstanding encumbrances other than the Rural Development lien(s) or previously approved prior liens, and the borrower is notified of the acceptance of the conveyance. For loans to organizations, OGC will be requested to review the case to verify that it was closed properly. The property will be assigned an ID number and entered into the Acquired Property Tracking System through the Automated Discrepancy Processing System (ADPS) terminal in the County Office.

(2) When costs incident to the completion of the transaction are to be paid by the Government, the servicing official will prepare and process the necessary documents as outlined in §1955.5(d) of this subpart and the costs will be charged to the borrower's account as recoverable costs. This includes taxes and assessments, water charges which protect the right to receive water, other liens, closing agent's fee, and any other costs related to the conveyance.

(h) *Actions to be taken after closing conveyance.* (1) When the Rural Development account is satisfied, the note(s) will be stamped "Satisfied by Surrender of Security and Borrower Released from Liability," and the statement must be signed by the servicing official.

(2) When the Rural Development account is not satisfied and the borrower is not released from liability, the note(s) will be retained by Rural Development.

(3) The servicing official will release the lien(s) of record, indicating that the debt was satisfied by surrender of security or that the lien is released but the debt not satisfied, whichever is applicable. If the lien is to be released but the debt not satisfied, OGC will provide the type of instrument required to comply with applicable State laws.

(4) After release of the lien(s), the servicing official will return the following to the borrower:

(i) If borrower is released from liability, the satisfied note(s) and a copy of Form RD 1955-1 showing acceptance by the Government; or

(ii) If borrower is not released from liability, a copy of Form RD 1955-1 showing acceptance by the Government.

(5) When the Rural Development account is not satisfied and the borrower not released from liability, the account balance, after deducting the "as is" market value and prior liens, if any, will be accelerated utilizing exhibit F of this subpart (available in any Rural Development office).

(6) For MFH loans, the State Director will cancel any interest credit and suspend any rental assistance. These actions will be accomplished by notifying the Finance Office unit which handles MFH accounts. In the interm the tenants will continue rental payments in accordance with their lease. Tenants will be informed of the pending liquidation action and the possible consequences of the action. If the project is to be removed from the Rural Development program, a minimum of 180 days' notice to the tenants is required. Letters of Priority Entitlement must be made available to any tenants that will be displaced.

(7) Actions outlined in § 1955.18 of this subpart will be taken, as applicable.

[50 FR 23904, June 7, 1985, as amended at 50 FR 45782, Nov. 1, 1985; 69 FR 69105, Nov. 26, 2004; 82 FR 19319, Apr. 27, 2017]

§ 1955.11 Conveyance of property to Rural Development by trustee in bankruptcy.

(a) *Authority.* With the advice of OGC (and prior approval of the National Office for MFH, Community Programs, and insured B&I loans), the State Director within his/her authority is authorized to accept a conveyance of property to the Government by the Trustee in Bankruptcy, provided:

(1) The Bankruptcy Court has approved the conveyance;

(2) The conveyance will permit a substantial recovery on the Rural Development debt; and

(3) Rural Development will acquire title free of all liens and encumbrances except Rural Development iens.

(b) *Fees and deed.* (1) Rural Development may pay any necessary and proper fees approved by the bankruptcy court in connection with the conveyance. Before paying a fee to a trustee for a Trustee's Deed in excess of $300 for any loan type(s) other than Farmer Programs or $1,000 for Farmer Program loans, prior approval of the Administrator must be obtained. The State Director will process the necessary documents as outlined in § 1955.5(d) of this subpart for payment of fees as recoverable costs.

(2) Conveyance may be by Trustee's Deed instead of a warranty deed. If upon advice of OGC it is determined a deed from any other person or entity (including the borrower) is necessary to obtain clear title, a deed from such person or entity will be obtained.

(c) *Acceptance.* The conveyance will be accepted for an amount of credit to the borrower's Rural Development account(s) as set forth in § 1955.18(e)(4) of this subpart.

(d) *Reporting.* Acquisition of property under this section will be reported in accordance with § 1955.18(a) of this subpart.

[50 FR 23904, June 7, 1985, as amended at 53 FR 27827, July 25, 1988]

§ 1955.12 Acquisition of property which served as security for a loan guarantee by Rural Development or at sale by another lienholder, bankruptcy trustee, or taxing authority.

When the servicing regulations for the type of loan(s) involved permit Rural Development to acquire property by one of these methods, the acquisition will be reported in accordance with § 1955.18(a) of this subpart.

§ 1955.13 Acquisition of property by exercise of Government redemption rights.

When the Government did not protect its interest in security property in a foreclosure by another lienholder, and if the Government has redemption rights, the State Director will determine whether to redeem the property. This determination will be based on all pertinent factors including the value of the property after the sale, and costs which may be incurred in acquiring and reselling the property. For Farmer Program loans, the County Supervisor will document the determination on exhibit G of this subpart. The decision must be made far enough in advance of expiration of the redemption period to permit exercise of the Government's rights. If the property is to be redeemed, complete information documenting the basis for not acquiring the property at the sale and factors which justify redemption of the property will be included in the case file. The assistance of OGC will be obtained in effecting the redemption. If the State Director decides not to redeem the property, the Government's right of redemption under Federal law (28 U.S.C. 2410) may be waived without consideration. If a State law right of redemption exists and may be sold, it will not be disposed of for less than its value.

[53 FR 35762, Sept. 14, 1988]

§ 1955.14 [Reserved]

§ 1955.15 Foreclosure by the Government of loans secured by real estate.

Foreclosure will be initiated when all reasonable efforts have failed to have the borrower voluntarily liquidate the loan through sale of the property, voluntary conveyance, or by entering into an accelerated repayment agreement when applicable servicing regulations permit; when either a net recovery can be made or when failure to foreclose would adversely affect Rural Development programs in the area. Also, in Farmer Program cases (except graduation cases under subpart F of part 1951 of this chapter), the borrower must have received exhibit A with attachments 1 and 2 of subpart S of part 1951 of this chapter, and any appeal must have been concluded. For real property located within the confines of a federally recognized Indian reservation and owned by a Native American borrower, proper notice of voluntary conveyance must be given as outlined in § 1955.9 (c)(1) of this subpart.

(a) *Authority*—(1) *Loans to individuals.* The District Director is authorized to approve or disapprove foreclosure and accelerate the account.

(2) *Loans to organizations.* (i) The State Director or District Director is authorized to approve or disapprove foreclosure of Rural Development secured debt does not exceed their respective loan approval authority. The State Director is authorized to approve or disapprove foreclosure of I&D, Shift-In-Land-Use (Grazing Association), loans to Indian Tribes and Tribal Corporations, and EOC loans, regardless of the amount of debt.

(ii) For all other organization loans, foreclosure will not be initiated without prior approval of the Administrator. The State Director will obtain OGC's opinion on the steps necessary to foreclose the loan, and forward the appropriate problem case report, a statement of essential facts, his/her recommendation, a copy of the OGC opinion, and the borrower's case file to the Administrator, Attn: Assistant Administrator (appropriate loan division) with a request for authorization to initiate foreclosure.

(b) *Problem case report.* When foreclosure is recommended, the servicing official will prepare Form RD 1955-2 for Farmer Program or SFH loans, exhibit A to this subpart for MFH loans, or exhibit A of RD Instruction 1951-E (available in any Rural Development office) for other organization loans. If chattel security is also involved, Forms RD 455-1, "Request for Legal Action"; 455-

2, "Evidence of Conversion"; and 455–22, "Information for Litigation"; as applicable to the case, will be prepared in accordance with the respective FMIs and made a part of the problem case submission. A statement must be included by the servicing official in the narrative that all servicing actions required by Rural Development loan servicing regulations have been taken and all required notices given to the borrower.

(1) *Appraisal.* The market value of the property may be estimated in completing the problem case report unless there are one or more prior liens other than current-year real estate taxes. Where such prior liens are involved, an appraisal report reflecting market value in existing condition will be included in the case file as a basis for determining the Government's prospects for financial recovery through foreclosure.

(2) *Recommendation for deficiency judgment.* If the debt will not be satisfied by the foreclosure, the borrower's financial situation will be assessed to determine if there is a possibility of further recovery on the account through a deficiency judgment. A summary of these determinations will be fully documented and appropriate recommendations made concerning deficiency judgment in the applicable problem case report.

(3) *Historic preservation.* If it is likely that Rural Development will acquire title to the property as a result of the foreclosure, and the structure(s) on the property will be in excess of 50 years old at the time of acquisition or meet any of the other criteria contained in § 1955.137(c) of subpart C of part 1955 of this chapter, steps should be initiated to meet the requirements of the National Historic Preservation Act as outlined in § 1955.137(c). Formal steps should not be initiated until the conclusion of all appeals. However, any such documentation required may be completed when the problem case report is prepared. This action should eliminate delays in selling the property after acquisition.

(c) *Submission of problem case.* The servicing official will submit the completed problem case docket to the official authorized to approve the fore-

closure (approval official). Before approval of foreclosure and acceleration of the account, the approval official is responsible for review of the problem case report to see that all items are complete and that all required servicing actions have been taken and all required notices given the borrower. The narrative portion of the report should provide complete information on the borrower's financial condition, deficiency judgment in case the debt is not satisfied by the foreclosure, and other pertinent background items. The approval official will approve or disapprove the foreclosure, or make a recommendation and refer the case to the National Office, if not within his/her approval authority. If foreclosure is not approved, the case will be returned to the originating office with instructions for further servicing. Problem case submission is as follows:

(1) *For loans to individuals.* The County Supervisors will submit the case to the District Director.

(2) *For loans to organizations.* The District Director will submit the case to the State Director along with a proposed liquidation and management plan covering the time the foreclosure is in process. The State Director will obtain the advice of OGC if required in connection with the type of loan being liquidated.

(d) *Approval of foreclosure.* When foreclosure is approved, it will be handled as follows:

(1) *Prior lien(s).* If there is a prior lien, all foreclosure alternatives should be explored including whether Rural Development will give the prior lienholder the opportunity to foreclose; join in the action if the prior lienholder wishes to foreclose; or foreclose the Rural Development loan(s), either settling the prior lien or foreclosing subject to it. The provisions of § 1965.11(c) of subpart A of part 1965 of this chapter must be followed for loans serviced under subpart A of part 1965. The assistance of OGC should be obtained in weighing the alternatives, with the objective being to pursue the course which will result in the greatest net recovery by the Government. After it is decided which option will be most advantageous to the Government, the approval official, either directly or

through a designee, will contact the prior lienholder to outline Rural Development's position. If State laws affect this action, a State Supplement will be issued with the advice of OGC to establish the procedure to be followed. For real property located within the confines of a federally recognized Indian reservation owned by a Native American borrower-owner, an analysis of whether Rural Development should acquire title must include facts which demonstrate the fair market value after considering the cost of clean-up of hazardous substances on the property.

(2) *Acceleration of account.* Subject to paragraphs (d)(2)(i), (d)(2)(ii), and (d)(2)(iii) of this section, the account will be accelerated using a notice substantially similar to exhibits B, C, D, or E of this subpart, or for multi-family housing, Guide Letters 1955–A–1 or 1955–A–2 (available in any Rural Development Office), as appropriate, to be signed by the official who approved the foreclosure. The accounts of borrowers with pending Chapter 12 and 13 cases which have not been discharged will be accelerated in accordance with instructions from OGC. Upon OGC approval, accounts of these borrowers may be accelerated using a notice substantially similar to exhibit D of this subpart. Loans secured by chattels must be accelerated at the same time as loans secured by real estate in accordance with §1965.26 (c) of subpart A of part 1965 of this chapter. The notice will be sent by certified mail, return receipt requested, to each obligor individually, addressed to the last known address. If different from the property address and/or the address the Finance Office uses, a copy of the notice will also be mailed to the property address and the address currently used by the Finance Office. (In chattel liquidation cases which have been referred for civil action under subpart A of part 1962 of this chapter, the Finance Office will be sent a copy of exhibits D, E, or E–1 (available in any Rural Development office) as applicable. County Office and Finance Office loan records will be adjusted to mature the entire debt in such cases). If a signed receipt for at least one of these acceleration notices sent by certified mail is received, no

further notice is required. If no receipt is received, a copy of the acceleration notice will be sent by regular mail to each address to which the certified notices were sent. This type mailing will be documented in the file. A State Supplement may be issued if OGC advises different or additional language or format is required to comply with State laws or if notice and mailing instructions are different from that outlined in this paragraph. A conformed copy of the acceleration notice will be forwarded to the servicing official. Farmer Program appeals will be concluded before acceleration. For MFH loans, a copy of the acceleration letter will also be forwarded to the National Office, ATTN: MFH Servicing and Property Management Division, for monitoring purposes. Accounts may be accelerated as follows:

(i) Where monetary default is involved, the account may be accelerated immediately after approval of foreclosure.

(ii) Where monetary default is *not* involved, the account will not be accelerated until the concurrence of OGC is obtained.

(iii) If borrower obtained the loan while a civilian, entered military service after the loan was closed, the Rural Development has not obtained a waiver of rights under the Soldiers and Sailors Relief Act, the account will not be accelerated until OGC has reviewed the case and given instructions.

(iv) If the decision is made to liquidate the farm loan(s) of a borrower who also has a SFH loan(s), and the dwelling was used as security for the farm loan(s) it will not be necessary to meet the requirements of 7 CFR part 3550 prior to accelerating the account. Except that, if the borrower is in default on his/her farm loan(s), the SFH account must have been considered for interest credit and/or moratorium at the time servicing options are being considered for the FP loan(s) prior to acceleration. If it is later determined the FP loan(s) are to receive additional servicing in lieu of liquidation, the RH loan will be reinstated simultaneously with the FP servicing actions and may be reamortized in accordance with 7 CFR part 3550. Accounts of a borrower who has both Farmer Program and

SFH loan(s) may be accelerated as follows:

(A) When the borrower's dwelling is financed with an SFH loan(s) is secured by and located on the same farm real estate as the Farmer Program loan(s) (dwelling located on the farm), the SFH loan(s) will be serviced in accordance with § 1965.26(c)(1) of subpart A of part 1965 of this chapter.

(B) When the borrower's dwelling is financed with an SFH loan(s) and is located on a nonfarm tract which also serves as additional security for the Farmer Program loan(s), the loans(s) will be serviced in accordance with § 1965.26 (c)(2) of subpart A of part 1965 of this chapter.

(C) When the borrower's dwelling is financed with an SFH loan(s) and is on a non-farm tract which does not serve as additional security for the Farmer Program loan(s), it will not be accelerated simultaneously with sending out attachments 5 and 6, or 5–A and 6–A, or attachment 9 and 10, or 9–A and 10–A, of exhibit A of subpart S of part 1951 of this chapter, as applicable, unless it is subject to liquidation based on provisions of 7 CFR part 3550, taking into consideration the prospects for success that may evolve when the borrower's livelihood is from a source other than the farming operation. If the SFH loan is in default and subject to liquidation based on provisions of 7 CFR part 3550, the SFH loan(s) must be accelerated at the same time the borrower is sent attachment 5 and 6, or 5–A and 6–A, or attachments 9 and 10, or 9–A and 10–A, to exhibit A of subpart S of part 1951 of this chapter, as applicable. For those borrowers who are in non-monetary default on their Farmer Programs loans and fail to return attachment 4 of exhibit A of subpart S of part 1951 of this chapter, the Farmer Programs loans and SFH loans will be accelerated at the same time. If the borrower appeals, one appeal hearing and one review will be held for both adverse actions.

(D) If a borrower's FP loan(s) were accelerated prior to May 7, 1987, and the SFH loan(s) is not accelerated, the SFH loan will be accelerated at the same time the borrower is sent attachments 5 and 6, or 5–A and 6–A, or attachments 7 and 8 to exhibit A of subpart S of 1951 of this chapter, as applicable, unless the requirements of § 1965.26 of subpart A of part 1965 of this chapter are met or the liquidation of the SFH loan is based on provisions of 7 CFR part 3550. If the borrower is sent attachments 5 and 6, or 5–A and 6–A to exhibit A of subpart S of 1951 of this chapter, as applicable, and requests an appeal, one hearing and one review will be held for both the adverse action on the FP loan restructuring request and SFH acceleration notices. If the borrower is sent attachments 7 and 8 to exhibit A of subpart S of 1951 of this chapter, there are no further appeals on the FP loans; but, the borrower is entitled to a hearing and a review on the SFH acceleration notice.

(v) For MFH loans, the acceleration notice will advise the borrower of all applicable prepayment requirements, in accordance with 7 CFR part 3560, subpart N. The requirements include the application of restrictive-use provisions to loans made on or after December 21, 1979, prepaid in response to acceleration notices and all tenant and agency notifications. The acceleration notice will also remind borrowers that rent levels cannot be raised during the acceleration without Rural Development approval, even after subsidies are canceled or suspended. Tenants are to be notified of the status of the project and of possible consequences of these actions. If the borrower wishes to prepay the project in response to the acceleration and Rural Development makes a determination that the housing is no longer needed, a minimum of 180 days' notice to tenants is required before the project can be removed from the Rural Development program. Letters of Priority Entitlement must be made available.

(3) *Offers by borrowers after acceleration of account*—(i) *Farmers Programs (FP) accelerations.* This category also includes non-FP loans to the same borrower which have been accelerated as part of the same action. After the account is accelerated, the borrower will have 30 days from the date of the acceleration notice to make payment in full to stop the acceleration, unless State or tribal law requires that the foreclosure be withdrawn if the account is

brought current and a State supplement is issued to specify the requirement.

(A) Payment in full [see exhibit D of this subpart (available in any Rural Development office)] may consist of the following means of fully satisfying the debt.

(1) Cash.

(2) Transfer and assumption.

(3) Sale of property.

(4) Voluntary conveyance.

(B) Payments which do not pay the account in full can be accepted subject to the following requirements:

(1) Payments will be accepted if there is no remaining security for the debt (real estate and chattel).

(2) If the borrower is in the process of selling security or nonsecurity, payments may be accepted unless State law would require the acceleration to be reversed. In States where payments cannot be accepted unless the acceleration is reversed, the payments will not be accepted. A State supplement will be issued to address State law on accepting payments after acceleration.

(3) If payments are mistakenly credited to the borrower's account, no waiver or prejudice to any rights which the United States may have for breach of any promissory note or convenant in the real estate instruments will result. Disposition of such payments will be made after consulting OGC.

(4) The servicing official will notify the approval official of any other offer. This includes a request by the borrower for an extension of time to accomplish voluntary liquidation or a proposal to cure the default(s). In all other cases, the approval official will decide whether an offer from a borrower will be accepted and servicing of the loan reinstated or whether foreclosure will be delayed to give the borrower additional time to voluntarily liquidate as authorized in servicing regulations for the type loan(s) involved. If an offer is received after the case has been referred to OGC, the approval official will consult OGC before accepting or rejecting the offer. The denial of an offer to stop foreclosure is not appealable. In all cases, the approval official will notify the servicing official of the decision made.

(ii) *All other accelerations.* After the account is accelerated, loan servicing ceases. For example, for SFH loans, the renewal or granting of interest credit or a moratorium is not authorized. The servicing official will accept no payment for less than the unpaid loan balance, unless State law requires that foreclosure be withdrawn if the account is brought current and a State supplement is issued to specify this requirement. If payments are mistakenly accepted and credited to the borrower's account, no waiver or prejudice to any rights which the United States may have for breach of any promissory note or covenants in the real estate instruments will result. Disposition of such payments will be made after consultation with OGC. The servicing official will notify the approval official of any offer received from the borrower. This includes a request by the borrower for an extension of time to accomplish voluntary liquidation or a written proposal to cure the default(s). The receipt of a payment with no proposal to cure the defaults is not considered a viable offer, and such payments will be returned to the borrower. The approval official will decide whether an offer from a borrower will be accepted and servicing of the loan reinstated or whether foreclosure will be delayed to give the borrower additional time to voluntarily liquidate as authorized in servicing regulations for the type loan involved. If an offer is received after the case has been referred to OGC, the approval official will consult OGC before accepting or rejecting the offer. The denial of an offer to stop foreclosure is not appealable. In all cases, the approval official will notify the servicing official of the decision made. For MFH loans, the National Office will be notified when foreclosure is withdrawn. When an account is reinstated under this section, the servicing official will grant or reinstate assistance for which the borrower qualifies, such as interest credit on an SFH loan. When granting interest credit in such a case:

(A) If an interest credit agreement expired after the account was accelerated, the effective date will be the date the previous agreement expired.

(B) If an interest credit agreement was not in effect when the account was accelerated, the effective date will be the date foreclosure action was withdrawn.

(C) For MFH loans with rental assistance, after acceleration and after any appeal or review has been concluded, rental assistance will be suspended if foreclosure is to continue. If the account is reinstated, the rental assistance will be reinstated retroactively to the date of suspension. In the interim, the tenants will continue rental payments in accordance with their leases, and all rental rates and lease renewals and provisions will be continued as if acceleration had not taken place.

(4) *Statement of account.* If a statement of account is required for foreclosure proceedings, Form RD 451–10, "Request for Statement of Account," will be processed in accordance with the FMI. When an official statement of account is not required, account balances and recapture information may be obtained from the field office terminal.

(5) *Appeals.* All appeals will be handled pursuant to subpart B of part 1900 of this chapter. Foreclosure actions will be held in abeyance while an appeal is pending. No case will be referred to OGC for processing of foreclosure until a borrower's appeal and appeal review have been concluded, or until the time has elapsed during which an appeal or a request for review may be made. In Farmer Programs cases, (except graduation cases under subpart F of part 1951 of this chapter), the borrower must have received the appropriate notices and consideration for primary loan servicing per subpart S of part 1951 of this chapter. Any Farmer Programs cases may be accelerated after all primary loan servicing options have been considered and all related appeals concluded, but will not be submitted to OGC for foreclosure action until all appeals related to any preservation rights have been concluded.

(6) *Petition in bankruptcy filed by borrower after acceleration of account.* (i) When bankruptcy is filed after an account has been accelerated, any foreclosure action initiated by Rural Development must be suspended until:

(A) The bankruptcy case is *dismissed* or *closed* (a discharge of debtor does not close the case);

(B) An Order lifting the automatic stay is obtained from the Bankruptcy Court; or

(C) The property is no longer property of the bankruptcy estate *and* the borrower has received a discharge.

(ii) The State Director will request the assistance of OGC in obtaining the Order(s) described in paragraph (c)(6)(i)(B) of this section.

(e) *Referral of case.* If the borrower fails to satisfy the account during the period of time specified in the acceleration notice, and no appeal is pending, the foreclosure process will continue:

(1) If the District Director is the approval official, he/she will forward the case file with all pertinent documents and information concerning the foreclosure action and appeal, if any, to the State Director for completion of the foreclosure.

(2) If the State Director is the approval official, or in cases referred by the District Director under paragraph (e)(1) of this section, the State Director will forward to OGC the case file and all documents needed by OGC to process the foreclosure. A State Supplement will be issued, with the advice and assistance of OGC, to reflect the make-up of the foreclosure docket. Since foreclosure processing varies widely from State to State, each State Supplement will be explicit in outlining step-by-step procedures. At the time indicated by OGC in the foreclosure instructions, Form RD 1951–6, "Borrower Account Description Flag," will be processed in accordance with the FMI. After referral to OGC, further actions will be in accordance with OGC's instructions for completion of the foreclosure. If prior approval of the Administrator is obtained, nonjudicial foreclosure for monetary default may be handled as outlined in a State Supplement approved by OGC without referral to OGC before foreclosure.

(f) *Completion of foreclosure*—(1) *Foreclosure advertisement for organization loans subject to title VI of the Civil Rights Act of 1964.* (i) The advertisement for foreclosure sale of property subject to title VI of the Civil Rights Act of 1964 will contain a statement substantially similar to the following: "The property described herein was purchased or improved with Federal financial assistance and is subject to the non-discrimination provisions of title VI of the Civil Rights Act of 1964, section 504 of the Rehabilitation Act of 1973 and other similarly worded Federal statutes and regulations issued pursuant thereto that prohibit discrimination on the basis of race, color, national origin, handicap, religion, age or sex in programs or activities receiving Federal financial assistance, for as long as the property continues to be used for the same or similar purposes for which the Federal assistance was extended or for so long as the purchaser owns it, whichever is later." At least 30 days before the foreclosure sale, the County Supervisor will notify, in writing, the Indian tribe which has jurisdiction over the reservation, and in which the real property is owned by a Native American member of said tribe that a foreclosure sale will be conducted to resolve this account, and will provide:

(A) Projected sale date and location;

(B) Fair market value of property;

(C) Amount Rural Development will bid on the property; and

(D) Amount of Rural Development debt against the property.

(ii) The purchaser will be required to sign Form RD 400–4, "Assurance Agreement," if the property will be used for its original or similar purposes.

(2) *Restrictive-use provisions for MFH loans.* For MFH loans, the advertisement will state the restrictive-use provisions which will be included in any deed used to transfer title.

(3) *Expenses.* Expenses which are incurred in connection with foreclosure, including legal fees, will be paid at the time recommended by OGC by processing the necessary documents as outlined in § 1955.5 (d) of this subpart. Costs will be charged as outlined in RD Instruction 2024–A (available in any Rural Development office).

(4) *Notice of judgment.* In states with judicial foreclosure, as soon as the foreclosure judgment is obtained, Form RD 1962–20, "Notice of Judgment," will be processed in accordance with the FMI. This will establish a judgment account to accrue interest at the rate stated in the judgment order so that an accurate account balance can be obtained for calculating the Government's foreclosure bid.

(5) *Gross investment.* The gross investment is the sum of the following:

(i) The unpaid balance of one of the following, as applicable:

(A) In States with nonjudicial foreclosure, the borrower's Rural Development account balance reflecting secured loan(s) and advances; and where State law permits, unsecured debts; or

(B) In States with judicial foreclosure, the judgment account established as a result of the foreclosure judgment in favor of Rural Development.

(ii) All recoverable costs charged (or to be charged) to the borrower's account in connection with the foreclosure action and other costs which OGC advises must be paid from proceeds of the sale before paying the Rural Development secured debt, including but not limited to payment of real estate taxes and assessments, prior liens, legal fees including U.S. Attorney's and U.S. Marshal's, and management fees; and

(iii) If a SFH loan subject to recapture of interest credit is involved, the total amount of subsidy granted and principal reduction attributed to subsidy.

(6) *Amount of Government's bid.* Except for FP loans and as modified by paragraph (f)(7)(ii) of this section, the Government's bid will be the amount of Rural Development's gross investment or the market value of the security, whichever is less. For real property located within the confines of a federally recognized Indian reservation and which is owned by an Rural Development borrower who is a member of the tribe with jurisdiction over the reservation, the Government's bid will be the greater of the fair market value or the Rural Development debt against the property, unless Rural Development determines that, because of the presence of hazardous substances on the property, it is not in the best interest of the Government to bid such amount, in which case there may be a deduction from the bid for the costs for hazardous material assessment and/or mitigation. For FP loans, except as modified by paragraph (f)(7)(ii) of this

section, the Government's bid will be the amount of Rural Development's gross investment or the amount determined by use of exhibit G–1 of this subpart, whichever is less. When the foreclosure sale is imminent, the State Director must request the servicing official to submit a current appraisal (in existing condition) as a basis for determining the Government's bid. Except for MFH properties, if an Rural Development appraiser is not available, the State Director may authorize an appraisal to be obtained by contract from a source outside Rural Development in accordance with RD Instruction 2024–A (available in any Rural Development office). For MFH properties, prior approval of the Assistant Administrator, Housing, is necessary to procure an outside appraisal.

(7) *Bidding.* The State Director will designate an individual to bid on behalf of the Government unless judicial proceedings or State nonjudicial foreclosure law provides for someone other than an Rural Development employee to enter the Government's bid. When the State Director determines attendance of an Rural Development employee at the sale might pose physical danger, a written bid may be submitted to the Marshal, Sheriff, or other party in charge of holding the sale. The Government's bid will be entered when no other party makes a bid or when the last bid will result in the property being sold for less than the bid authorized in paragraph (f)(6) of this section.

(i) When Rural Development is the senior lienholder, only one bid will be entered, and that will be for the amount authorized by the State Director.

(ii) When Rural Development is not the senior lienholder *and* OGC advises that the borrower has no redemption rights or if a deficiency judgment will be obtained, the State Director may authorize the person who will bid for the Government to make incremental bids in competition with other bidders. If incremental bidding is desired, the State Director's instructions to the bidder will state the initial bid, bidding increments, and the maximum bid.

(g) *Reports on sale and finalizing foreclosure.* Immediately after a foreclosure sale at which the State Director has

designated a person to bid on behalf of the Government, the servicing official will furnish the State Director a report on the sale. The State Director will forward a copy of this report to OGC and, for MFH loans, to the National Office. Based on OGC's instructions, a State supplement will provide a detailed outline of actions necessary to complete the foreclosure.

[50 FR 23904, June 7, 1985, as amended at 80 FR 9895, Feb. 24, 2015]

EDITORIAL NOTE: For FEDERAL REGISTER citations affecting § 1955.15, see the List of CFR Sections Affected, which appears in the Finding Aids section of the printed volume and at *www.govinfo.gov.*

§§ 1955.16–1955.17 [Reserved]

§ 1955.18 Actions required after acquisition of property.

The approval official may employ the services of local designated attorneys, of a case by case basis, to process all legal procedures necessary to clear the title of foreclosure properties. Such attorneys shall be compensated at not more than their usual and customary charges for such work. Contracting for such attorneys shall be accomplished pursuant to the Federal acquisition regulations and related procurement regulations and guidance.

(a)–(d) [Reserved]

(e) *Credit to the borrower's account or foreclosure judgment account*—(1) *For SFH accounts.* When Rural Development acquired the property, the account will be satisfied unless:

(i) In a voluntary conveyance case where the debt exceeds the market value of the property and the borrower is *not* released from liability, in which case the account credit will be the market value (less outstanding liens if any); or

(ii) In a foreclosure where the bid is less than the account balance and a deficiency judgment will be sought for the difference, in which case the account credit will be the amount of Rural Development's bid.

(2) *For all types of accounts other than SFH.* When Rural Development acquired the property, the account credit will be as follows:

(i) In a voluntary conveyance case:

(A) Where the market value of the property equals or exceeds the debt or where the borrower is released from liability for any difference, the account will be satisfied.

(B) Where the debt exceeds the market value of the property and the borrower is *not* released from liability, the account credit will be the market value (less outstanding liens, if any).

(ii) In a foreclosure, the account credit will be the amount of Rural Development's bid *except* when incremental bidding as provided for in § 1955.15(f)(7)(ii) of this subpart was used, in which case the account credit will be the maximum bid that was authorized by the State Director.

(3) *For all types of accounts when Rural Development did not acquire the property.* The sale proceeds will be handled in accordance with applicable State laws with the advice and assistance of OGC, including remittance of funds, application of the borrower's account credit, and disbursement of any funds in excess of the amount due Rural Development.

(4) *In cases where Rural Development acquired security property by means other than voluntary conveyance or foreclosure.* In these cases, such as conveyance by a bankruptcy trustee or by Court Order, the account credit will be as follows:

(i) If the market value of the acquired property equals or exceeds the debt, the account will be satisfied.

(ii) If the debt exceeds the market value of the acquired property, the account credit will be the market value.

(f)–(1) [Reserved]

[50 FR 23904, June 7, 1985, as amended at 52 FR 41957, Nov. 2, 1987; 53 FR 27827, July 25, 1988; 53 FR 35764 Sept. 14, 1988; 55 FR 35295, Aug. 29, 1990; 56 FR 10147, Mar. 11, 1991; 56 FR 29402, June 27, 1991; 58 FR 38927, July 21, 1993; 58 FR 68725, Dec. 29, 1993; 60 FR 34455, July 3, 1995]

§ 1955.19 [Reserved]

§ 1955.20 Acquisition of chattel property.

Every effort will be made to avoid acquiring chattel property by having the borrower or Rural Development liquidate the property according to subpart A of part 1962 of this chapter and apply the proceeds to the borrower's account(s). Methods of acquisition authorized are:

(a) *Purchase at the following types of sale:* (1) Execution sale conducted by the U.S. Marshal, sheriff or other party acting under Court order to satisfy judgment liens.

(2) Rural Development foreclosure sale conducted by the U.S. Marshal or sheriff in States where a State Supplement provides for sales to be conducted by them.

(3) Sale by trustee in bankruptcy.

(4) Public sale by prior lienholder.

(5) Public sale conducted under the terms of Form RD 455–4, "Agreement for Voluntary Liquidation of Chattel Security," the power of sale in security agreements or crop and chattel mortgage, or similar instrument, if authorized by State Supplement.

(b) *Voluntary conveyance.* Voluntary conveyance of chattels will be accepted only when the borrower can convey ownership free of other liens *and* the borrower can be released from liability under the conditions set forth in § 1955.10(f)(2) of this subpart. Payment of other lienholders' debts by Rural Development in order to accept voluntary conveyance of chattels is not authorized. Before a voluntary conveyance from a Farmer Program loan borrower can be accepted, the borrower must be sent Exhibit A with Attachments 1 and 2 of subpart S of part 1951 of this chapter.

(1) *Offer.* The borrower's offer of voluntary conveyance will be made on Form RD 1955–1. If it is determined the conveyance offer can be accepted, the borrower will execute a bill of sale itemizing each item of chattel property being conveyed and will provide titles to vehicles or other equipment, where applicable.

(2) *Acceptance of offer release from liability.* Before accepting an offer to convey chattels to Rural Development, the concurrence of the State Director must be obtained. When chattel security is voluntarily conveyed to the Government and the borrower and co-signer(s), if any, are to be released from liability, the servicing official will stamp the note(s) "Satisfied by Surrender of Security and Borrower Released from Liability." When the Agency debt less the market value and

prior liens is $1 million or more (including principal, interest and other charges), release of liability must be approved by the Administrator or designee; otherwise, the State Director must approve the release of liability. All cases requiring a release of liability will be submitted in accordance with Exhibit A of Subpart B of Part 1956 of this chapter (available in any Rural Development office). Form RD 1955-1 will be executed by the servicing official showing acceptance by the Government, and the satisfied note(s) and a copy of Form RD 1955-1 will be furnished to the borrower.

(3) *Release of lien(s).* When an offer has been accepted as outlined in paragraph (b)(2) of this section, the servicing official will release any liens of record which secured the satisfied indebtedness.

(4) *Rejection of offer.* If it is determined an offer of voluntary conveyance will not be accepted, the servicing official will indicate on Form RD 1955-1 that the offer is rejected, execute the form, and furnish a copy to the borrower.

(c) *Attending sales.* The servicing official will:

(1) Attend all sales described in paragraph (a)(5) of this section unless an exception is authorized by the State Director because of physical danger to the Rural Development employee or adverse publicity would be likely.

(2) Attend public sales by prior lienholders when the market value of the chattel property is significantly more than the amount of the prior lien(s).

(3) Obtain the advice of the State Director on attending sales described in paragraphs (a) (1), (2), and (3) of this section.

(d) *Appraising chattel property.* Prior to the sale, the servicing official will appraise chattel property using Form RD 440-21, "Appraisal of Chattel Property." If a qualified appraiser is not available to appraise chattel property, the State Director may obtain an appraisal from a qualified source outside Rural Development by contract in accordance with Rural Development Instruction 2024-A (available in any Rural Development office).

(e) *Abandonment of security interest.* The State Director may authorize abandonment of the Government's security interest when chattel property, considering costs of moving or rehabilitation, has no market value and obtaining title would not be in the best interest of the Government.

(f) *Bidding at sale.* (1) The servicing official is authorized to bid at sales described in paragraph (a) of this section. Ordinarily, only one bid will be made on items of chattel security unless the State Director authorizes incremental bidding. Bids will be made only when no other party bids or when it appears bidding will stop and the property will be sold for less than the amount of the Government's authorized bid. When the State Director determines attendance of an Rural Development employee might pose physical danger, a written bid may be submitted to the party holding the sale. The bid(s) will be the lesser of:

(i) The market value of the item(s) less the estimated costs involved in the acquisition, care, and sale of the item(s) of security; or

(ii) The unpaid balance of the borrower's secured Rural Development debt plus prior liens, if any.

(2) Bids will not be made in the following situations unless authorized by the State Director:

(i) When chattel property under prior lien has a market value which is not significantly more than the amount owed the prior lienholder. If Rural Development holds a junior lien on several items of chattel property, advice should be obtained from the State Director on bidding.

(ii) After sufficient chattel property has been bid in by Rural Development to satisfy the Rural Development debt; prior liens, and cost of the sale.

(iii) When the sale is being conducted by a lienholder junior to Rural Development.

(iv) At a private sale.

(v) When the sale is being conducted under the terms of Form RD 455-3, "Agreement for Sale by Borrower (Chattels and/or Real Estate)".

(g) *Payment of costs.* Costs to be paid by Rural Development in connection with acquisition of chattel property will be paid as outlined in § 1955.5(d) of this subpart as recoverable costs.

NOTE: Payment of other lienholders' debts in connection with voluntary conveyance of chattels is not authorized.

(h) *Reporting acquisition of chattel property.* Acquisition of chattel property will be reported by use of Form RD 1955-3 prepared and distributed in accordance with the FMI.

[50 FR 23904, June 7, 1985, as amended at 50 FR 45783, Nov. 1, 1985; 51 FR 45433, Dec. 18, 1986; 53 FR 27828 July 25, 1988; 53 FR 35764, Sept. 14, 1988; 60 FR 28320, May 31, 1995]

§ 1955.21 Exception authority.

The Administrator may, in individual cases, make an exception to any requirement or provision of this subpart or address any omission of this subpart which is not inconsistent with the authorizing statute or other applicable law if the Administrator determines that the Government's interest would be adversely affected or the immediate health and/or safety of tenants or the community are endangered if there is no adverse effect on the Government's interest. The Administrator will exercise this authority upon the request of the State Director with recommendation of the appropriate program Assistant Administrator; or upon request initiated by the appropriate program Assistant Administrator. Requests for exceptions must be made in writing and supported with documentation to explain the adverse effect, propose alternative courses of action, and show how the adverse effect will be eliminated or minimized if the exception is granted.

§ 1955.22 State supplements.

State Supplements will be prepared with the assistance of OGC as necessary to comply with State laws or only as specifically authorized in this regulation to provide guidance to Rural Development officials. State supplements will be submitted to the National Office for post approval in accordance with RD Instruction 2006-B (available in any Rural Development office).

§§ 1955.23-1955.49 [Reserved]

§ 1955.50 OMB control number.

The collection of information requirements contained in this regulation have been approved by the Office of Management and Budget (OMB) and have been assigned OMB control number 0575-0109. Public reporting burden for this collection of information is estimated to vary from 5 minutes to 5 hours per response, with an average of .56 hours per response including time for reviewing instructions, searching existing data sources, gathering and maintaining the data needed, and completing and reviewing the collection of information. Send comments regarding this burden estimate or any other aspect of this collection of information, including suggestions for reducing this burden, to Department of Agriculture, Clearance Officer, OIRM, room 404-W, Washington, DC 20250; and to the Office of Management and Budget, Paperwork Reduction Project (OMB #0575-0109), Washington, DC 20503.

[57 FR 1372, Jan. 14, 1992]

EXHIBITS A-F TO SUBPART A OF PART 1955 [RESERVED]

Subpart B—Management of Property

SOURCE: 53 FR 35765, Sept. 14, 1988, unless otherwise noted.

§ 1955.51 Purpose.

This subpart delegates authority and prescribes policies and procedures for the Rural Housing Service (RHS), Rural Business-Cooperative Service (RBS) andherein referred to as "Agency." This subpart does not apply to Farm Service Agency, Farm Loan Programs, or to RHS single family housing loans or community program loans sold without insurance to the private sector. These community program loans will be serviced by the private sector, and future revisions to this subpart no longer apply to such loans. This subpart does not apply to the Rural Rental Housing, Rural Cooperative Housing, or Farm Labor Housing Program of RHS. In addition, this subpart does not apply to Water and Waste

Programs of the Rural Utilities Service, Watershed loans, ·and Resource Conservation and Development loans, which are serviced under part 1782 of this title. This subpart covers:

(a) Management of real property which has been taken into custody by the respective Agency after abandonment by the borrower;

(b) Management of real and chattel property which is in Agency inventory; and

(c) Management of real and chattel property which is security for a guaranteed loan liquidated by an Agency (or which the Agency is in the process of liquidating).

[61 FR 59778, Nov. 22, 1996, as amended at 69 FR 69106, Nov. 26, 2004; 72 FR 55019, Sept. 28, 2007; 72 FR 64123, Nov. 15, 2007]

§ 1955.52 Policy.

Inventory and custodial real property will be effectively managed to preserve its value and protect the Government's financial interests. Properties owned or controlled by Rural Development will be maintained so that they are not a detriment to the surrounding area and they comply with State and local codes. Generally, Rural Development will continue operation of Multiple Family Housing (MFH) projects which are acquired or taken into custody. Servicing of repossessed or abandoned chattel property is covered in subpart A of part 1962 of this chapter, and management of inventory chattel property is covered in § 1955.80 of this subpart.

§ 1955.53 Definitions.

As used in this subpart, the following definitions apply:

CONACT or CONACT property. Property acquired or sold pursuant to the Consolidated Farm and Rural Development Act (CONACT). Within this subpart, it shall also be construed to cover property which secured loans made pursuant to the Agriculture Credit Act of 1978; the Emergency Agricultural Credit Adjustment Act of 1978; the Emergency Agricultural Credit Act of 1984; the Food Security Act of 1985; and other statutes giving agricultural lending authority to Rural Development.

Contracting Officer (CO). CO means a person with the authority to enter into, administer, and/or terminate contracts and make related determinations and findings. The term includes authorized representatives of the CO acting within the limits of their authority as delegated by the CO.

Custodial property. Borrower-owned real property and improvements which serve as security for an Rural Development loan, have been abandoned by the borrower, and of which the respective Agency has taken possession.

Farmer program loans. This includes Farm Ownership (FO), Soil and Water (SW), Recreation (RL), Economic Opportunity (EO), Operating (OL), Emergency (EM), Economic Emergency (EE), Special Livestock (SL), Softwood Timber (ST) loans, and Rural Housing loans for farm service buildings (RHF).

Government. The United States of America, acting through the respective agency, U.S. Department of Agriculture.

Indian reservation. All land located within the limits of any Indian reservation under the jurisdiction of the United States notwithstanding the issuance of any patent, and including rights-of-way running through the reservation; trust or restricted land located within the boundaries of a former reservation of a federally recognized Indian tribe in the State of Oklahoma; or all Indian allotments the Indian titles to which have not been extinguished if such allotments are subject to the jurisdiction of a federally recognized Indian tribe.

Inventory property. Real and chattel property and related rights to which the Government has acquired title.

Loans to individuals. Farmer Program loans, as defined above, whether to individuals or entities; Land Conservation and Development (LCD); and Single-Family Housing (SFH), including both Sections 502 and 504 loans.

Loans to organizations. Community Facility (CF), Water and Waste Disposal (WWD), Association Recreation, Watershed (WS), Resource Conservation and Development (RC&D), loans to associations for Irrigation and Drainage and other Soil and Water Conservation measures, loans to Indian Tribes and Tribal Corporations, Shift-in-Land-Use (Grazing Associations) Business and Industrial (B&I) to both individuals and groups, Rural Development

Loan Fund (RDLF), Intermediary Relending Program (IRP), Nonprofit National Corporation (NNC), Economic Opportunity Cooperative (EOC), Rural Housing Site (RHS), Rural Cooperative Housing (RCH), and Rural Rental Housing (RRH) and Labor Housing (LH) to both individuals and groups. The housing-type loans identified here are referred to in this subpart collectively as MFH loans.

Nonprogram (NP) property. SFH and MFH property acquired pursuant to the Housing Act of 1949, as amended, that cannot be used by a borrower to effectively carry out the objectives of the respective loan program; for example, a dwelling that cannot be feasibly repaired to meet the requirements for existing housing as described in 7 CFR part 3550. It may contain a structure which would meet program standards; however, is so remotely located it would not serve as an adequate residential unit or an older house which is excessively expensive to heat and/or maintain for a very-low or low-income homeowner.

Nonrecoverable cost is a contractual or noncontractual program loan cost expense not chargeable to a borrower, property account, or part of the loan subsidy.

Office of the General Counsel (OGC). The OGC, U.S. Department of Agriculture, refers to the Regional Attorney or Attorney-in-Charge in an OGC field office unless otherwise indicated.

Program property. SFH and MFH inventory property that can be used to effectively carry out the objectives of their respective loan programs with financing through that program. Inventory property located in an area where the designation has been changed from rural to nonrural will be considered as if it were still in a rural area.

Recoverable cost is a contractual or noncontractual program loan expense chargeable to a borrower, property account, or part of the loan subsidy.

Servicing official. For loans to individuals as defined in this section, the servicing official is the County Supervisor. For insured B&I loans, the servicing official is the State Director. For Rural Development Loan Fund and Intermediary Relending Program loans, the servicing official is the Director, Business and Industry Division. For Nonprofit National Corporations loans, the servicing official is Director, Community Facility Division. For all other types of loans, the servicing official is the District Director.

Suitable property. For FSA inventory property, real property that can be used for agricultural purposes, including those farm properties that may be used as a start up or add-on parcel of farmland. It also includes a residence or other off-farm site that could be used as a basis for a farming operation. For agencies other than FSA, real property that could be used to carry out the objectives of the Agency's loan program with financing provided through that program.

Surplus property. For FSA inventory property, real property that cannot be used for agricultural purposes including nonfarm properties. For other agencies, property that cannot be used to carry out the objectives of financing available through the applicable loan program.

[53 FR 35765, Sept. 14, 1988, as amended at 56 FR 29402, June 27, 1991; 57 FR 19525, 19528, May 7, 1992; 58 FR 58648, Nov. 3, 1993; 62 FR 44396, Aug. 21, 1997; 63 FR 41716, Aug. 5, 1998; 67 FR 78329, Dec. 24, 2002]

§ 1955.54 Redelegation of authority.

Authorities will be redelegated to the extent possible, consistent with program objectives and available resources.

(a) Any authority in this subpart which is specifically provided to the Administrator or to an Assistant Administrator may only be delegated to a State Director. The State Director cannot redelegate such authority.

(b) Except as provided in paragraph (a) of this section, the State Director may redelegate, in writing, any authority delegated to the State Director in this subpart, unless specifically excluded, to a Program Chief, Program Specialist, or Property Management Specialist on the State Office staff.

(c) The District Director may redelegate, in writing, any authority delegated to the District Director in this subpart to an Assistant District Director or District Loan Specialist. Authority of District Directors in this subpart applies to Area Loan Specialists in Alaska and the Director for the Western Pacific Territories.

(d) The County Supervisor may redelegate, in writing, any authority delegated to the County Supervisor in this subpart to an Assistant County Supervisor, GS–7 or above, who is determined by the County Supervisor to be qualified. Authority of County Supervisors in this subpart applies to Area Loan Specialists in Alaska, Island Directors in Hawaii, the Director for the Western Pacific Territories, and Area Supervisors in the Western Pacific Territories and American Samoa.

§ 1955.55 Taking abandoned real or chattel property into custody and related actions.

(a) *Determination of abandonment.* (Multi-family housing type loans will be handled in accordance with 7 CFR part 3560, subpart J.) When it appears a borrower has abandoned security property, the servicing official shall make a diligent attempt to locate the borrower to determine what the borrower's intentions are concerning the property. This includes making inquiries of neighbors, checking with the Postal Service, utility companies, employer(s), if known, and schools, if the borrower has children, to see if the borrower's whereabouts can be determined and an address obtained. A State supplement may be issued if necessary to further define "abandonment" based on State law. If the borrower is not occupying or is not in possession of the property but has it listed for sale with a real estate broker or has made other arrangements for its care or sale, it will not be considered abandoned so long as it is adequately secured and maintained. Except for borrowers with Farmers Program loans, if the borrower has made no effort to sell the property and can be located, an opportunity to voluntarily convey the property to the Government will be offered the borrower in accordance with § 1955.10 of Subpart A of this part. In farmer program cases, borrowers must receive Attachments 1 and 2 of Exhibit A of Subpart S of Part 1951 of this chapter and any appeal must be concluded before any adverse action can be taken. The County Supervisor will send

these forms to the borrower's last known address as soon as it is determined that the borrower has abandoned security property.

(b) *Taking security property into Rural Development custody.* When security property is determined to be abandoned, the running record in the borrower's file will be fully documented with the facts substantiating the determination of abandonment, and the servicing official shall proceed as follows without delay:

(1) For loans to individuals (except those with Farmer Program loans), if there are no prior liens, or if a prior lienholder will not take the measures necessary to protect the property, the County Supervisor shall take custody of the property, and a problem case report will be prepared recommending foreclosure in accordance with § 1955.15 of Subpart A of this part, unless the borrower can be located and voluntary liquidation accomplished. Farmer Program loan borrowers will be sent the forms listed in paragraph (a) of this section and the provisions of § 1965.26 of Subpart A of Part 1965 of this chapter will be followed.

(2) For MFH loans, if there are no prior liens, the District Director will immediately notify the State Director, who will request guidance from OGC and may also request advice from the National Office. The State Director, with the advice of OGC, will advise the borrower by writing a letter, certified mail, return receipt requested, at the address currently used by Finance Office, outlining proposed actions by Rural Development to secure, maintain, and operate the project.

(i) If the unpaid loan balance plus recoverable costs do not exceed the State Director's loan approval authority, the State Director will authorize the District Director to take custody of the property, make emergency repairs if necessary to protect the Government's interest, and will advise how the property is to be managed in accordance with 7 CFR part 3560.

(ii) If the unpaid loan balance plus recoverable costs exceeds the State Director's loan approval authority, the State Director will refer the case to the National Office for advice on emergency actions to be taken. The docket will be forwarded to the National Office with detailed recommendations for immediate review and authorization for further action, if requested by the MFH staff.

(iii) Costs incurred in connection with procurement of such things as management services will be handled in accordance with RD Instruction 2024–A (available in any Rural Development office).

(iv) The District Director will prepare a problem case report to initiate foreclosure in accordance with §1955.15 of Subpart A of this part and submit the report to the State Director along with a proposed plan for managing the project while liquidation is pending.

(3) For organization loans other than MFH, if there are no prior liens, the District Director will immediately notify the State Director that the property has been abandoned and recommend action which should be taken to protect the Government's interest. After obtaining the advice of OGC and the appropriate staff in the National Office, the State Director may authorize the District Director to take custody of the property and give instructions for immediate actions to be taken as necessary. The District Director will prepare a Report on Servicing Action (Exhibit A of Subpart E of Part 1951 of this chapter) recommending that foreclosure be initiated in accordance with §1955.15 of Subpart A of this part and submit the report to the State Director, along with a proposed plan for management and/or operation of the project while liquidation is pending.

(c) *Protecting custodial property.* The Rural Development official who takes custody of abandoned property shall take the actions necessary to secure, maintain, preserve, lease, manage, or operate the property.

(1) *Nonsecurity personal property on premises.* If a property has been abandoned by a borrower who left nonsecurity personal property on the premises, the personal property will not be removed and disposed of *before* the real property is acquired by the Government. If the premises are in a condition which presents a fire, health or safety hazard, but also contains items of value, only the trash and debris presenting the hazard will be removed. The servicing official may request advice from the State Director as necessary. The servicing official shall check for liens on nonsecurity personal property left on abandoned premises. If there is a known lienholder(s), the lienholder(s) will be notified by certified mail, return receipt requested, that the borrower has abandoned the property and that Rural Development has taken the real property into custody.

Actions by Rural Development must not damage or jeopardize livestock, growing crops, stored agricultural products, or any other personal property which is not Rural Development security.

(2) *Repairs to custodial property.* Repairs to custodial property will be limited to those which are essential to prevent further deterioration of the property. Expenditures in excess of an aggregate of $1,000 per property must have prior approval of the state Director.

(d) *Emergency advances where liquidation is pending.* Although security property may not be defined as abandoned in accordance with paragraph (a) of this section, if the borrower is not occupying the property and refuses or is unable to protect the security property, the servicing official is authorized to make expenditures necessary to protect the Government's interest. This would include, but is not limited to, securing or winterizing the property or making emergency repairs to prevent deterioration. Expenditures will be handled in accordance with paragraph (e) of this section. Situations where this authority may be used include, but are not limited to, where a borrower has a sale pending or when a voluntary conveyance is in process.

(e) *Income and costs.* Income received from the property will be applied to the borrower's account as an extra payment. Expenditures will be charged to the borrower's account as a recoverable cost.

(f) *Off-site procurements.* Circumstances may require off-site procurement action(s) to be taken by Rural Development to protect custodial, security or inventory property from damage or destruction and/or protect the Government's investment in the property. Such procurements may include, but are not limited to construction or reconstruction of roads, sewers, drainage work or utility lines. This type work may be accomplished either through Rural Development procurement or cooperative agreement. However, if Rural Development is obtaining a service or product for itself only, it must be a procurement and any such actions will be in accordance with RD Instruction 2024-A (available in any Rural Development office). Funding will come from the appropriate insurance fund.

(1) *Conditions for procurement.* Such expenditures may be made only when all of the following conditions are met:

(i) A determination is made that failure to procure work would likely result in a property loss greater than the expenditure;

(ii) There are no other feasible means (including cooperative agreements) to accomplish the same result;

(iii) The recovery of such advance(s) is not authorized by security instruments in the case of security or custodial property (no such limitation exists for inventory property);

(iv) Written documentation supporting subparagraphs (i), (ii) and (iii) has been obtained from the authorized program official;

(v) Approval has been obtained from the appropriate Assistant Administrator.

(2) *Direct procurement action.* Where direct procurement action is contemplated, an opinion must be obtained from the Regional Attorney that:

(i) Rural Development has the authority to enter the off-site property to accomplish the contemplated work, or

(ii) A specific legal entity has authority to grant an easement (right-of-way) to Rural Development for the contemplated work and such an easement, in a form approved by the Regional Attorney, has been obtained.

(3) *Cooperative agreements.* Cooperative agreements between Rural Development and other entities may be made to accomplish the requirement where the principal purpose is to provide money, property, services or items of value to state or local governments or other recipients to accomplish a public purpose. Exhibit C of this subpart (available in any Rural Development office) is an example of a typical cooperative agreement. A USDA handbook providing detailed guidance for all parties is available from the USDA—Office of Operations and Finance. Although cooperative agreements are not a contracting action, the authority, responsibility and administration of these agreements will be handled consistent with contracting actions.

(4) *Consideration of maintenance agreements.* Maintenance requirements must be considered in evaluating the economic benefits of off-site procurements. Where feasible, arrangements or agreements should be made with state, local governments or other entities to ensure continued maintenance by dedication or acceptance, letter agreements, or other applicable statutes.

[53 FR 35765, Sept. 14, 1988, as amended at 54 FR 20521, May 12, 1989; 57 FR 36591, Aug. 14, 1992; 68 FR 61331, Oct. 28, 2003; 69 FR 69106, Nov. 26, 2004]

§ 1955.56 **Real property located in Coastal Barrier Resources System (CBRS).**

(a) *Approval official's scope of authority.* Any action that is not in conflict with the limitations in paragraphs (a)(1), (a)(2) or (a)(3) of this section shall not be undertaken until the approval official has consulted with the appropriate Regional Director of the U.S. Fish and Wildlife Service. The Regional Director may or may not concur that the proposed action does or does not violate the provisions of the Coastal Barrier Resources Act (CBRA). Pursuant to the requirements of the CBRA, and except as specified in paragraphs (b) and (c) of this section, no maintenance or repair action may be taken for property located within a CBRS where:

(1) The action goes beyond maintenance, replacement-in-kind, reconstruction, or repair and would result in the expansion of any roads, structures or facilities. Water and waste disposal

facilities as well as community facilities may be improved to the extent required to meet health and safety requirements but may not be improved or expanded to serve additional users, patients, or residents;

(2) The action is inconsistent with the purposes of the CBRA; or

(3) The property to be repaired or maintained was initially the subject of a financial transaction that violated the CBRA.

(b) *Administrator's review.* Any proposed maintenance or repair action that does not conform to the requirements of paragraph (a) of this section must be forwarded to the Administrator for review and approval. Approval will not be granted unless the Administrator determines, through consultation with the Department of the Interior, that the proposed action does not violate the provisions of the CBRA.

(c) *Emergency provisions.* In emergency situations to prevent imminent loss of life, imminent substantial damage to the inventory property or the disruption of utility service, the approval official may take whatever minimum steps are necessary to prevent such loss or damage without first consulting with the appropriate Regional Director of the U.S. Fish and Wildlife Service. However, the Regional Director must be immediately notified of any such emergency action.

§1955.57 Real property containing underground storage tanks.

Within 30 days of acquisition of real property into inventory, Rural Development must report certain underground storage tanks to the State agency identified by the Environmental Protection Agency (EPA) to receive such reports. Notification will be accomplished by completing an appropriate EPA or alternate State form, if approved by EPA. A State supplement will be issued providing the appropriate forms required by EPA and instructions on processing same.

(a) Underground storage tanks which meet the following criteria must be reported:

(1) It is a tank, or combination of tanks (including pipes which are connected thereto) the volume of which is

ten percent or more beneath the surface of the ground, including the volume of the underground pipes; and

(2) It is not exempt from the reporting requirements as outlined in paragraph (b) of this section; and

(3) The tank contains petroleum or substances defined as hazardous under section 101(14) of the Comprehensive Environmental Response Compensation and Liability Act, 42 U.S.C. 9601. The State Environmental Coordinator should be consulted whenever there is a question regarding the presence of a regulated substance; or

(4) The tank contained a regulated substance, was taken out of operation by Rural Development since January 1, 1974, and remains in the ground. Extensive research of records of inventory property sold before the effective date of this section is not required.

(b) The following underground storage tanks are *exempt* from the EPA reporting requirements:

(1) Farm or residential tanks of 1,100 gallons or less capacity used for storing motor fuel for noncommercial purposes;

(2) Tanks used for storing heating oil for consumptive use on the premises where stored;

(3) Septic tanks;

(4) Pipeline facilities (including gathering lines) regulated under; (i) The Natural Gas Pipeline Safety Act of 1968; (ii) the Hazardous Liquid Pipeline Safety Act of 1979; or (iii) for an intrastate pipeline facility, regulated under State laws comparable to the provisions of law referred to in (b)(4) (i) or (ii) of this section;

(5) Surface impoundments, pits, ponds, or lagoons;

(6) Storm water or wastewater collection systems;

(7) Flow-through process tanks;

(8) Liquid traps or associated gathering lines directly related to oil or gas production and gathering operations; or

(9) Storage tanks situated in an underground area (such as a basement, cellar, mineworking, drift, shaft, or tunnel) if the tank is situated upon or above the surface of the floor.

(c) A copy of each report filed with the designated State agency will be forwarded to and maintained in the State Office by program area.

(d) Prospective purchasers of Rural Development inventory property with a reportable underground storage tank will be informed of the reporting requirement, and provided a copy of the form filed by Rural Development.

(e) In a State which has promulgated additional underground storage tank reporting requirements, Rural DevelopmentRural Development will comply with such requirements and a State supplement will be issued to provide necessary guidance.

(f) Regardless of whether an underground storage tank must be reported under the requirements of this section, if Rural Development personnel detect or believe there has been a release of petroleum or other regulated substance from an underground storage tank on an inventory property, the incident will be reported to the appropriate State Agency, the State Environmental Coordinator and appropriate program chief. These parties will collectively inform the servicing official of the appropriate response action.

§§ 1955.58–1955.59 [Reserved]

§ 1955.60 Inventory property subject to redemption by the borrower.

If inventory property is subject to redemption rights, the State Director, with prior approval of OGC, will issue a State Supplement giving guidance concerning the former borrower's rights, whether or not the property may be leased or sold by the Government, payment of taxes, maintenance, and any other items OGC deems necessary to comply with State laws. Routine care and maintenance will be provided according to § 1955.64 of this subpart to preserve and protect the property. Repairs are limited to those essential to prevent further deterioration of the property or to remove a health or safety hazard to the community in accordance with § 1955.64(a) of this subpart unless State law permits full recovery of cost of repairs in which case usual policy on repairs is applicable. If the former borrower with redemption rights has possession of the property or has a right to lease proceeds, Rural Development will not rent the property

until the redemption period has expired unless the State Director obtains prior authorization from OGC. Further guidance on sale subject to redemption rights is set forth in § 1955.138 of Subpart C of this part.

§ 1955.61 Eviction of persons occupying inventory real property or dispossession of persons in possession of chattel property.

Advice and assistance will be obtained from OGC where eviction from realty or dispossession of chattel property is necessary. Where OGC has given written authorization, eviction may be effected through State courts rather than Federal courts when the former borrower is involved, or through local courts instead of Federal/State courts when the party occupying/possessing the Rural Development property is not the former borrower. In those cases, a State Supplement will be issued to provide explicit instructions. For MFH, eviction of tenants will be handled in accordance with 7 CFR part 3560, subpart D and with the terms of the tenant's lease. If no written lease exists, the State Director will obtain advice from OGC.

[54 FR 20522, May 12, 1989, as amended at 69 FR 69106, Nov. 26, 2004]

§ 1955.62 Removal and disposition of nonsecurity personal property from inventory real property.

If the former borrower has vacated the inventory property but left items of value which do not customarily pass with title to the real estate, such as furniture, personal effects, and chattels not covered by an Rural Development lien, the personal property will be handled as outlined below unless otherwise directed by a State supplement approved by OGC which is necessary to comply with State law. For MFH, the removal and disposition of nonsecurity personal property will be handled in accordance with the tenant's lease or advice from OGC. When property is deemed to have no value, it is recommended that it be photographed for documentation before it is disposed of. The Rural Development official having custody of the property may request advice from the State Office staff as necessary. Actions to effect removal of items of value from inventory property shall be as follows:

(a) *Notification to owner or lienholder.* The servicing official will check the public records to see if there is a lien on any of the personal property.

(1) If there is a lien(s) of record, the servicing official will notify the lienholder(s) by certified mail, return receipt requested, that the personal property will be disposed of by Rural Development unless it is removed from the premises within 7 days from the date of the letter.

(2) If there are no liens of record, or if a lienholder notified in accordance with paragraph (a)(1) of this section fails to remove the property within the time specified, the servicing official will notify the former borrower at the last known address by certified mail, return receipt requested, that the personal property remaining on the premises will be disposed of by Rural Development unless it is removed within 7 days from the date of the letter. If no address can be determined, a copy of the letter should be posted on the front door of the property and documentation entered in the running record of the Rural Development file.

(b) *Disposal of unclaimed personal property.* If the property is not removed by the former borrower or a lienholder after notification as outlined in paragraphs (a)(1) and (a)(2) of this section, the servicing official shall list the items with clear description, estimated value, and indication of which are covered by a lien, if any, and submit the list to the State Director with a request for authorization to have the items removed and disposed of. Based on advice from OGC, the State Director will give authorization and provide instructions for removal and disposal of the personal property. If approved by OGC, the property may be disposed of as follows:

(1) If a reasonable amount can likely be realized by the agency from sale of the personal property, it may be sold at public sale. Items under lien will be sold first and the proceeds up to the amount of the lien paid to the lienholders less a pro rata share of the sale expenses. Proceeds from sale of items not under lien and proceeds in excess of the amount due a lienholder

will be remitted and applied in the following order:

(i) To the inventory account up to the amount of expenses incurred by the Government in connection with sale of the personal property (such as advertising and auctioneer, if used).

(ii) To an unsatisfied balance on the Rural Development loan account, if any.

(iii) To the borrower, if whereabouts are known.

(2) If personal property is not sold, a mover or hauler may be authorized to take the items for moving costs. Refer to RD Instruction 2024-A (available in any Rural Development office) for guidance.

(c) *Payment of costs.* Upon payment of all expenses incurred by the Government in connection with the personal property, Rural Development will allow the former borrower or a lienholder access to the property to reclaim the personal property at any time prior to its disposal.

(d) *Removal of abandoned motor vehicles from inventory property.* Since State laws vary concerning disposal of abandoned motor vehicles, the State Director shall, with the advice of OGC, issue a State supplement outlining the method to be followed which will comply with applicable State laws.

[53 FR 35765, Sept. 14, 1988, as amended at 68 FR 61332, Oct. 28, 2003]

§ 1955.63 Suitability determination.

As soon as real property is acquired, a determination must be made as to whether or not the property can be used for program purposes. The suitability determination will be recorded in the running record of the case file.

(a) *Determination.* The Agency will classify property that secured loans or was acquired under the CONACT as "suitable property" or "surplus property" in accordance with the definitions found in § 1955.53.

(b) *Grouping and subdividing farm properties.* To the maximum extent practicable, the Agency will maximize the opportunity for beginning farmers and ranchers to purchase inventory properties. Farm properties may be subdivided or grouped according to § 1955.140, as feasible, to carry out the .

objectives of the applicable loan program. Properties may also be subdivided to facilitate the granting or selling of a conservation easement or the fee title transfer of portions of a property for conservation purposes. The environmental effects of such actions will be considered pursuant to part 1970 of this chapter.

(c) *Housing property.* Property which secured housing loans will be classified as "program" or "nonprogram (NP)." After a determination of whether the property is suited for retention in the respective program, the repair policy outlined in § 1955.64(a) of this subpart will be followed. In determining whether a property is suited for retention in the program, items such as size, design, possible health and/or safety hazards and obsolescence due to functional, economic, or locational conditions must carefully be considered. Generally, program property will meet, or can be realistically repaired to meet, the standards for existing housing outlined in Subpart A of Part 1944 of this chapter provided the property is typical of modest homes in the area. The cost of repairs will generally not be considered in determining suitability. Since houses, sites and locations vary widely throughout the country, discretion and sound judgment must be used in determining suitability. The majority of houses RHS acquires will be suited for retention and classified as program property. In some instances, property will not be suited for retention in the program and will be classified as "nonprogram (NP)" property. Situations of this type include, but are not limited to:

(1) A dwelling which has been enlarged or improved to the point where it is clearly above modest.

(2) When a determination is made that the property should not have been financed originally.

(3) A dwelling brought into the program as an existing dwelling which met program standards at the time it was originally financed by the Agency but which does not conform to current policies. This includes older and/or larger houses of a type which have proven to create excessive energy and/or maintenance costs to very-low and low-income borrowers.

(4) A dwelling which is obsolete due to location, design, construction or age.

(5) A dwelling which requires major redesign/renovation to be brought to program standards.

(d) [Reserved]

[53 FR 35765, Sept. 14, 1988, as amended at 54 FR 20522, May 12, 1989; 58 FR 58648, Nov. 3, 1993; 60 FR 34455, July 3, 1995; 60 FR 55147, Oct. 27, 1995; 62 FR 44396, Aug. 21, 1997; 68 FR 7700, Feb. 18, 2003; 82 FR 19319, Apr. 27, 2017]

§ 1955.64   [Reserved]

§ 1955.65   Management of inventory and/or custodial real property.

(a) *Authority*—(1) *County Supervisor.* The County Supervisor, with the assistance of the District Director and State Office program staff as necessary, will select the management method(s) used for property which secures (or secured) loans to individuals as defined in this subpart.

(2) *State Director.* The State Director will select the management method to be used for property which secures (or secured) loans to organizations as defined in this subpart. The State Director shall also provide guidance and assistance to County Supervisors and District Directors as necessary to insure that property under their jurisdiction is effectively managed.

(b) *Management methods.* Management methods and requirements will vary depending on such things as the number of properties involved, their density of location, and market conditions. Management tools which may be used effectively range from contracts to secure individual property, have the grass cut, or winterize a dwelling; a simple management contract to provide maintenance and other services on a group of properties (including but not limited to specification writing, inspection of repairs, and yard and directional signs and their installation), or manage an MFH project; blanket-purchase arrangement contracts to obtain services for more than one property; to a broad-scope management contract with a real estate broker or management agent which may include inspection and specification-writing services, making simple repairs, obtaining lessees, collecting rents, coordination with listing brokers in marketing the properties and effecting eviction of tenants when necessary. A contractor may handle evictions only where State laws permit the contractor to do so in his/her own name; a contractor may not pursue eviction in the name of the Government. Custodial property may be managed in the same manner as inventory property except that it may be leased only if it is habitable without repairs in excess of those authorized in §1955.55(c) of this subpart. Farm or organization property, such as rental housing and community facilities, may be operated under a management contract if the State Director has determined it is approporiate to have the property in operation. In any case, the primary consideration in selecting the method of management to be used is to protect the Government's interest. If property to be operated or leased under a management contract is located in an area identified by the Federal Insurance Administration as a special flood or mudslide hazard area, lessees or tenants must be notified to that effect in accordance with §1955.66(e) of this subpart. A management contract which covers property in such a hazard area may provide for the contractor to issue the required notices.

(c) *Obtaining services for management and/or operation of properties.* Services for management, repair, and/or operation of properties will be obtained by contract in accordance with RD Instruction 2024–A (available in any Rural Development office).

(1) *Management contracts.* Management contracts are flexible instruments which may be tailored to meet the specific needs of almost any situation involving custodial or inventory property. This type of contract may be used to manage and maintain SFH properties, farms, and any other type of facility for which Rural Development is responsible. Organization-type properties will be secured, maintained, repaired, and operated if authorized, in accordance with a management plan prepared by the District Director and approved by the State Director if the amount of total debt does not exceed the State Director's loan approval authority, or by the Administrator. For MFH projects, tenant occupancy and selection will be in accordance with the occupancy standards set forth in 7 CFR part 3560, subpart D. Tenants will be required to sign a written lease if one does not exist when the property is acquired or taken into custody. If a contract involves management of an MFH project with 5 or more units, or 5 or more single-family dwellings located in the same subdivision, the contractor must furnish Form HUD 935.2, "Affirmative Fair Housing Marketing Plan," subject to Rural Development's approval. Contracts for management of farm inventory property will be offered on a competitive bid basis, giving preference to persons who live in, and own and operate qualified small businesses in the area where the property is located in accordance with the provisions in RD Instruction 2024–Q (available in any Rural Development office).

(2) *Authority to enter into management contracts.* (i) The County Supervisor may enter into a management contract for basic services involving farms or not more than 25 single-family dwellings; however, the aggregate amount paid under a contract may not exceed the contracting authority limitation for County Supervisors outlined in RD Instruction 2024–A (available in any Rural Development office).

(ii) A District Director may enter into a management contract for basic maintenance and management services for an MFH project within the contracting authority outlined in RD Instruction 2024–A (available in any Rural Development office). The aggregate amount of any contract may not exceed that contracting authority.

(iii) A CO in the State Office may enter into a management contract for basic services involving more than 25 single-family dwellings, a more complex management contract for SFH property, or an appropriate contract for management or operation of farm or organization-type property. The aggregate amount paid under a contract may not exceed the contracting authority limitation for State Office staff outlined in RD Instruction 2024–A (available in any Rural Development office).

(iv) If a proposed management contract will exceed the contracting authority for State Office staff within a short time, a request for contract action will be forwarded to the Administrator, to the attention of the appropriate program division.

(3) *Specification of services.* All management contracts will provide for termination by either the contractor or the Government upon 30 days written notice. Contracts providing for management of multiple properties will also provide for properties to be added or removed from the contractor's assignment whenever necessary, such as when a property is acquired or taken into custody during the period of a contract or when a property is sold from inventory. If a contractor prepares repair specifications, that contractor will be excluded from the solicitation for making the repairs to avoid a conflict of interest.

If a management contract calls for specification writing services, a clause must be inserted in the contract prohibiting the preparer or his/her associates from doing the repair work.

(4) *Costs.* Costs incurred with the management of property will be paid according to RD Instruction 2024-A (available in any Rural Development office). For management of custodial property, costs will be charged to the borrower's account as recoverable; and for management of inventory property as nonrecoverable. Except for management fees, costs of managing MFH inventory property when tenants are still in residence will be paid to the extent possible with rental income. Management fees will be paid to the manager in accordance with RD Instruction 2024-A (available in any Rural Development Office).

(d) *Additional management services.* Additional types of management services and supplies for which the State Director may authorize acquisition include: Appraisal services (except for MFH), security services, newspaper copy preparation services, market data and comparable list acquisition, and tax data acquisition. If the State Director believes there is a need to acquire other services not listed in this paragraph or authorized elsewhere in this subpart, the State Director should make a written request to the Assistant Administrator (appropriate program) for consideration and/or authorization.

[53 FR 35765, Sept. 14, 1988, as amended at 57 FR 36591, Aug. 14, 1992; 69 FR 69106, Nov. 26, 2004; 70 FR 20704, Apr. 21, 2005]

§ 1955.66 Lease of real property.

When inventory real property, except for FSA and MFH properties, cannot be sold promptly, or when custodial property is subject to lengthy liquidation proceedings, leasing may be used as a management tool when it is clearly in the best interest of the Government. Leasing will not be used as a means of deferring other actions which should be taken, such as liquidation of loans in abandonment cases or repair and sale of inventory property. Leases will provide for cancellation by the lessee or the Agency on 30-day written notice unless Special Stipulations in an individual lease for good reason provide otherwise. If extensive repairs are needed to render a custodial property suitable for occupancy, this will preclude its being leased since repairs must be limited to those essential to prevent further deterioration of the security in accordance with § 1955.55(c) of this subpart. The requirements of part 1970 of this chapter will be met for all leases.

(a) *Authority to approve lease of property*—(1) *Custodial property.* Custodial property may be leased pending foreclosure with the servicing official approving the lease on behalf of the Agency.

(2) *Inventory property.* Inventory property may be leased under the following conditions. Except for farm property proposed for a lease under the Homestead Protection Program, any property that is listed or eligible for listing on the National Register of Historic Places may be leased only after the servicing official and the State Historic Preservation Officer determine that the lease will adequately ensure the property's condition and historic character.

(i) *SFH.* SFH inventory will generally not be leased; however, if unusual circumstances indicate leasing may be prudent, the county official is authorized to approve the lease.

(ii) *MFH.* MFH projects will generally not be leased, although individual living units may be leased under a management agreement. After the property is placed under a management contract, the contractor will be responsible for leasing the individual units in accordance with 7 CFR part 3560. In cases where an acceptable management contract cannot be obtained, the District Director may execute individual leases.

(iii) *Organization property other than MFH.* Only the State Director, with the advice of appropriate National Office staff, may approve the lease of organization property other than MFH, such as community facilities, recreation projects, and businesses. A lease of utilities may require approval by State regulatory agencies.

(b) *Selection of lessees for other than farm property.* When the property to be leased is residential, a special effort will be made to reach prospective lessees who might not otherwise apply because of existing community patterns. A lessee will be selected considering the potential as a program applicant for purchase of the property (if property is suited for program purposes) and ability to preserve the property. The leasing official may require verification of income or a credit report (to be paid for by the prospective lessee) as he or she deems necessary to assure payment ability and creditworthiness of the prospective lessee.

(c) *Selection of lessees for FSA property.* FSA inventory property may only be leased to an eligible beginning farmer or rancher who was selected to purchase the property through the random selection process in accordance with §1955.107(a)(2)(ii) of subpart C of this part. The applicant must have been able to demonstrate a feasible farm plan and Agency funds must have been unavailable at the time of the sale. Any applicant determined not to be a beginning farmer or rancher may request that the State Executive Director conduct an expedited review in accordance with §1955.107(a)(2)(ii) of subpart C of this part.

(d) *Property securing Farm Credit Programs loans located within an Indian Reservation.* (1) State Executive Directors will contact the Bureau of Indian Affairs Agency supervisor to determine the boundaries of Indian Reservations and Indian allotments.

(2) Not later than 90 days after acquiring a property, FSA will afford the Indian tribe having jurisdiction over the Indian reservation within which the inventory property is located an opportunity to purchase the property. The purchase shall be in accordance with the priority rights as follows:

(i) To a member of the Indian tribe that has jurisdiction over the reservation within which the real property is located;

(ii) To an Indian corporate entity;

(iii) To the Indian tribe.

(3) The Indian tribe having jurisdiction over the Indian reservation may revise the order of priority and may restrict the eligibility for purchase to:

(i) Persons who are members of such Indian tribe;

(ii) Indian corporate entities that are authorized by such Indian tribe to purchase lands within the boundaries of the reservation; or

(iii) The Indian tribe itself.

(4) If any individual, Indian corporate entity, or Indian tribe covered in paragraphs (d)(1) and (d)(2) of this section wishes to purchase the property, the county official must determine the prospective purchaser has the financial resources and management skills and experience that is sufficient to assure a reasonable prospect that the terms of the purchase agreement can be fulfilled.

(5) If the real property is not purchased by any individual, Indian corporate entity or Indian tribe pursuant to paragraphs (d)(1) and (d)(2) of this section and all appeals have concluded, the State Executive Director shall transfer the property to the Secretary of the Interior if they are agreeable. If present on the property being transferred, important resources will be protected as outlined in §§1955.137 and 1955.139 of subpart C of this part.

(6) Properties within a reservation formerly owned by entities and non-tribal members will be treated as regular inventory that is not located on an Indian Reservation and disposed of pursuant to this part.

(e) *Lease amount.* Inventory property will be leased for an amount equal to that for which similar properties in the area are being leased or rented (market rent). Inventory property will not be leased for a token amount.

(1) *Farm property.* To arrive at a market rent amount, the county official will make a survey of lease amounts of farms in the immediate area with similar soils, capabilities, and income potential. The income-producing capability of the property during the term of the lease must also be considered. This rental data will be maintained in an operational file as well as in the running records of case files for leased inventory properties. While cash rent is preferred, the lease of a farm on a crop-share basis may be approved if this is the customary method in the area. The lessee will market the crops, provide FSA with documented evidence of crop income, and pay the pro rata share of the income to FSA.

(2) *SFH property.* The lease amount will be the market rent unless the lessee is a potential program applicant, in which case the lease amount may be set at an amount approximating the monthly payment if a loan were made (reflecting payment assistance, if any) calculated on the basis of the price of the house and income of the lessee, plus $1/12$ of the estimated real estate taxes, property insurance, and maintenance which would be payable by a homeowner.

(3) *Property other than farm or SFH.* Any inventory property other than a farm or single-family dwelling will generally be leased for market rent for that type property in the area. However, such property may be leased for less than market rent with prior approval of the Administrator.

(f) *Property containing wetlands or located in a floodplain or mudslide hazard area.* Inventory property located in areas identified by the Federal Insurance Administration as special flood or mudslide hazard areas will not be leased or operated under a management contract without prior written notice of the hazard to the prospective lessee or tenant. If property is leased by FSA, the servicing official will provide the notice, and if property is leased under a management contract, the contractor must provide the notice in compliance with a provision to that effect included in the contract. The notice must be in writing, signed by the servicing official or the contractor, and delivered to the prospective lessee or tenant at least one day before the lease is signed. A copy of the notice will be attached to the original and each copy of the lease. Property containing floodplains and wetlands will be leased subject to the same use restrictions as contained in § 1955.137(a)(1) of subpart C of this part.

(g) *Highly erodible land.* If farm inventory property contains "highly erodible land," as determined by the NRCS, the lease must include conservation practices specified by the NRCS and approved by FSA as a condition for leasing.

(h) *Lease of FSA property with option to purchase.* A beginning farmer or rancher lessee will be given an option to purchase farm property. Terms of the option will be set forth as part of the lease as a special stipulation.

(1) The lease payments will not be applied toward the purchase price.

(2) The purchase price (option price) will be the advertised sales price as determined by an appraisal prepared in accordance with § 761.7 of this title.

(3) For inventory properties leased to a beginning farmer or rancher applicant, the term of the lease shall be the earlier of:

(i) A period not to exceed 18 months from the date that the applicant was selected to purchase the inventory farm, or

(ii) The date that direct, guaranteed, credit sale or other Agency funds become available for the beginning farmer or rancher to close the sale.

(4) Indian tribes or tribal corporations which utilize the Indian Land Acquisition program will be allowed to purchase the property for its market value less the contributory value of the buildings, in accordance with subpart N of part 1823 of this chapter.

(i) *Costs.* The costs of repairs to leased property will be paid by the Government. However, the Government will not pay costs of utilities or any other costs of operation of the property by the lessee. Repairs will be obtained pursuant to subpart B of part 1924 of this chapter. Expenditures on custodial property as limited in § 1955.55 (c) (2) of this subpart will be charged to the borrower's account as recoverable costs.

(j) *Security deposit.* A security deposit in at least the amount of one month's rent will be required from all lessees of SFH properties. The security deposit for farm property should be determined by considering only the improvements or facilities which might be subject to misuse or abuse during the term of the lease. For all other types of property, the leasing official may determine whether or not a security deposit will be required and the amount of the deposit.

(k) *Lease form.* Form RD 1955–20 approved by OGC will be used by the agency to lease property.

(l) *Lease income.* Lease proceeds will be applied as follows:

(1) *Custodial property.* The proceeds from a lease of custodial property will be applied to the borrower's account as an extra payment unless foreclosure proceedings require that such payments be held in suspense.

(2) *Inventory property.* The proceeds from a lease of inventory property will be applied to the lease account.

[62 FR 44397, Aug. 21, 1997, as amended at 64 FR 62568, Nov. 17, 1999; 68 FR 61332, Oct. 28, 2003; 69 FR 69106, Nov. 26, 2004; 82 FR 19319, Apr. 27, 2017]

§§ 1955.67–1955.71 [Reserved]

§ 1955.72 Utilization of inventory housing by Federal Emergency Management Agency (FEMA) or under a Memorandum of Understanding between the Agency and the Department of Health and Human Services (HHS) for transitional housing for the homeless.

(a) *FEMA.* By a Memorandum of Understanding between the Agency and FEMA, inventory housing property not under lease or sales agreement may be made available to shelter victims in an area designated as a major disaster area by the President. See Exhibit A of this subpart. Authority is hereby delegated to the State Director to implement this Memorandum of Understanding; and the State Director may redelegate this authority to County Supervisors or District Directors.

(b) *HHS.* By a Memorandum of Understanding between the Agency and HHS, inventory housing property not under lease or sales agreement may be made available by lease to public bodies and nonprofit organizations to provide transitional housing for the homeless. See Exhibit D of this subpart. Authority is hereby delegated to the State Director to implement this Memorandum of Understanding; and the State Director may redelegate this authority to County Supervisors or District Directors. Copies of all executed leases and/or questions regarding this program should be referred by State Offices to the Single Family Housing Servicing and Property Management (SFH/SPM) Division in the National Office.

[54 FR 20523, May 12, 1989, as amended at 60 FR 34455, July 3, 1995]

§§ 1955.73–1955.80 [Reserved]

§ 1955.81 Exception authority.

The Administrator may, in individual cases, make an exception to any requirement or provision of this subpart, or address any omission of this subpart which is not inconsistent with the authorizing statute or other applicable law, if the Administrator determines that the Government's interest would be adversely affected or the immediate health and/or safety of tenants or the community are endangered if there is no adverse effect on the Government's interest. The Administrator will exercise this authority upon request of the State Director with the recommendation of the appropriate program Assistant Administrator or upon a request initiated by the appropriate program Assistant Administrator. Requests for exceptions must be made in writing and supported with documentation to explain the adverse effect, propose alternative courses of action, and show how the adverse effect will be eliminated or minimized if the exception is granted.

[53 FR 35765, Sept. 14, 1988, as amended at 58 FR 58649, Nov. 3, 1993]

§ 1955.82 State supplements.

State supplements will be prepared with the assistance of OGC as necessary to comply with State laws or only as specifically authorized in this regulation to provide guidance to Rural Development officials. State supplements applicable to MFH must have prior approval of the National Office; others may receive post approval. Requests for approval for those affecting MFH must include complete justification, citations of State law, and an opinion from OGC.

§§ 1955.83–1955.99 [Reserved]

§ 1955.100 OMB control number.

The collection of information requirements in this regulation have been approved by the Office of Management and Budget and assigned OMB control number 0575–0110.

EXHIBIT A TO SUBPART B OF PART 1955— MEMORANDUM OF UNDERSTANDING BETWEEN THE FEDERAL EMERGENCY MANAGEMENT AGENCY AND RURAL DEVELOPMENT

EDITORIAL NOTE: Exhibit A is not published in the Code of Federal Regulations. It is available in any Rural Development County Office.

[53 FR 35765, Sept. 14, 1988, as amended at 80 FR 9897, Feb. 24, 2015]

EXHIBIT B TO SUBPART B OF PART 1955— NOTIFICATION OF TRIBE OF AVAILABILITY OF FARM PROPERTY FOR PURCHASE

(To Be Used By Farm Service Agency To Notify Tribe)

From: County official
To: (Name of Tribe and address)
Subject: Availability of Farm Property for Purchase
[To be Used within 90 days of acquisition]
Recently the Farm Service Agency (FSA) acquired title to _____ acres of farm real property located within the boundaries of your Reservation. The previous owner of this property was _____. The property is available for purchase by persons who are members of your tribe, an Indian Corporate entity, or the tribe itself. Our regulations provide for those three distinct priority categories which may be eligible; however, you may revise the order of the priority categories and may restrict the eligibility to one or any combination of categories. Following is a more detailed description of these categories:
1. Persons who are members of your Tribe. Individuals so selected must be able to meet the eligibility criteria for the purchase of Government inventory property and be able to carry on a family farming operation. Those persons not eligible for FSA's regular programs may also purchase this property as a Non-Program loan on ineligible rates and terms.
2. Indian corporate entities. You may restrict eligible Indian corporate entities to those authorized by your Tribe to purchase lands within the boundaries of your Reservation. These entities also must meet the basic eligibility criteria established for the type of assistance granted.
3. The Tribe itself is also considered eligible to exercise their right to purchase the property. If available, Indian Land Acquisition funds may be used or the property financed as a Non-Program loan on ineligible rates and terms.
We are requesting that you notify the local FSA county office of your selection or intentions within 45 days of receipt of this letter, regarding the purchase of this real estate. If you have questions regarding eligibility for any of the groups mentioned above, please contact our office. If the Tribe wishes to purchase the property, but is unable to do so at this time, contact with the FSA county office should be made.

Sincerely,

County official

[62 FR 44399, Aug. 21, 1997]

EXHIBIT C TO SUBPART B OF PART 1955— COOPERATIVE AGREEMENT (EXAMPLE)

EDITORIAL NOTE: Exhibit C is not published in the Code of Federal Regulations. It is available in any Rural Development County Office.

EXHIBIT D TO SUBPART B OF PART 1955— FACT SHEET—THE FEDERAL INTERAGENCY TASK FORCE ON FOOD AND SHELTER FOR THE HOMELESS

EDITORIAL NOTE: Exhibit D is not published in the Code of Federal Regulations. It is available in any Rural Development County Office.

Subpart C—Disposal of Inventory Property

INTRODUCTION

§ 1955.101 Purpose.

This subpart delegates program authority and prescribes policies and procedures for the sale of inventory property including real estate, related real estate rights, and chattels. It also covers the granting of easements and rights-of-way on inventory property. Credit sales of inventory property to ineligible (non-program (NP)) purchasers will be handled in accordance with Subpart J of Part 1951 of this chapter, except Community and Business Programs (C&BP) and Multi-Family Housing (MFH) which will be handled in accordance with this Subpart. In addition, credit sales of Single Family Housing (SFH) properties converted to MFH will be handled in accordance with this Subpart.This subpart does not apply to Farm Service Agency, Farm Loan Programs, Single Family Housing (SFH) inventory property, or to the Rural Rental Housing, Rural Cooperative Housing, and Farm Labor Housing Programs. In addition, this subpart does not apply to Water and Waste Programs of the Rural Utilities Service, Watershed loans, and Resource Conservation and Development loans, which are serviced under part 1782 of this title.

[72 FR 55019, Sept. 28, 2007, as amended at 72 FR 64123, Nov. 15, 2007]

§1955.102 Policy.

The terms "nonprogram (NP)" and "ineligible" may be used interchangeably throughout this subpart, but are identical in their meaning. Sales efforts will be initiated as soon as property is acquired in order to effect sale at the earliest practicable time. When a property is of a nature that will enable a qualified applicant for one of the applicable loan programs to meet the objectives of that loan program, preference will be given to the program applicants. Sales are authorized for program purposes which differ from the purposes of the loan the property formerly secured, and property which secured more than one type loan may be sold under the program most appropriate for the specific property and community needs as long as the price is not diminished. Examples are: (RH) property; detached Labor Housing or Rural Rental Housing units may be sold as SFH units; or SFH units may be sold as a Rural Rental Housing project. All such properties and applicants must meet the requirements for the loan program under which the sale is proposed.

[53 FR 35776, Sept. 14, 1988, as amended at 58 FR 52652, Oct. 12, 1993; 62 FR 44399, Aug. 21, 1997]

§1955.103 Definitions.

As used in this subpart, the following apply:

Approval official. The Rural Development official having loan and grant approval authority auhorized under Subpart A of Part 1901 of this chapter.

Auction sale. A public sale in which property is sold to the highest bidder in open verbal competition.

Beginning farmer or rancher. A beginning farmer or rancher is an individual or entity who:

(1) Is an eligible applicant for FO loan assistance in accordance with §1943.12 of subpart A of part 1943 of this chapter or §1980.180 of subpart B of part 1980 of this chapter.

(2) Has not operated a farm or ranch, or who has operated a farm or ranch for not more than 10 years. This requirement applies to all members of an entity.

(3) Will materially and substantially participate in the operation of the farm or ranch.

(i) In the case of a loan made to an individual, individually or with the immediate family, material and substantial participation requires that the individual provide substantial day-to-day labor and management of the farm or ranch, consistent with the practices in the county or State where the farm is located.

(ii) In the case of a loan made to an entity, all members must materially and substantially participate in the operation of the farm or ranch. Material and substantial participation requires that the individual provides some amount of the management, or labor and management necessary for day-to-day activities, such that if the individual did not provide these inputs, operation of the farm or ranch would be seriously impaired.

(4) Agrees to participate in any loan assessment, borrower training, and financial management programs required by Rural Development regulations.

(5) Does not own real farm or ranch property or who, directly or through interests in family farm entities, owns real farm or ranch property, the aggregate acreage of which does not exceed 30 percent of the average farm or ranch acreage of the farms or ranches in the

county where the property is located. If the farm is located in more than one county, the average farm acreage of the county where the applicant's residence is located will be used in the calculation. If the applicant's residence is not located on the farm or if the applicant is an entity, the average farm acreage of the county where the major portion of the farm is located will be used. The average county farm or ranch acreage will be determined from the most recent Census of Agriculture developed by the U.S. Department of Commerce, Bureau of the Census. State Directors will publish State supplements containing the average farm or ranch acreage by county.

(6) Demonstrates that the available resources of the applicant and spouse (if any) are not sufficient to enable the applicant to enter or continue farming or ranching on a viable scale.

(7) In the case of an entity:

(i) All the members are related by blood or marriage.

(ii) All the stockholders in a corporation are qualified beginning farmers or ranchers.

Borrower. An individual or entity which has outstanding obligations to the Rural Development under any Farmer Programs loan(s), without regard to whether the loan has been accelerated. A borrower includes all parties liable for the Rural Development debt, including collection-only borrowers, except for debtors whose total loans and accounts have been voluntarily or involuntarily foreclosed or liquidated, or who have been discharged of all Rural Development debt.

Capitalization value. The value determined in accordance with subpart E of part 1922 of this chapter.

Closing agent. An attorney or title insurance company which is approved as a loan closing agent in accordance with subpart B of part 1927 of this chapter.

CONACT or CONACT property. Property acquired or sold pursuant to the Consolidated Farm and Rural Development Act (CONACT). Within this subpart, it shall also be construed to cover property which secured loans made pursuant to the Emergency Agricultural Credit Act of 1984; the Food Security Act of 1985; and other statutes giving agricultural lending authority to the respective Agency.

Credit sale. A sale in which financing is provided to an applicant for the purchase of inventory property.

Decent, safe and sanitary (DSS) housing. Standards required for the sale of Government acquired SFH, MFH and LH structures acquired pursuant to the Housing Act of 1949, as amended. "DSS" housing unit(s) are structures which meet the requirements of Rural Development as described in Subpart A of Part 1924 of this chapter for existing construction or if not meeting the requirements:

(1) Are structurally sound and habitable,

(2) Have a potable water supply,

(3) Have functionally adequate, safe and operable heating, plumbing, electrical and sewage disposal systems,

(4) Meet the Thermal Performance Standards as outlined in exhibit D of subpart A of part 1924 of this chapter, and

(5) Are safe; that is, a hazard does not exist that would endanger the safety of dwelling occupants.

Eligible terms. Credit terms, for other than SFH or MFH property sales, prescribed in Rural Development program regulations for its various loan programs; available only to persons/entities meeting eligibility requirements set forth for the respective loan program. For SFH and MFH properties, see the definition of "Program terms."

Farmer program loans. This includes Farm Ownership (FO), Soil and Water (SW), Recreation (RL), Economic Opportunity (EO), Operating (OL), Emergency (EM), Economic Emergency (EE), Special Livestock (SL), Softwood Timber (ST) and Rural Housing loans for farm service buildings (RHF).

Homestead protection (FP only). The program which permits former Farmer Program borrowers to lease their former principal residence with an option to buy. See subpart S of part 1951 of this chapter.

Indian Reservation. All land located within the limits of any Indian reservation under the jurisdiction of the United States notwithstanding the issuance of any patent and including rights-of-way running through the reservation; trust or restricted land located within the boundaries of a former reservation of a federally recognized Indian Tribe in the State of Oklahoma; or all Indian allotments the Indian titles to which have not been extinguished if such allotments are subject to the jurisdiction of a federally recognized Indian Tribe.

Ineligible terms. Credit terms, for other than SFH or MFH property sales, offered for the convenience of the Government to facilitate sales; more stringent than terms offered under Rural Development's loan programs. Applicable when the purchaser does not meet program eligibility requirements or when the property is classified as surplus. Loans made on ineligible terms are classified as Nonprogram (NP) loans and are serviced accordingly. For SFH and MFH properties, see the definition of "Nonprogram (NP) terms."

Inventory property. Property for which title is vested in the Government and which secured an a Rural Development loan loan or which was acquired from another Agency for program purposes.

Market value. The most probable price which property should bring, as of a specific date, in a competitive and open market, assuming the buyer and seller are prudent and knowledgeable, and the price is not affected by undue stimulus such as forced sale or loan interest subsidy.

Negotiated sale. A sale in which there is a bargaining of price and/or terms.

Nonprogram (NP) property. SFH and MFH property acquired pursuant to the Housing Act of 1949, as amended, that cannot be used by a borrower to effectively carry out the objectives of the respective loan program; for example, a dwelling that cannot be feasibly repaired to meet the requirements for existing housing as described in subpart A of part 1944 of this chapter. It may contain a structure which would meet program standards, however is so remotely located it would not serve as an adequate residential unit or be an older house which is excessively expensive to heat and/or maintain for a very-low or low-income homeowner.

Nonprogram (NP) terms. Credit terms for SFH or MFH property sales, offered for the convenience of the Government to facilitate sales; more stringent than terms offered under Rural Development's loan programs. Applicable when the purchaser does not meet program eligibility requirements or when the property is classified as nonprogram (NP). Loans made on NP terms are classified as NP loans and are serviced accordingly. For property other than SFH and MFH, see the definition of "Ineligible terms."

Organization property. Property for which the following loans were made is considered organization property. Community Facility (CF); Water and Waste Disposal (WWD); Association Recreation; Watershed (WS); Resource Conservation and Development (RC&D); loans to associations for Shift-In-Land Use (Grazing Association); loans to associations for Irrigation and Drainage and other soil and water conservation measures; loans to Indian Tribes and Tribal corporations; Rural Rental Housing (RRH) to both groups and individuals; Rural Cooperative Housing (RCH); Rural Housing Site (RHS); Labor Housing (LH) to both groups and individuals; Business and Industry (B&I) to both individuals and groups or corporations; Rural Development Loan Fund (RDLF); Intermediary Relending Program (IRP); Nonprofit National Corporations (NNC); and Economic Opportunity Cooperative (EOC). Housing-type (RHS, RCH, RRH and LH) organization property is referred to collectively in this subpart as Multiple Family Housing (MFH) property.

Owner. An individual or an entity which owned the farm but who may or may not have been operating the farm at the time the farm was taken into inventory.

Participating broker. A duly licensed real estate broker who has executed a listing agreement with Rural Development.

Program property. SFH and MFH inventory property that can be used to effectively carry out the objectives of their respective loan programs with financing through that program. Inventory property located in an area where the designation has been changed from rural to nonrural will be considered as if it were still in a rural area.

Program terms. Credit terms for SFH or MFH property sales, prescribed in Rural Development program regulations for its various loan programs; available only to persons/entities meeting eligibility requirements set forth for the respective loan program. For property sales other than SFH and MFH, see the definition of "Eligible terms."

Regular Agency sale. Sale made by Rural Deveopment employees or real estate brokers other than by sealed bid, auction, or negotiation.

Regular sale. Sale by Rural Development employees or real estate brokers other than by sealed bid, auction or negotiation.

Safe. No hazard exists on property which would likely endanger the health or safety of occupants or users.

Sealed bid sale. A public sale in which property is offered to the highest bidder by prior written bid submitted in a sealed envelope.

Servicing official. For loans to individuals, as defined in § 1955.53 of subpart B of part 1955 of this chapter, the servicing official is the County Supervisor. For all other loans, excluding insured B&I, the servicing official is the District Director. For insured B&I loans, the servicing official is the State Director.

Socially disadvantaged applicant (SDA). An applicant who is a member of a socially disadvantaged group whose members have been subjected to racial, ethnic, or gender prejudice because of their identity as a member of a group, without regard to their individual qualities. For entity SDA applicants, the majority interest in the entity must be held by socially disadvantaged individuals. The Agency has identified socially disadvantaged groups as Women, Blacks, American Indians, Alaskan Natives, Hispanics, Asians, and Pacific Islanders.

Suitable property. Real property that could be used to carry out the objectives of Rural Development's loan programs with financing provided through that program.

Surplus property. Property that cannot be used to carry out the objectives of financing available through the applicable loan program.

[50 FR 23904, June 7, 1985]

EDITORIAL NOTE: For FEDERAL REGISTER citations affecting § 1955.103, see the List of CFR Sections Affected, which appears in the Finding Aids section of the printed volume and at *www.govinfo.gov.*

§ 1955.104 Authorities and responsibilities.

(a) *Redelegation of authority.* Rural Development officials will redelegate authorities to the maximum extent possible, consistent with program objectives and available resources.

(1) Any authority in this subpart which is specifically provided to the Administrator or to an Assistant Administrator may only be delegated to a State Director. The State Director cannot redelegate such authority.

(2) Except as provided in paragraph (a)(1) of this section, the State Director may redelegate, in writing, any authority delegated to the State Director in this subpart, unless specifically excluded, to a Program Chief, Program Specialist, or Property Management Specialist on the State Office staff.

(3) The District Director may redelegate, in writing, any authority delegated to the District Director in this subpart to an Assistant District Director or District Loan Specialist. Authority of District Directors in this subpart applies to Area Loan Specialists in Alaska and the Director for the Western Pacific Territories.

(4) The County Supervisor may redelegate, in writing, any authority delegated to the County Supervisor in this subpart to an Assistant County Supervisor, GS-7 or above, who is determined by the County Supervisor to be qualified. Authority of County Supervisors in this subpart applies to Area Loan Specialists in Alaska, Island Directors in Hawaii, the Director for the Western Pacific Territories, and Area Supervisors in the Western Pacific Territories and American Samoa.

(b) *Responsibility.* (1) National Office program directors are responsible for reviewing and providing guidance to State, District and County Offices in disposing of inventory property.

(2) The State Director is responsible for establishing an effective program and insuring compliance with Rural Development regulations.

(3) District Directors are responsible for disposal actions for programs under

their supervision and for monitoring County Office compliance with Rural Development regulations and State Supplements.

(4) County Supervisors are responsible for timely disposal of inventory property for programs under their supervision.

[53 FR 27830, July 25, 1988, as amended at 66 FR 7568, Jan. 24, 2001]

CONSOLIDATED FARM AND RURAL DEVELOPMENT ACT (CONACT) REAL PROPERTY

§ 1955.105 Real property affected (CONACT).

(a) *Loan types.* Sections 1955.106–1955.109 of this subpart prescribe procedures for the sale of inventory real property which secured any of the following type of loans (referred to as CONACT property in this subpart): Farm Ownership (FO); Recreation (RL); Soil and Water (SW); Operating (OL); Emergency (EM); Economic Opportunity (EO); Economic Emergency (EE); Softwood Timber (ST); Community Facility (CF); Water and Waste Disposal (WWD); Reserve Conservation and Development (RC&D); Watershed (WS); Association Recreation; EOC: Rural Renewal; Water Facility; Business and Industry (B&I); Rural Development Loan Fund (RDLF); Intermediary Relending Program (IRP); Nonprofit National Corporation (NNC); Irrigation and Drainage; Shift-in-Land Use (Grazing Association); and loans to Indian Tribes and Tribal Corporations. Homestead Protection, as set forth in Subpart S of Part 1951 of this chapter, is only applicable to Farmer Program loans as defined in § 1955.103 of this subpart.

(b) *Controlled substance conviction.* In accordance with the Food Security Act of 1985 (Pub. L. 99–198), after December 23, 1985, if an individual or any member, stockholder, partner, or joint operator of an entity is convicted under Federal or State law of planting, cultivating, growing, producing, harvesting, or storing a controlled substance (see 21 CFR Part 1308, which is Exhibit C to Subpart A of Part 1941 of this chapter and is available in any Rural Development office, for the definition of "controlled substance") prior to a credit sale approval in any crop year, the individual or entity shall be ineligible for a credit sale for the crop year in which the individual or member, stockholder, partner, or joint operator of the entity was convicted and the four succeeding crop years. Applicants will attest on Form RD 410–1, "Application for RD Services," that as individuals or that its members, if an entity, have not been convicted of such crime after December 23, 1985.

(c) *Effects of farm property sales on farm values.* State Directors will analyze farm real estate market conditions within the geographic areas of their jurisdiction and determine whether or not the sale of the Rural Development farm inventory properties will have a detrimental effect on the value of farms within these areas. Such analysis will be carried out in January of each year and as often throughout the year as necessary to reflect changing farm real estate conditions. If the analyses of farm real estate conditions indicate that such sales would put downward pressure on farm real estate values in any area, all farm properties within the area affected will be withheld from the market and managed in accordance with the provisions of Subpart B of this Part until such time that a subsequent analysis indicates otherwise. The State Director will notify, in writing, the County Supervisor(s) servicing those areas that are restricted from selling farm inventory property. State Directors in consultation with other lenders, real estate agents, auctioneers, and others in the community will analyze all available information such as:

(1) The number of farms and acres that Rural Development expects to acquire in inventory.

(2) The number of farms and acres other lenders expect to acquire in inventory.

(3) The number of farms and acres that Rural Development currently has in inventory.

(4) The number of farms and acres other lenders currently have in inventory.

(5) The number of farms not included in paragraphs (c)(3) and (c)(4) of this section which are currently listed for sale.

111

(6) Published real estate values and trend reports such as those available from the Economic Research Service or professional appraisal organizations.

(d) *Highly erodible land.* If farm inventory property contains "highly erodible land," as determined by the SCS, the lease must include conservation practices specified by the SCS and approved by Rural Development as a condition for leasing. Refer to § 1955.137(d) of this subpart for implementation requirements.

[53 FR 35777, Sept. 14, 1988, as amended at 57 FR 19528, May 7, 1992; 58 FR 58649, Nov. 3, 1993; 62 FR 44399, Aug. 21, 1997]

§ 1955.106 Disposition of farm property.

(a) *Rights of previous owner and notification.* Before property which secured a Farm Credit Programs loan is taken into inventory, the FSA county official will advise the borrower-owner of Homestead Protection rights (see subpart S of part 1951 of this chapter.)

(b) *Racial, ethnic, and gender consideration.* The County Supervisor will make a special effort to insure that prospective purchasers, who traditionally would not be expected to apply for farm ownership loan assistance because of existing racial, ethnic, or gender prejudice, are informed of the availability of the Socially Disadvantaged Program. Emphasis will be placed on providing assistance to such socially disadvantaged applicants in accordance with the applicable sections of subpart A of part 1943 of this chapter.

(c) *Nonprogram (NP) borrowers.* Nonprogram (NP) borrowers are not eligible for Homestead Protection provisions as set forth in subpart S of part 1951 of this chapter. When it is determined that all conditions of § 1951.558(b) of subpart L of part 1951 of this chapter have been met, loans for unauthorized assistance will be treated as authorized loans and will be eligible for homestead protection.

[53 FR 35777, Sept. 14, 1988, as amended at 58 FR 58649, Nov. 3, 1993; 62 FR 44399, Aug. 21, 1997]

§ 1955.107 Sale of FSA property (CONACT).

FSA inventory property will be advertised for sale in accordance with the provisions of this subpart. If a request is received from a Federal or State agency for transfer of a property for conservation purposes, the advertisement should be conditional on that possibility. Real property will be managed in accordance with the provisions of subpart B of this part until sold.

(a) *Suitable Property.* Not later than 15 days from the date of acquisition, the Agency will advertise suitable property for sale. For properties currently under a lease, except leases to beginning farmers and ranchers under § 1955.66(a)(2)(iii) of subpart B of this part, the property will be advertised for sale not later than 60 days after the lease expires or is terminated. There will be a preference for beginning farmers or ranchers. The advertisement will contain a provision to lease the property to a beginning farmer or rancher for up to 18 months should FSA credit assistance not be available at the time of sale. The first advertisement will not be required to contain the sales price but it should inform potential beginning farmer or rancher applicants that applications will be accepted pending completion of the advertisement process. When possible, the sale of suitable FSA property should be handled by county officials. Farm property will be advertised for sale by publishing, as a minimum, two weekly advertisements in at least two newspapers that are widely circulated in the area in which the farm is located. Consideration will be given to advertising inventory properties in major farm publications. Either Form RD 1955-40 or Form RD 1955-41, "Notice of Sale," will be posted in a prominent place in the county. Maximum publicity should be given to the sale under guidance provided by § 1955.146 of this subpart and care should be taken to spell out eligibility criteria. Tribal Councils or other recognized Indian governing bodies having jurisdiction over Indian reservations (see § 1955.103 of this subpart) shall be responsible for notifying those parties in § 1955.66(d)(2) of subpart B of this part.

(1) *Price.* Property will be advertised for sale for its appraised market value based on the condition of the property at the time it is made available for

sale. The market value will be determined by an appraisal made in accordance with §761.7 of this title. Property contaminated with hazardous waste will be appraised "as improved" which will be used as the sale price for advertisement to beginning farmers or ranchers.

(2) *Selection of purchaser*. After homestead protection rights have expired, suitable farmland must be sold in the priority outlined in this paragraph. When farm inventory property is larger than family size, the property will be subdivided into suitable family size farms pursuant to §1955.140 of this subpart. ·

(i) *Sale to beginning farmers/ranchers*. Not later than 135 days from the date of acquisition, FSA will sell suitable farm property, with a priority given to applicants who are classified as beginning farmers or ranchers, as defined in §1955.103, at the time of sale.

(ii) *Random selection*. The county official will first determine whether applicants meet the eligibility requirements of a beginning farmer or rancher. For applicants who are not determined to be beginning farmers or ranchers, they may request that the State Executive Director provide an expedited review and determination of whether the applicant is a beginning farmer or rancher for the purpose of acquiring inventory property. This review shall take place not later than 30 days after denial of the application. The State Executive Director's review decision shall be final and is not administratively appealable. When there is more than one beginning farmer or rancher applicant, the Agency will select by lot by placing the names in a receptacle and drawing names sequentially. Drawn offers will be numbered and those drawn after the first drawn name will be held in suspense pending sale to the successful applicant. The random selection drawing will be open to the public, and applicants will be advised of the time and place.

(iii) *Notification of applicants not selected to purchase suitable farmland*. When the Agency selects an applicant to purchase suitable farmland, in accordance with this paragraph, all applicants not selected will be notified in writing that they were not selected.

The outcome of the random selection by lot is not appealable if such selection is conducted in accordance with this subpart.

(3) *Credit sale procedure*. Subject to the availability of funds, credit sale to program applicants will be processed as follows:

(i) The interest rate charged by the Agency will be the lower of the interest rates in effect at the time of loan approval or closing.

(ii) The loan limits for the requested type of assistance are applicable to a credit sale to an eligible applicant.

(iii) Title clearance and loan closing for a credit sale and any subsequent loan to be closed simultaneously must be the same as for an initial loan except that:

(A) Form RD 1955–49, "Quitclaim Deed," or other form of nonwarranty deed approved by the Office of the General Counsel (OGC) will be used.

(B) The buyer will pay attorney's fees and title insurance costs, recording fees, and other customary fees unless they are included in a subsequent loan. A subsequent loan may not be made for the primary purpose of paying closing costs and fees.

(iv) Property sold on credit sale may not be used for any purpose that will contribute to excessive erosion of highly erodible land or to the conversion of wetlands to produce an agricultural commodity. All prospective buyers will be notified in writing as a part of the property advertisement of the presence of highly erodible land and wetlands on inventory property.

(b) *Surplus property and suitable property not sold to a beginning farmer or rancher*. Except where a lessee is exercising the option to purchase under the Homestead Protection provision of subpart S of part 1951 of this chapter, surplus property will be offered for public sale by sealed bid or auction within 15 days from the date of acquisition in accordance with §1955.147 or §1955.148. Suitable farm property which has been advertised for sale to a beginning farmer or rancher in accordance with paragraph (a) of this section, but has not sold within 135 days from the date of acquisition will be offered for public sale by sealed bid or auction to the highest bidder as provided in paragraph (b)(1) of this section. All prospective buyers will be notified in writing as part of the property advertisement of the presence of any highly erodible land, converted wetlands, floodplains, wetlands, or other special characteristics of the property that may limit its use or cause an easement to be placed on the property.

(1) *Advertising surplus property.* FSA will advertise surplus property for sale by sealed bid or auction within 15 days from the date of acquisition or, for those suitable properties not sold to beginning farmers or ranchers in accordance with this section, within 135 days of the date of acquisition.

(2) *Sale by sealed bid or auction.* Surplus real estate must be offered for public sale by sealed bid or auction and must be sold no later than 165 days from the date of acquisition to the highest bidder. Preference will be given to a cash offer which is at least *percent of the highest offer requiring credit. (*Refer to Exhibit B of RD Instruction 440.1 (available in any Agency office) for the current percentage.) Equally acceptable sealed bid offers will be decided by lot.

(3) *Negotiated sale.* If no acceptable bid is received through the sealed bid or auction process, the State Executive Director will sell surplus property at the maximum price obtainable without further public notice by negotiation with interested parties, including all previous bidders. The rates and terms offered for a credit sale through negotiation will be within the limitations established in paragraph (b) (4) of this section. A sale made through negotiation will require a bid deposit of not less than 10 percent of the negotiated price in the form of a cashier's check, certified check, postal or bank money order, or bank draft payable to FSA. Preference will be given to a cash offer which is at least * percent of the highest offer requiring credit. [*Refer to Exhibit B of RD Instruction 440.1 (available in any Agency office) for the current percentage.] Equally acceptable offers will be decided by lot.

(4) *Rates and terms.* Subject to the availability of funds, rates and terms for Homestead Protection will be in accordance with subpart S of part 1951 of this chapter. Sales of suitable property offered to program eligible applicants will be on rates and terms provided in subpart A of part 1943 of this chapter. Surplus property and suitable property, which has not been sold to program eligible applicants will be offered for cash or on ineligible terms in accordance with subpart J of part 1951 of this chapter. The State Executive Director will determine the loan terms for surplus property within these limitations. A credit sale made on ineligible terms will be closed at the interest rate in effect at the time the credit sale was approved. After extensive sales efforts where no acceptable offer has been received, the State Executive Director may request the Administrator to permit offering surplus property for sale on more favorable rates and terms; however, the terms may not be more favorable than those legally permissible for eligible borrowers. Surplus property will be offered for sale for cash or terms that will provide the best net return for the Government. The term of financing extended may not be longer than the period for which the property will serve as adequate security. All credit sales on ineligible terms will be identified as NP loans.

[62 FR 44399, Aug. 21, 1997, as amended at 64 FR 62569, Nov. 17, 1999; 68 FR 7700, Feb. 18, 2003; 82 FR 19320, Apr. 27, 2017]

§ 1955.108   Sale of (CONACT) property other than FSA property.

Program officials will immediately contact the National Office whenever they acquire real property to obtain further instructions on the time frames and procedures for advertising and disposing of such property.

[62 FR 44401, Aug. 21, 1997]

§ 1955.109   Processing and closing (CONACT).

(a) *Determining repayment ability and creditworthiness.* If a credit sale is involved, the applicant must furnish necessary financial information to assist in determining repayment ability and creditworthiness. Information regarding eligibility, planned development and total operations will be provided the same as for the respective type of FSA loan. Purchasers requesting credit on ineligible terms, except for C&BP,

will be handled in accordance with subpart J of part 1951 of this chapter. For C&BP, information will be provided which is similar to an application including financial information required for the respective loan program to establish financial stability, creditworthiness and repayment ability.

(b) [Reserved]

(c) *Form of payment.* Payments at closing will be in the form of cash, cashier's check, certified check, postal or bank money order, or bank draft made payable to the Agency.

(d)–(e) [Reserved]

(f) *Earnest money.* Earnest money, if any, will be used to pay purchaser's closing costs with any balance of the costs being paid by the purchaser. Any excess earnest money will be credited to the purchase price or recognized as a part of the purchaser's downpayment.

(g) *Closing and reporting sales.* Title clearance, loan closing and property insurance requirements for a credit sale will be the same as for a program loan, except the property will be conveyed by Form RD 1955–49, in accordance with § 1955.141(a) of this subpart.

(h) *Classification.* Credit sales on ineligible terms for C&BP will be classified as NP loans and serviced accordingly.

(i) [Reserved]

(j) *Form RD 1910–11, "Applicant Certification, Federal Collection Policies for Consumer or Commercial Debts."* The County Supervisor or District Director must review Form RD 1910–11 "Applicant Certification, Federal Collection Policies for Consumer or Commercial Debts," with the applicant, and the form must be signed by the applicant.

[53 FR 35780, Sept. 14, 1988, as amended at 54 FR 29333, July 12, 1989; 58 FR 52652, Oct. 12, 1993; 60 FR 34455, July 3, 1995; 62 FR 44401, Aug. 21, 1997; 68 FR 61332, Oct. 28, 2003]

RURAL HOUSING (RH) REAL PROPERTY

§ 1955.110 [Reserved]

§ 1955.111 Sale of real estate for RH purposes (housing).

Sections 1955.112 through 1955.120 of this subpart pertain to the sale of acquired property pursuant to the Housing Act of 1949, as amended, (RH property). Single family units (generally which secured loans made under sec-

tion 502 or 504 of the Housing Act of 1949, as amended) are referred to as SFH property. All other property is referred to as MFH property. Notwithstanding the provisions of §§ 1955.112 through 1955.118 of this subpart, § 1955.119 is the governing section for the sale of SFH inventory property to a public body or nonprofit organization to use for transitional housing for the homeless.

[55 FR 3942, Feb. 6, 1990]

§ 1955.112 Method of sale (housing).

(a) *Sales by Rural Development* . Sales customarily will be made by Rural Development personnel in accordance with §§ 1955.114 and 1955.115 of this subpart (as appropriate) when staffing and workload permit and inventory levels do not exceed those outlined in paragraph (b) of this section. Adequate and timely advertising in accordance with § 1955.146 of this subpart is of utmost importance when this method is used. No earnest money will be collected in connection with sales by Rural Development. For MFH, this method will always be used unless another method is authorized by the Assistant Administrator, Housing.

(b) *Real estate brokers.* The County Office will utilize the services of real estate brokers for regular sales when there are five or more properties in inventory at any one time during the calendar year. When real estate brokers are used, first consideration will be given to utilizing such services under an exclusive broker contract as provided for in § 1955.130 of this subpart. Only when it is determined that an exclusive broker contract is not practicable, will the services of real estate brokers under an open listing agreement be utilized. The use of real estate brokers in offices having less than five properties in inventory at any one time during the calendar year is optional provided staffing and workload permit diligent and timely sales by Rural Development. When broker services for SFH are utilized, the Rural Development office will not conduct direct sales, but will refer inquiries to the broker or list of participating brokers. However, if Rural Development has been approached by a potential buyer desiring to purchase a specific property and a sales contract has been accepted, the property will not be listed for sale with real estate brokers. Earnest money held by real estate brokers will be used to pay the purchaser's closing costs with any balance of the costs to be paid by the purchaser. Any required earnest money deposit is exclusive of any required credit report fee. Brokers may only be used for MFH with authorization of the Assistant Administrator, Housing.

(c) *Sealed bid or auction.* The use of sealed bids or auctions is an effective method by which to sell inventory property. If the State Director determines that NP SFH property has been given adequate market exposure and that diligent sales efforts have not produced buyers, or under unusual circumstances as outlined in § 1955.115(a)(1) of this subpart, he/she will authorize sale by sealed bid or auction unless additional sales methods appear more prudent. Program SFH property will be sold by regular sale *only*, unless the Assistant Administrator, Housing, authorizes sale by sealed bid or auction. The State Director will request such authorization when all reasonable marketing efforts fail to produce buyers and the conditions of § 1955.114(a)(6) of this subpart have been met. The case file, including documentation of all marketing efforts, will be forwarded to the Assistant Administrator, Housing, ATTN: Single Family Housing Servicing and Property Management (SFH/SPM) Division, to request authority to sell program property by sealed bid or auction. The decision to utilize a sealed bid or auction must be carefully weighed when the property is located in a subdivision, since the resultant sale may have an adverse effect on surrounding property values. Detailed guidance for conducting sealed bid sales is provided in § 1955.147 of this subpart and for conducting auction sales in §§ 1955.131 and 1955.148 of this subpart.

[53 FR 27831, July 25, 1988]

§ 1955.113 Price (housing).

Real property will be offered or listed for its present market value, as adjusted by any administrative price reductions provided for in this section. Market value will be based upon the condition of the property at the time it is made available for sale. However, when a section 515 RRH credit sale is being made to a nonprofit organization or public body to utilize former single family dwellings as a rental or cooperative project for very-low-income residents, the price will be the lesser of the Government's investment or market value, less administrative price reductions, if any. Market value for multifamily housing projects will be determined through an appraisal conducted in accordance with subpart B to part 1922 of this chapter. Multi-family housing appraisals conducted shall reflect the impact of any restrictive-use provisions attached to the project as part of the credit sale.

(a) *SFH price reduction.* SFH property will be appraised at any time additional market data indicates this action is warranted. If SFH inventory has not sold after being actively marketed, the price will be administratively reduced. An administrative price reduction will be made without changing the SFH appraisal. For ease in computing dates for administrative price reductions, each month is assumed to have thirty days. The following schedule of administrative price reductions will be followed:

(1) *Program property.* If program property has not sold after being actively marketed at the current appraised value for 45 days during which time program applicants have exclusive rights to purchase the property, plus an additional 30 days to any offeror, the price will be administratively reduced by 10 percent of the appraised value. During the first 45 days after the price reduction, the property will be actively marketed with program applicants having exclusive rights to purchase the property, and at the expiration of this 45-day period, the property may be sold to any offeror. If at the end of this 75-day period the property remains unsold, a second price reduction of 10 percent of the appraised value will be made. During the first 45 days after the second price reduction, the property will be actively marketed with program applicants having exclusive rights to purchase the property, and at the expiration of this 45-day period, the property may be sold to any

offeror. If the property does not sell within 75 days of the second price reduction, further guidance is provided in § 1955.114(a)(6) and Exhibit D (available in any Rural Development office) of this subpart.

(2) *Nonprogram (NP) property.* If NP property has not been sold after being actively marketed for 45 days, the price will be administratively reduced by 10 percent of the appraised value. If the property remains unsold after an additional 45-day period of active marketing, one further price reduction of 10 percent of the appraised value will be made. If the property does not sell within 45 days of the second price reduction, further guidance is provided in § 1955.115(a)(1) and Exhibit D (available in any Rural Development office) of this subpart.

(b) *MFH price reduction.* For multiple-family property, the sale price will only be reduced to the extent that the market value has decreased as shown in a current market appraisal. The District Director will not reduce the price without the prior written approval of the State Director. The State Director must request National Office authorization on reductions in price for multiple-family property if the inventory value at the time of acquisition exceeded the State Director's loan approval authority.

[53 FR 27831, July 25, 1988; 54 FR 6875, Feb. 15, 1989, as amended at 58 FR 38927, July 21, 1993]

§ 1955.114 Sales steps for program property (housing).

Program property will be sold by regular sale unless the Assistant Administrator, Housing, authorizes another method. If the State Director determines that program property has been given adequate market exposure and that diligent sales efforts including the use of real estate brokers has not produced purchasers, the State Director may request the Assistant Administrator, Housing, to authorize sale by sealed bid or public auction as specified in § 1955.112(c) of this subpart.

(a) *Single family housing (SFH).* Sale prices will be established in accordance with § 1955.113 of this subpart. The County Supervisor will either offer the property or list it with real estate brokers for regular sale under the provisions of § 1955.112 of this subpart. See Exhibit D of this subpart (available in any Rural Development office) which outlines chronologically the sales steps for program property.

(1) The following provisions apply to all offers to purchase SFH inventory property:

(i) Program property will be available for purchase only by program applicants for the first 45 days from the date of the initial offering or listing, and for the first 45 days following the date of any reduction in price. During these 45-day period(s), offers from others may be received and held until the first business day following the 45-day period (the 46th day) when any such offer(s) will be considered as received on the 46th day along with offers received on that same (46th) day. After the expiration of each 45-day exclusive period for program applicants, program property may be purchased by offerors requesting credit on program terms, nonprogram (NP) terms or for cash in the order of priority set forth in paragraph (a)(3) of this section.

(ii) In regular sales, an acceptable offer *must* be for at least the sale price. No offer for less than the sale price will be considered, accepted or held. Offers will be considered as acceptable or unacceptable independent of any accompanying credit request (on program or NP terms).

(iii) All offers will be date-stamped when received. Selection of equally acceptable offers, considering offers in the category order outlined in paragraph (a)(3) of this section, received on the same business day will be made by lot by placing the names in a receptacle and drawing names sequentially. Drawn offers will be numbered and those drawn after the first drawn offer will be held as back-up offers pending sale to the successful offeror, unless the offeror has specifically noted on the offer that it may not be held as a back-up offer.

(iv) An offer may be submitted any time after the effective date the property is available for sale or any price reduction; however, it is not considered until five business days after the effective date. An offer received during the five business day period is considered on the 6th day, at the same time as any offer received on the 6th day.

(v) If an offer subject to Rural Development financing is accepted, and the offeror's credit request is later denied, the next offer (if any) will be accepted regardless of whether the rejected applicant appeals the adverse decision (NP applicants do not receive appeal rights). In cases involving program property, if no back-up offers are on hand, the property will be reoffered/relisted for sale utilizing the balance of any outstanding retention period. Property will not be held off the market pending the outcome of an appeal.

(2) *Effective date and method of offering.* When ready for sale, each property will be offered for sale by use of Form RD 1955-43 unless Rural Development has on hand a signed offer from a program applicant to purchase a specific program property or an offer from any offeror to purchase a specific NP property. The date the form is posted or mailed to real estate brokers is the effective date the offer for sale has begun.

Listings will provide for sales on program and NP terms, as appropriate.

(3) *Priority of offers.* For program properties, acceptable offers received after the 45-day retention period specified in paragraph (a)(1)(i) of this section have priority in the order given in paragraphs (a)(3) (i), (ii), (iii) and (iv) of this section. For NP properties, acceptable offers have priority in the order given in paragraphs (a)(3) (ii), (iii) and (iv) of this section. Program applicants may purchase NP property, however, credit may only be extended on NP terms.

(i) Offers with requests for credit on program terms. An offer from an applicant requesting credit on program terms in excess of the sale price will be considered as equally acceptable with other acceptable offers from program applicants and will be sold for the sale price.

(ii) Cash offers, in descending order from highest to lowest, provided the cash offer is higher than any other offer which falls into the parameters of paragraph (a)(3)(iii) of this section multiplied by the current cash preference percentage listed in exhibit B of

RD Instruction 440.1 (available in any Rural Development office).

(iii) Offers with requests for credit on NP terms in descending order from highest to lowest, for *more* than the sale price. An offer with a request for credit in excess of the market value of the property will not be accepted. If an offer of this type is received, the offeror will be given the opportunity to reduce the credit request to the market value (or lower) with no change to be made in the offered price.

(iv) Offers with requests for credit on NP terms for the sale price.

(4) *Back-up offers and notification to offerors.* Back-up offers will be taken in accordance with paragraph (a)(1)(iii) of this section. County offices utilizing the services of real estate brokers will advise the brokers of changes in the status of the property. County offices not utilizing real estate brokers will advise offerors of changes in the status of the property utilizing exhibit E of this subpart (available in any Rural Development office) or similar format. Use of exhibit E is optional in offices utilizing real estate brokers.

(5) *Finalizing sales.* Credit sales on program terms will be made in accordance with § 1955.117 of this subpart and 7 CFR part 3550. Cash sales will be handled in accordance with § 1955.118 of this subpart and credit sales on NP terms will be made in accordance with subpart J of part 1951 of this chapter.

(6) *Unsold property.* If program property remains unsold after eight months of active marketing, the case file, with documentation of all marketing efforts, will be forwarded to the State Office for review with a recommendation of future sales efforts. The State Director will determine whether a request should be made to the Assistant Administrator, Housing, to sell the property by sealed bid or auction, or whether additional guidance such as, but not limited to advertising, reappraisal, offering a special effort sales bonus, or 20-year amortization factor (with balloon after 10 years) on NP financing may facilitate a sale.

(b) *Multiple family housing.* The sale price will be established in accordance with § 1955.113 of this subpart. Notification of known interested prospective offerors and advertising should be handled as set forth in §1955.146 of this subpart. The sale information will include a sale price, any restrictive-use provisions the project will be subject to and made part of the title, a date/time/location when offers will be drawn, and require all offerors to submit an application package comparable to that required by the respective loan program, which will be reviewed by the State Director or designee. The sale/time/location will be established by the District Director and will allow adequate time for advertising and review of applications to determine eligibility in accordance with MFH program requirements. Offerors whose applications are rejected by by Rural Development will be notified in writing by the approval official, and for program applicants, given appeal rights in accordance with subpart B of part 1900 of this chapter. If an application is rejected, the sale will continue regardless of whether the rejected applicant appeals the adverse decision. Property will not be held pending the outcome of an appeal. An offeror may withdraw an offer prior to the sale date, but not on the sale date. All offers from applicants determined eligible for the type loan being offered will be considered. The District Director, or delegate, and one other Rural Development employee will conduct the drawing at which time the public may be present. Offers will be placed in a receptacle and drawn sequentially. Drawn offers will be numbered and those drawn after the first drawn will be held as back-up offers, unless the offeror has indicated that the offer may not be held as back-up. Award will be made to the first offer drawn provided the offer is acceptable as to the terms and conditions set forth in the sale notice. The successful offeror will be notified immediately in writing by the approval official, return receipt requested, that the successful offeror's offer has been accepted even if the successful offeror was present at the sale. The remaining offerors will each be notified by letter, return receipt requested, that their offer was not successful, but will be held as a back-up offer. The selection of the offeror was by lot and is therefore not appealable. If an unsuccessful offeror was not

present at the sale and requests the name of the successful offeror, the name may be released. If the MFH property has been listed with real estate brokers after receiving authorization from the Assistant Administrator, Housing, Form RD 1955–40, or another appropriate form designated for MFH property, will be used and the property sold to the first eligible program applicant. Any other method of sale must receive prior written authorization from the Assistant Administrator, Housing. Cash sales of program property will remain subject to restrictive-use provisions determined needed and included in the advertisement. The deed will contain the applicable restrictive-use provisions. Tenants and prospective tenants will receive the applicable protections for the specific restrictive-use provision contained in 7 CFR part 3560, subpart N.

(c) *Single family inventory converted to MFH.* Written offers by nonprofit organizations, public bodies or for-profit entities, which have good records of providing low income housing under section 515, will be considered by Rural Development for the purchase of multiple SFH units for conversion to MFH. Section 514 credit sale mortgages may contain repayment terms up to 33 years and section 515 credit sale mortgage terms may be up to 50 years.

(1) The price provisions of §1955.113 and the processing provisions for MFH in §1955.117 of this subpart apply to such a conversion.

(2) The provisions of §1955.130 of this subpart pertaining to real estate brokers apply, as applicable, and a commission will be due in the normal manner on units which were listed with the broker(s).

(3) Prior approval of the National Office is required before issuance of Form AD–622, "Notice of Preapplication Review Action." A preapplication with documentation as required by the Agency, along with the State Director's recommendation, will be forwarded to the National Office, Attention: Assistant Administrator, Housing, for a determination and further guidance.

(4) A credit sale for this purpose will be made according to the provisions of 7 CFR part 3560, as modified by

§ 1955.117 of this subpart, except the units need not be contiguous, but they must be located in close enough proximity so that management costs are not increased nor management capabilities diminished because of distance.

(5) An additional loan may be made simultaneously with the credit sale, or later, only when the property involved meets the requirements of 7 CFR part 3560, subpart K.

(d) *CONACT residential property suitable for the SFH program.* When a single family house acquired under the CONACT is determined to be suited for the SFH program, it may be offered for sale as a SHF unit as though it had been acquired under the SFH program. It may, however, be sold in this manner to a program RH applicant on *program terms only*—not for cash or on NP terms. When a house is offered for sale under this paragraph, the listing notices and any advertising (whether being sold by Rural Development or through real estate brokers) must state this restriction.

[53 FR 27832, July 25, 1988, as amended at 55 FR 3942, Feb. 6, 1990; 56 FR 2257, Jan. 22, 1991; 58 FR 38927, July 21, 1993; 58 FR 38949, July 21, 1993; 58 FR 52652, Oct. 12, 1993; 67 FR 78329, Dec. 24, 2002; 69 FR 69106, Nov. 26, 2004]

§ 1955.115 Sales steps for nonprogram (NP) property (housing).

The appropriate Rural Development office will take the following steps after repairs, if economically feasible, are completed. The appraisal will be updated to reflect changes in market conditions, repairs and improvements, if any. Form RD 1955–43 for SFH and 1955–40 for MFH will be completed to offer the property for sale. The advertising requirements and deed restrictions in § 1955.116 of this subpart apply if the property does not meet Rural Development DSS standards.

(a) *Single Family Housing.* Sales steps will be the same as for program properties as provided in § 1955.114(a) of this subpart, except that sales must be for cash in accordance with § 1955.118 or credit on NP terms as provided in subpart J of part 1951 of this chapter. See exhibit D of this subpart (available in any Rural Development office) which outlines chronologically the sales steps for NP properties.

(1) *Sale by sealed bid or auction.* If a NP property has not sold within 150 days after being offered for sale, the inventory case file with documentation of marketing efforts will be submitted to the State Director. The State Director will authorize sale by sealed bid or auction in accordance with § 1955.112(c) of this subpart unless additional sales methods appear more prudent. Use of the sealed bid or auction method may be considered as an initial sales effort under special or unusual circumstances such as, but not limited to, structures which have been substantially destroyed by fire or other causes.

(2) *Sale as chattel.* If efforts to sell NP property by sealed bid or auction prove unsuccessful, the structure(s) may be sold as chattel (for chattel or salvage value, as appropriate) when authorized by the State Director. When the structure is to be sold as chattel (exclusive of land) further guidance is provided in §§ 1955.121, 1955.122 and 1955.141(b) of this subpart. If no offer is received, the structure(s) may be demolished and removed from the site and then the site offered for sale. If this method is utilized, Rural Development will attempt to have the structure removed in exchange for the salvageable materials by contract, otherwise, will solicit for contracts to have the structure removed in accordance with Rural Development Instruction 2024–A (available in any Rural Development office).

(3) *Sale of vacant land.* When Rural Development has vacant land in inventory which was security for an SFH loan, the land will be sold in accordance with this subparagraph. When the lot meets the requirements of 7 CFR part 3550, and a program applicant desires to purchase the lot and construct a dwelling, a credit sale will not be made. Instead, one section 502 loan will be made which will include funds for the purchase of the lot and construction of a dwelling. Otherwise, the lot will be sold for cash or on NP terms with a loan not to exceed ten years in term and amortization.

(b) *Multiple family housing.* Sales steps will be the same as for program MFH property as provided in § 1955.114(b) of this subpart except that sales must be for cash or on NP terms as set forth in § 1955.118 of this subpart.

Additionally, if cash offers are received, they will be given first preference by drawing from the cash offers only. If the State Director determines an auction sale should be used to sell NP MFH property, authority to use that method of sale must be requested from the Assistant Administrator, Housing. Inventory files, including information on the acquisition, marketing efforts made, management of the property, other pertinent information, a memorandum covering the facts of the case, and recommendations of the State Director must be submitted for review. If the housing is sold out of the Rural Development program as NP property, the closing of the sale may not take place until tenants have received all notifications and benefits afforded to tenants in prepaying projects in accordance with 7 CFR part 3560, subpart N.

[53 FR 27833, July 25, 1988, as amended at 58 FR 38928, July 21, 1993; 58 FR 52652, Oct. 12, 1993; 67 FR 78329, Dec. 24, 2002; 69 FR 69106, Nov. 26, 2004]

§ 1955.116 Requirements for sale of property not meeting decent, safe and sanitary (DSS) standards (housing).

For real property (exclusive of improvements) which is unsafe, refer to § 1955.137(e) of this subpart for further guidance. For all other housing inventory property which does not meet decent, safe and sanitary (DSS) standards, the provisions of this section apply.

(a) *Notices and advertising.* If the inventory property has a single family dwelling or MFH unit thereon which does not meet DSS standards as defined in § 1955.103 of this subpart, but which could meet such standards through the repair or renovation activities of the future owner, any "Notice of Real Property For Sale," "Notice of Sale," or other advertisement used in conjunction with advertising the property for sale must include the following language which is contained in Form RD 1955-44, "Notice of Residential Occupancy Restriction":

This property contains a dwelling unit or units which Rural Development has deemed to be inadequate for residential occupancy. The Quitclaim Deed by which this property

will be conveyed will contain a covenant restricting the residential unit(s) on the property from being used for residential occupancy until the dwelling unit(s) is repaired, renovated or razed. This restriction is imposed pursuant to section 510(e) of the Housing Act of 1949, as amended, 42 U.S.C. 1480. The property must be repaired and/or renovated as follows:*

* For advertisements, the sentence preceding the asterisk may be deleted and replaced with the following, or similar sentence: "Contact Rural Development (or any real estate broker/name of exclusive broker) for a list of items which must be repaired/renovated." For notices other than advertising, insert those items which are necessary to make the dwelling unit(s) meet DSS standards. Examples are:

—Replace flooring and floor joists in kitchen and bathroom.

—Drill new well to provide for an adequate and potable water supply.

—Hook-up to community water and sewage system now being installed.

—Provide a functionally adequate, safe and operable * system. * Insert heating, plumbing, electrical and/or sewage disposal, etc., as appropriate.

—Install *. * Insert new roof, foundation, sump pump, bathroom fixtures, etc., as appropriate.

—Install R-* insulation in basement walls or ceiling, R-* insulation in attic, and storm windows/doors throughout. * Insert appropriate R-Values to meet Thermal Performance Standards.

(b) *Sale agreements.* If a housing structure in inventory does not meet DSS standards, Form RD 1955-44 must be attached to Forms RD 1955-45 or RD1955-46, as appropriate, to provide notification of the deed restriction and required repairs/renovations before the dwelling can be used for residential purposes.

(c) *Quitclaim Deed.* The following, the original of Form RD 1955-44, or similar restrictive clause adapted for use in an individual State pursuant to a State Supplement approved by OGC must be added to the Quitclaim Deed for properties which do not meet DSS standards at the time of sale but which could through the repair/renovation activities of the future owner:

Pursuant to section 510(e) of the Housing Act of 1949, as amended, 42 U.S.C. 1480(e), the purchaser ("Grantee" herein) of the above-described real property (the "subject property" herein) covenants and agrees with the United States acting by and through Rural Development (the "Grantor" herein) that the dwelling unit(s) located on the subject property as of the date of this Quitclaim Deed will not be occupied or used for residential purposes until the item(s) listed at the end of this paragraph have been accomplished. This covenant shall be binding on Grantee and Grantee's heirs, assigns and successors and will be construed as both a covenant running with the subject property and as equitable servitude. This covenant will be enforceable by the United States in any court of competent jurisdiction. When the existing dwelling unit(s) on the subject property complies with the aforementioned standards of Rural Development or the unit(s) has been completely razed, upon application to Rural Development in accordance with its regulations, the subject property may be released from the effect of this covenant and the covenant will thereafter be of no further force or effect. The property must be repaired and/or renovated as follows: *.'' * Insert the same items referenced in the listing notice(s) and sale agreement which are necessary to make the dwelling unit(s) meet DSS standards.

(d) *Release of restrictive covenant.* Upon request of the property owner for a release of the restrictive covenant, Rural Development will inspect the property to ensure that the repairs/renovations outlined in the restrictive covenant have been properly completed or the structure(s) razed. A State Supplement outlining the procedure for releasing the restrictive covenant will be issued with the advice of OGC.

[53 FR 27834, July 25, 1988]

§ 1955.117 Processing credit sales on program terms (housing).

The following provisions apply to all credit sales on program terms:

(a) *Offers.* Form RD 1955-45 will be used to document the offer and acceptance for regular Rural Development sales. The contract is accepted prior to processing Form RD 410-4, "Application for Rural Housing Assistance (Non-Farm Tract)," for SFH property with the provision that acceptance is subject to program approval. MFH property sales require an application package comparable to that submitted for the respective loan program application.

(b) *Processing.* Rural Development regulations pertaining to the type of credit being extended will be followed in making credit sales on program terms except as modified by the provi-

sions of this section. All MFH credit sales may be made for up to 100 percent of the current market value of the security, less any prior lien. However, if a profit or limited profit applicant desires to earn a return, the applicant will be required to contribute at least 3 percent of the purchase price as a cash downpayment. All credit sales of RRH, RCH, and LH properties will be subject to prepayment and restrictive-use provisions specified by the respective program requirements.

(c) *Approval.* Forms RD 1940-1 or RD 3560-51, as appropriate, will be used to approve a credit sale even though no obligation of funds is required.

(d) *Downpayment.* When a downpayment is made, it will be collected at closing.

(e) *Interest rate.* Upon request of the applicant, the interest rate charged by Rural Development will be the lower of the interest rate in effect at the time of loan approval or closing. If the applicant does not indicate a choice, the loan will be closed at the rate in effect at the time of loan approval.

(f) *Closing costs.* MFH purchasers will pay closing costs from their own funds. Where necessary, SFH purchasers who qualify may be made a subsequent loan to pay closing costs in an amount not to exceed 1 percent of the sale price of the dwelling. Any closing costs which are legally or customarily paid by the seller will be paid by Rural Development and charged to the inventory account as a nonrecoverable cost items.

(g) *Closing sale.* Title clearance, loan closing and property insurance requirements for a credit sale, and any loan closed simultaneously with the credit sale, are the same as for a program loan of the same type except:

(1) The property will be conveyed in accordance with § 1955.141(a) of this subpart.

(2) Earnest money, if any, will be used to pay purchaser's closing costs with any balance of closing costs being paid from the purchaser's personal funds except as provided in paragraph (f) of this section. For SFH credit sales and MFH credit sales to nonprofit organizations or public bodies, any excess deposit will be refunded to the purchaser. For MFH credit sales to profit or limited profit buyers, any excess

earnest money deposit will be credited to the purchase price and recognized as a part of the purchaser's initial investment.

(3) The County Supervisor or District Director will provide the closing agent with the necessary information for closing the sale. The assistance of OGC will be requested to provide closing instructions in exceptional or complex cases and for all MFH sales.

(h) *Reporting.* After the sale is closed, it will be reported according to §1955.142 of this subpart.

[53 FR 27834, July 25, 1988; 54 FR 6875, Feb. 15, 1989, as amended at 58 FR 38928, July 21, 1993; 68 FR 61332, Oct. 28, 2003; 69 FR 69106, Nov. 26, 2004]

§1955.118 Processing cash sales or MFH credit sales on NP terms.

(a) *Cash sales.* Cash sales will be closed by the servicing official collecting the purchase price (less any earnest money deposit or bid deposit) and delivering the deed to the purchaser.

(b) *Credit sales.* The following provisions apply to MFH credit sales on NP terms:

(1) *Offers.* Form RD 1955–45 or RD 1955–46, as appropriate, will be used to document the offer and acceptance. Contract acceptance is made prior to processing a request for credit on NP terms.

(2)' *Processing.* Purchasers requesting credit on NP terms will be required to submit documentation to establish financial stability, repayment ability, and creditworthiness. Standard forms used to process program applications may be utilized or comparable documentation may be accepted from the purchaser with the servicing official having the discretion to determine what information is required to support loan approval for the type property involved. Individual credit reports will be ordered for each individual applicant and each principal within an applicant entity in accordance with subpart B of part 1910 of this chapter. Commercial credit reports will be ordered for profit corporations and partnerships, and organizations with a substantial interest in the applicant entity in accordance with subpart C of part 1910 of this chapter.

(3) *Approval.* Form RD 3560–51 will be used to approve a credit sale even though no obligation of funds is involved. Special instructions on the FMI pertaining to NP credit sales will be followed.

(4) *Downpayment.* A downpayment of not less than 10 percent of the purchase price is required at closing.

(5) *Interest rate.* The Section 515 RRH interest rate plus ½ percent will be charged on all types of housing credit sales, except SFH. Refer to exhibit B of RD Instruction 440.1 (available in any Rural Development office) for interest rates. Loans made on NP terms will be closed at the interest rate which was in effect at the time the loan was approved.

(6) *Term of note.* The note amount will be amortized over a period not to exceed 10 years. If the State Director determines more favorable terms are necessary to facilitate the sale, the note amount may be amortized using a 30-year factor with payment in full (balloon payment) due not later than 10 years from the date of closing. In no case will the term be longer than the period for which the property will serve as adequate security.

(7) *Modification of security instruments.* If applicable to the type property being sold, modification of security instruments may be made. On the promissory note and/or security instrument (mortgage or deed of trust) any covenants relating to graduation to other credit, restrictive-use provisions on MFH projects, personal occupancy, inability to secure other financing, and restrictions on leasing may be deleted. Deletions are made by lining through only the specific inapplicable language with both the NP borrower and Rural Development initialing the changes.

(8) *Closing sale.* Title clearance, loan closing and property insurance requirements for a credit sale are the same as for a program loan except:

(i) The property will be conveyed in accordance with §1955.141(a) of this subpart.

(ii) The purchaser will pay his/her own closing costs. Earnest money, if any, will be used to pay purchaser's closing costs with any balance of closing costs being paid by the purchaser. Any closing costs which are legally or customarily paid by the seller will be paid by Rural Development from the downpayment.

(iii) The County Supervisor or District Director will provide the closing agent with the necessary information for closing the sale. The assistance of OGC will be requested to provide closing instructions for all MFH sales.

(iv) When more than one property is bought by the same buyer and the transactions are closed at the same time, a separate promissory note will be prepared for each property, but one mortgage will cover all the properties.

(9) *Reporting.* After the sale is closed, it will be reported according to § 1955.142 of this subpart.

(10) *Classification.* MFH credit sales on NP terms will be classified as NP loans and serviced accordingly.

(11) *Form RD 1910–11, "Applicant Certification, Federal Collection Policies for Consumer or Commercial Debts."* The County Supervisor or District Director must review Form RD 1910–11, "Applicant Certification, Federal Collection Policies for Consumer or Commercial Debts," with the applicant, and the form must be signed by the applicant.

[53 FR 27835, July 25, 1988, as amended at 54 FR 29333, July 12, 1989; 55 FR 3942, Feb. 6, 1990; 58 FR 38928, July 21, 1993; 58 FR 52653, Oct. 12, 1993; 68 FR 61332, Oct. 28, 2003; 69 FR 69106, Nov. 26, 2004]

§ 1955.119 Sale of SFH inventory property to a public body or nonprofit organization.

Notwithstanding the provisions of § 1955.111 through § 1955.118 of this subpart, this section contains provisions for the sale of SFH inventory property to a public body or nonprofit organization to use for transitional housing for the homeless. A public body or nonprofit organization is a nonprogram applicant. All other SFH credit sales on nonprogram terms will be handled in accordance with subpart J of part 1951 of this chapter.

(a) *Method of sale.* The method of sale is according to § 1955.112 of this subpart. Upon request from a public body or nonprofit organization, Rural Development will provide a list of all SFH inventory property, regardless of whether it is listed for sale with real estate brokers. The list will indicate whether the property is program or nonprogram. Upon written notice of the organization's intent to buy a specific property, if it is not under a sale contract, Rural Development will withdraw the property from the market for a period not to exceed 30 days to provide the organization sufficient time to execute Form RD 1955–45.

(b) *Price.* The price of the property will be established according to § 1955.113 of this subpart; however, a 10 percent discount of the listed price is authorized on nonprogram property. No discount is authorized on program property.

(c) *Decent, safe and sanitary (DSS) standards.* If an organization wants to buy a property which does not meet DSS standards, Rural Development will repair it to meet those standards, including thermal performance standards, unless Rural Development determines it is not feasible to do so according to § 1955.64(a)(1)(ii) of subpart B of part 1955 of this chapter. The price will be adjusted to reflect any resulting change in value. Cosmetic repairs, if needed, such as painting, floor covering, landscaping, etc., are the responsibility of the organization. Form RD 1955–44, itemizing the required repairs and Rural Developments agreement to complete them before closing will be made a part of Form RD 1955–45, the sales contract, before it is signed. Required repairs must be completed before closing so DSS restrictions will not be required in the deed.

(d) *Approval and closing.* Processing cash sales or MFH credit sales on nonprogram terms is according to § 1955.118 of this subpart, except as follows:

(1) *Earnest money deposit.* No earnest money deposit is required.

(2) *Downpayment.* No downpayment is required.

(3) *Term of note.* The term of the note may not exceed 30 years.

[55 FR 3942, Feb. 6, 1990, as amended at 58 FR 52653, Oct. 12, 1993]

§ 1955.120 Payment of points (housing).

To effect regular sale of inventory SFH property to a purchaser who is financing the purchase of the property with a non-Rural Development loan, the County Supervisor may authorize the payment by Rural Development of not more than three points. The payment must be a customary requirement of the lender for the seller within the community where the property is located. Terms of payment will be incorporated in Form RD 1955–45 and will be fixed as of the date the form is signed by the appropriate Rural Development official. Points will *not* be paid to reduce the purchaser's interest rate. The payment will be deducted from the funds to be received by Rural Development at closing.

[53 FR 27836, July 25, 1988. Redesignated at 55 FR 3942, Feb. 6, 1990, as amended at 58 FR 52653, Oct. 12, 1993; 68 FR 61332, Oct. 28, 2003]

CHATTEL PROPERTY

§1955.121 Sale of acquired chattels (chattel).

Sections 1955.122 through 1955.124 of this subpart prescribe procedures for the sale of all acquired chattel property except real property rights. The State Director is authorized to sell acquired chattels by auction, sealed bid, regular sale or, for perishable items and crops, by negotiated sale. The State Director may redelegate authority to any qualified Rural Development employee.

§1955.122 Method of sale (chattel).

Acquired chattels will be sold as expeditiously as possible using the method(s) considered most appropriate. If the chattel is not sold within 180 days after acquisition, assistance will be requested as outlined in §1955.143 of this subpart.

(a) *Sale to beginning farmers or ranchers.* Beginning farmers or ranchers obtaining special OL loan assistance under §1941.15 of subpart A of part 1941 of this chapter will receive priority in the purchase of farm equipment held in government inventory during the commitment period. The County Supervisor will notify such applicants/borrowers of any farm equipment held in government inventory within the service area of the Rural Development County Office. These applicants/borrowers will be given 10 working days to respond that they are interested in purchasing any or all items of equipment at the appraised fair market value established by Rural Development. Rural Development Form Letter 1955–C–1 will be used to notify applicants/borrowers of the availability of farm equipment in Rural Development inventory. The equipment must be essential to the success of the operation described in the loan application in order for the applicant to have an opportunity to purchase such equipment. The County Supervisor will determine what equipment is essential.

(b) *Regular sale.* Chattels will be sold by Rural Development employees at market value to program applicants. Form RD 440–21, "Appraisal of Chattel Property," will be used when appraising chattels for regular sale.

(c) *Auctions.* Section 1955.148 of this subpart provides detailed guidance on auctions applicable to the sale of chattels, as supplemented by this section.

(1) *Established public auction.* An established public auction is an auction that is widely advertised and held on a regularly scheduled basis at the same facility. This method of sale is particularly suited for the sale of commodities, farm machinery and livestock. No additional public notice of sale is required other than that commonly used by the facility. This is the preferred method of disposal.

(2) *Other auctions.* Other auctions, whether conducted by Rural Development employees or fee auctioneers, are suitable for on-premises sales, for sale of dissimilar chattels, and for the sale of chattels in conjunction with the auction of real property. A minimum of 5 days public notice will be given prior to the date of auction.

(d) *Sealed bid sales.* Section 1955.147 of this subpart provides detailed guidance on sealed bid sales applicable to the sale of chattels. When it is believed that financing will have to be provided through a credit sale, this method has advantages over auction sales. It requires, however, additional steps in the event any established minimum price is not obtained. Preference will be given to a cash offer which is at least ____* percent of the highest offer requiring credit.

[* Refer to exhibit B of RD Instruction 440.1 (available in any Rural Development office) for the current percentage.]

(e) *Negotiated sale.* Perishable acquired items and crops (except timber) and chattels for which no acceptable bid was received from auction or sealed bid methods may be sold by direct negotiation for the best price obtainable. No public notice is required to negotiate with interested parties including prior bidders. Justification for the use of this method of sale will be documented.

(f) *Notification.* In many States the original owner of the chattel property must personally be notified of the sale date and method of sale within a certain time prior to the sale. The State Director then will issue a State supplement clearly stating what notices are to be sent, if any. County Supervisor will review State supplements to determine what notices must be sent to the previous owner of the chattel property prior to Rural Development taking action to sell the property.

No public notice is required to negotiate with interested parties including prior bidders. Justification for the use of this method of sale will be documented. A copy of the sale instrument (Form RD 1955–47, "Bill of Sale 'A'—Sale of Government Property") will be kept in the County or District Office inventory file. Sale proceeds will be remitted according to RD Instruction 1951–B (available in any Rural Development office). A State Supplement, when needed, will be prepared with the assistance of OGC to provide additional guidance on negotiated sales and to insure compliance with State laws.

[50 FR 23904, June 7, 1985, as amended at 53 FR 35780, Sept. 14, 1988; 58 FR 48290, Sept. 15, 1993; 58 FR 58650, Nov. 3, 1993; 62 FR 44401, Aug. 21, 1997; 68 FR 61332, Oct. 28, 2003]

§ 1955.123 Sale procedures (chattel).

(a) *Credit sales.* Although cash sales are preferred in the sale of chattel, credit sales may be used advantageously in the sale of chattels to eligible purchasers and to facilitate sales of high-priced chattels. Credit sales to eligible purchasers will be in accordance with the provisions of this chapter for the appropriate program for which a loan would otherwise be made including eligibility determinations. Preference will be given to a cash offer that is at least * percent of the higher

offer requiring credit. [*Refer to exhibit B of RD Instruction 440.1 (available in any Rural Development office) for the current percentage.] Credit sales made to ineligible purchasers will require not less than a 10 percent downpayment with the remaining balance amortized over a period not to exceed 5 years. The interest rate for ineligible purchasers of C&BP chattel will be the current ineligible interest rate for C&BP property set forth in Exhibit B of RD Instruction 440.1 (available in any Rural Development office). District Directors and State Directors are authorized to approve or disapprove sale of C&BP chattel on ineligible terms in accordance with the respective type of program approval authorities in Exhibit E of Subpart A of Part 1901 of this chapter (available in any Rural Development office). For other than C&BP, credit sales to NP purchasers will be handled in accordance with Subpart J of Part 1951 of this chapter.

(b) *Receipt of payment.* Payment will be by cashier's check, certified check, postal or bank money order or personal check (not in excess of $500) made payable to the agency. Cash may be accepted if it is not possible for one of these forms of payment to be used. Third party checks are not acceptable. If full payment is not received at the time of sale, the offer will be documented by Form RD 1955–45 or Form RD 1955–46 where the chattel is sold jointly with real estate by regular sale.

(c) *Transfer of title.* Title will be transferred to a purchaser in accordance with § 1955.141(b) of this subpart.

(d) *Reporting sale.* Sales will be reported in accordance with § 1955.142 of this subpart.

(e) *Reporting and disposal of inventory property not sold.* Refer to §§ 1955.143 and 1955.144 of this subpart for additional guidance in disposing of problem property.

[50 FR 23904, June 7, 1985, as amended at 58 FR 52653, Oct. 12, 1993; 58 FR 58650, Nov. 3, 1993; 68 FR 61332, Oct. 28, 2003; 80 FR 9899, Feb. 24, 2015]

§ 1955.124 Sale with inventory real estate (chattel).

Inventory chattel property may be sold with inventory real estate if a higher aggregate price can be obtained. Proceeds from a joint sale will be applied to the respective inventory accounts based on the value of the property sold. Form RD 440–21 will be used to determine the value of the chattel property. The offer for the sale of the chattels will be documented by incorporating the terms and conditions of the sale of Form RD 1955–45 or Form RD 1955–46, and may be accepted by the appropriate approval official based upon the combined final sale price.

§§ 1955.125–1955.126 [Reserved]

USE OF CONTRACTORS TO DISPOSE OF INVENTORY PROPERTY

§ 1955.127 Selection and use of contractors to dispose of inventory property.

Sections 1955.128 through 1955.131 prescribe procedures for contracting for services to facilitate disposal of inventory property. RD Instruction 2024–A (available in any Rural Development office) is applicable for procurement of nonpersonal services.

[53 FR 27836, July 25, 1988]

§ 1955.128 Appraisers.

(a) *Real property.* The State Director may authorize the County Supervisor or District Director to procure fee appraisals of inventory property, except MFH properties, to expedite the sale of inventory real or chattel property. (Fee appraisals of MFH properties will only be authorized by the Assistant Administrator, Housing, when unusual circumstances preclude the use of a qualified Rural Development MFH appraiser.) The decision will be based on the availability of comparables, the capability and availability of personnel, and the number and type of properties (such as large farms and business property) requiring valuation. For Farmer Programs real estate properties, all contract (fee) appraisers should include the sales comparison, income (when applicable), and the cost approach to value. All Rural Development real estate contract appraisers must be certified as State-Certified General Appraisers.

(b) *Chattel property.* For Farmer Programs chattel appraisals, the contractor/appraiser completing the report must meet at least one of the following qualifications:

(1) Certification by a National or State appraisal society.

(2) If the contractor is not a certified appraiser and a certified appraiser is not available, the contractor may qualify or may use other qualified appraisers, if the contractor can establish that he/she or that the appraiser meets the criteria for a certification in a National or State appraisal society.

(3) The appraiser has recent, relevant, documented appraisal experience or training, or other factors clearly establish the appraiser's qualifications.

[58 FR 58650, Nov. 3, 1993]

§ 1955.129 Business brokers.

The services of business brokers or business opportunity brokers may be authorized by the appropriate Assistant Administrator in lieu of or in addition to real estate brokers for the sale of businesses as a whole, including goodwill and chattel, when:

(a) The primary use of the structure included in the sale is other than residential;

(b) The business broker is duly licensed by the respective state; and

(c) The primary function of the business is other than farming or ranching.

§ 1955.130 Real estate brokers.

Contracting authority for the use of real estate brokers is prescribed in Exhibit D of RD Instruction 2024–A (available in any Rural Development office). Brokers who are managing custodial or inventory property may also participate in sales activities under the same conditions offered other brokers. Brokers must be properly licensed in the State in which they do business.

(a) *Type of listings.* The State Director may authorize use of exclusive listings during any calendar year. Since the Agency receives many more marketing services for its commission dollar and saves time listing the property with only one broker, it is strongly recommended that all County Offices be authorized the use of exclusive brokers.

(1) *Exclusive broker contract.* An exclusive broker contract provides for the

selection of one broker by competitive negotiation who will be the only authorized broker for the Rural Development office awarding the contract within a defined area and for specific property or type of property. Criteria will be specified in the solicitation together with a numerical weighting system to be used (usually 1–100). Responses will be calculated on the basis of the criteria such as personal qualifications, membership in Multiple Listing Service (MLS), previous experience with Rural Development sales, advertising plans, proposed innovative promotion methods, and financial capability. The responsibilities of the broker under an exclusive broker contract exceed those of the open listing agreement and therefore, an exclusive broker contract is the preferred method of listing properties.

(2) *Open listing.* Open listing agreements provide for any licensed real estate broker to provide sales services for any property listed under the terms and conditions of Form RD 1955–42, "Open Real Property Master Listing Agreement." If this method is used, a newspaper advertisement will be published at least once yearly, or a notice sent to all real estate brokers in the counties served by the Rural Development office, informing brokers that sales services are being requested. The advertising will be substantially similar to the example given in Exhibit B of this subpart (available in any Rural Development office). An open listing agreement may be executed at any time during the year, but must be effective prior to the broker showing the property. When this method is used, the Rural Development office is responsible for ensuring that adequate advertising is performed to effectively market the property.

(b) *Listing notices.* Forms RD 1955–40 or RD 1955–43, as appropriate, will be used to provide brokers with notice of initial listing, withdrawal, price change, terms change, relisting, sale cancellation, restrictions on sale, etc.

(c) *Priority of offers.* All offers received during the same business day will be considered as having been received at the same time. The successful offer from among equally acceptable offers within each category will be de-

termined by lot by Rural Development. Priority rules for specific categories of property are:

(1) *Program SFH.* See § 1955.114(a) of this subpart.

(2) *Program MFH.* Offers will be considered from program applicants only.

(3) *NP SFH.* See § 1955.115(a) of this subpart.

(4) *NP MFH.* See § 1955.115(b) of this subpart.

(5) *Suitable and surplus FSA CONACT.* See § 1955.107 of this subpart.

(6) *Suitable and Surplus Non-FSA CONACT.* See § 1955.108 of this subpart.

(d) *Price.* No offer for less than the listed price will be accepted during the period of regular sale.

(e) *Earnest money.* The broker will collect earnest money in the amount specified in paragraph (e)(1) of this section when a sale contract is executed. The earnest money will be retained by the broker until contract closing, withdrawal, cancellation, or rejection by Rural Development. When a contract is cancelled because Rural Development rejects the offeror's application for credit, the earnest money will be returned to the offeror. When a contract closes, the broker will make the earnest money available to be used toward closing costs, or in the case of a cash sale it may be returned to the purchaser. For MFH sales to profit or limited profit buyers, any excess earnest money deposit will be credited to the purchaser's initial investment.

(1) *Amount.* The amount of earnest money collected will be:

(i) For single family properties or MFH projects of 2 to 5 units, $50.

(ii) For all property other than that covered in paragraph (e)(1)(i) of this section, the *greater* of the estimated closing costs shown on the notice of listing (Form RD 1955–40) or ½ of 1 percent of the purchase price.

(2) *Offeror default.* When a contract is cancelled due to offeror default, the earnest money will be delivered to and retained by the agency as full liquidated damages.

(f) *Commission—*(1) *Amount—*(i) *Exclusive broker contract.* Rural Development may not set the commission rate in an exclusive broker solicitation/contract. The rate of commission will be one of the evaluation criteria in the solicitation. However, any broker who submits an offer with a commission rate lower than the typical rate for such services in the area must provide documentation that they have successfully sold properties at the lower rate with no compromise in services. The solicitation/contract will explicitly detail this policy.

(ii) *Open listing agreement.* A uniform fee or commission schedule, by property type, will be established by the servicing official within a given sales area. The commission rate to be paid will be the typical rate for such services in the sales area and will not exceed or be lower than commissions paid for similar types of services provided by the broker to other sellers of similar property.

(2) *Special effort sales bonuses.* The servicing official may request authorization from the State Director to pay fixed amount bonuses for special effort property, such as a property with a value so low that the commission alone does not warrant broker interest or property that has been held in inventory for an extended period of time where it is believed that an added bonus will create additional efforts by the broker to sell the property. The State Director may authorize use of short-term (not to exceed three months) special effort sales bonuses on a group, county, district or state-wide basis, if it appears necessary to facilitate the sale of nonprogram property.

(3) *Payment of commission.* Payment of a broker's commission is contingent on the closing of the sale and will not be paid until the sale has closed and title has passed to the purchaser. No commission will be paid where the sale is to the broker, broker's salesperson(s), to persons living in his/her or salesperson(s) immediate household or to legal entities in which the broker or salesperson(s) have an interest if the sale is contingent upon receiving Rural Development credit. If credit is not being extended in these instances (a cash sale), a commission will be paid. Under an exclusive broker contract, if a cooperating broker purchases the property and is receiving Rural Development credit, one-half the respective commission will be paid to the exclusive broker. Commissions will be paid at closing if sufficient cash to cover the commission is paid by the purchaser. Otherwise, the commission will be paid by the appropriate Rural Development official by completing Form AD–838 and processing Form RD 838–B for payment in accordance with the respective FMI's, and charged to the inventory account as a nonrecoverable cost.

(g) *Nondiscrimination.* Brokers who execute listing agreements with Rural Development shall certify to nondiscrimination practices as provided in Form RD 1955–42. In addition, all brokers participating in the sale of property shall sign the nondiscrimination certification on Form RD 1955–45.

[53 FR 27836, July 25, 1988, as amended at 55 FR 3943, Feb. 6, 1990; 62 FR 44401, Aug. 21, 1997; 68 FR 61332, Oct. 28, 2003]

§1955.131 Auctioneers.

The services of licensed auctioneers, if required, may be used to conduct auction sales as described in §1955.148 of this subpart and procured by competitive negotiation under the contracting authority of Exhibit C to RD Instruction 2024–A (available in any Rural Development office).

(a) *Selection criteria.* The auctioneer should be selected by evaluating criteria such as proposed sales dates, location, advertising, broker cooperation, innovations, mechanics of sale, sample advertising, personal qualifications, financial capability, private sector financing and license/bonding.

(b) *Commission.* Rural Development may not set the commission rate in an auctioneer solicitation/contract. The rate of commission will be one of the evaluation criteria in the solicitation. However, any offeror that submits an offer with a commission rate lower than the typical rate for such services in the area must include documentation that they have successfully sold properties at the lower rate with no compromise in services. The solicitation/contract will explicitly detail this policy. Commissions will be paid at closing if sufficient cash to cover the commission is paid by the purchaser. Otherwise, the commission will be paid by the appropriate Rural Development official completing Form AD–838 and processing Form RD 838–B for payment in accordance with the respective FMI's, and charged to the inventory account as a nonrecoverable cost.

(c) *Auctioneer restriction.* The auctioneer, his/her sales agents, cooperating brokers or persons living in his, her or their immediate household are restricted from bidding or from subsequent purchase of any property sold or offered at the auctioneer's sale for a period of one year from the auction date.

[50 FR 23904, June 7, 1985, as amended at 53 FR 27837, July 25, 1988]

GENERAL

§ 1955.132 Pilot projects.

Rural Development may conduct pilot projects to test policies and procedures for the management and disposition of inventory property which deviate from the provisions of this subpart, but are not inconsistent with the provisions of the authorizing statute or other applicable Acts. A pilot project may be conducted by Rural Development employees or by contract with individuals, organizations or other entities. Prior to initiation of a pilot project, Rural Development will publish notice in the FEDERAL REGISTER of its nature, scope, and duration.

[55 FR 3943, Feb. 6, 1990]

§ 1955.133 Nondiscrimination.

(a) *Title VI provisions.* If the inventory real property to be sold secured a loan that was subject to Title VI of the Civil Rights Act of 1964, and the property will be used for its original or similar purpose, or if Rural Development extends credit and the property then becomes subject to Title VI, the buyer will sign Form RD 400–4. "Assurance Agreement." The instrument of conveyance will contain the following statement:

The property described herein was obtained or improved through Federal financial assistance. This property is subject to the provisions of Title VI of the Civil Rights Act of 1964 and the regulations issued pursuant thereto for so long as the property continues to be used for the same or similar purposes for which the Federal financial assistance was extended.

(b) *Affirmative Fair Housing Marketing Plan.* Exclusive listing brokers or auctioneers selling SFH properties having 5 or more properties in the same subdivision listed or offered for sale at the same time will prepare and submit to Rural Development an acceptable Form HUD 935.2, "Affirmative Fair Housing Marketing Plan," for each such subdivision in accordance with § 1901.203(c) of Subpart E of Part 1901 of this chapter.

(c) *Equal Housing Opportunity logo.* All Rural Development and contractor sale advertisements will contain the Equal Housing Opportunity logo.

§ 1955.134 Loss, damage, or existing defects in inventory real property.

(a) *Property under contract.* If a bid or offer has been accepted by the Rural Development and through no fault of either party, the property is lost or damaged as a result of fire, vandalism, or an act of God between the time of acceptance of the bid or offer and the time the title of the property is conveyed by Rural Development, FmHA or its successor agency under Public Law 103–354 will reappraise the property. The reappraised value of the property will serve as the amount Rural Development will accept from the purchaser. However, if the actual loss based on the reduction in market value of the property as determined by Rural Development is less than $500, payment of the full purchase price is required. In the event the two parties cannot agree upon an adjusted price, either party, by mailing notice in writing to the other, may terminate the contract of sale, and the bid deposit or earnest money, if any, will be returned to the offeror.

(b) *Existing defects.* Rural Development does not provide any warranty on property sold from inventory. Subsequent loans may be made, in accordance with applicable loan making regulations for the respective loan program, to correct defects.

[50 FR 23904, June 7, 1985, as amended at 53 FR 27837, July 25, 1988]

§ 1955.135 Taxes on inventory real property.

Where Rural Development owned property is subject to taxation, taxes and assessment installments will be prorated between Rural Development and the purchaser as of the date the title is conveyed in accordance with the conditions of Forms RD 1955–45 or RD 1955–46. The purchaser will be responsible for paying all taxes and assessment installments accruing after the title is conveyed. The County Supervisor or District Director will advise the taxing authority of the sale, the purchaser's name, and the description of the property sold. Only the prorata share of assessment installments for property improvements (water, sewer, curb and gutter, etc.) accrued as of the date property is sold will be paid by Rural Development for inventory property. At the closing, payment of taxes and assessment installments due to be paid by Rural Development will be paid from cash proceeds Rural Development is to receive as a result of the sale or by voucher and will be accomplished by one of the following:

(a) For purchasers receiving Rural Development credit and required to escrow, Rural Development's share of accrued taxes and assessment installments will be deposited in the purchaser's escrow account.

(b) For purchasers not required to escrow, accrued taxes and assessment installments may be:

(i) Paid to the local taxing authority if they will accept payment at that time; or

(ii) Paid to the purchaser. If appropriate, for program purchasers, the funds can be deposited in a supervised bank account until the taxes can be paid.

(c) Except for SFH, deducted from the sale price (which may result in a promissory note less than the sale price), if acceptable to the purchaser.

[56 FR 6953, Feb. 21, 1991]

§ 1955.136 Environmental review requirements.

(a) Prior to a final decision on some disposal actions, the action must comply with the environmental review requirements in accordance with the agency's environmental policies and procedures found in 7 CFR part 1970.

(1) The conveyance is controversial for environmental reasons and/or is qualified within those categories described in § 1955.137 of this subpart.

(2) The Rural Development approval official has reason to believe that conveyance would result in a change in use of the real property. For example, farmland would be converted to a nonfarm use; or an industrial facility would be changed to a different industrial use that would produce increased gaseous, liquid or solid wastes over the former use or changes in the type or contents of such wastes. Assessments are not required for conveyance where the real property would be retained in its former use within the reasonably foreseeable future.

(b) When an EA or EIS is prepared it shall address the requirements of Departmental Regulation 9500–3, "Land Use Policy," in connection with the conversion to other uses of prime and unique farmlands, farmlands of statewide or local importance, prime forest and prime rangelands, the alteration of wetlands or flood plains, or the creation of nonfarm uses beyond the boundaries of existing settlements.

[50 FR 23904, June 7, 1985, as amended at 81 FR 11032, Mar. 2, 2016; 81 FR 26667, May 4, 2016; 82 FR 19320, Apr. 27, 2017]

§ 1955.137 Real property located in special areas or having special characteristics.

(a) *Real property located in flood, mudslide hazard, wetland or Coastal Barrier Resources System (CBRS)—(1) Use restrictions.* Executive Order 11988, "Floodplain Management," and Executive Order 11990, "Protection of Wetlands," require the conveyance instrument for inventory property containing floodplains or wetlands which is proposed for lease or sale to specify those uses that are restricted under identified Federal, State and local floodplains or wetlands regulations as well as other appropriate restrictions. The restrictions shall be to the uses of the property by the lessee or purchaser and any successors, except where prohibited by law. Applicable restrictions will be incorporated into quitclaim deeds in a format similar to that contained in Exhibits H and I of RD Instruction 1955–C (available in any

Agency office). A listing of all restrictions will be included in the notices required in paragraph (a)(2) of this section.

(2) *Notice of hazards.* Acquired real property located in an identified special flood or mudslide hazard area as defined in, subpart B of part 1806 of this chapter will not be sold for residential purposes unless determined by the county official or district director to be safe (that is, any hazard that exists would not likely endanger the safety of dwelling occupants).

(3) *Limitations placed on financial assistance.* (i) Financial assistance is limited to property located in areas where flood insurance is available. Flood insurance must be provided at closing of loans on program-eligible and non-program (NP)-ineligible terms. Appraisals of property in flood or mudslide hazard areas will reflect this condition and any restrictions on use. Financial assistance for substantial improvement or repair of property located in a flood or mudslide hazard area is subject to the limitations outlined in 7 CFR part 1970 for Rural Development programs.

(ii) Pursuant to the requirements of the Coastal Barrier Resources Act (CBRA) and except as specified in paragraph (a)(3)(v) of this section, no credit sales will be provided for property located within a CBRS where:

(A) It is known that the purchaser plans to further develop the property;

(B) A subsequent loan or any other type of Federal financial assistance as defined by the CBRA has been requested for additional development of the property;

(C) The sale is inconsistent with the purpose of the CBRA; or

(D) The property to be sold was the subject of a previous financial transaction that violated the CBRA.

(iii) For purposes of this section, additional development means the expansion, but not maintenance, replacement-in-kind, reconstruction, or repair of any roads, structures or facilities. Water and waste disposal facilities as well as community facilities may be repaired to the extent required to meet health and safety requirements, but may not be improved or expanded to serve new users, patients or residents.

(iv) A sale which is not in conflict with the limitations in paragraph (a)(3)(ii) of this section shall not be completed until the approval official has consulted with the appropriate Regional Director of the U.S. Fish and Wildlife Service and the Regional Director concurs that the proposed sale does not violate the provisions of the CBRA.

(v) Any proposed sale that does not conform to the requirements of paragraph (a)(3)(ii) of this section must be forwarded to the Administrator for review. Approval will not be granted unless the Administrator determines, through consultation with the Department of Interior, that the proposed sale does not violate the provisions of the CBRA.

(b) *Wetlands located on FSA inventory property.* Perpetual wetland conservation easements (encumbrances in deeds) to protect and restore wetlands or converted wetlands that exist on suitable or surplus inventory property will be established prior to sale of such property. The provisions of paragraphs (a) (2) and (3) of this section also apply, as does paragraph (a)(1) of this section insofar as floodplains are concerned. This requirement applies to either cash or credit sales. Similar restrictions will be included in leases of inventory properties to beginning farmers or ranchers. Wetland conservation easements will be established as follows:

(1) All wetlands or converted wetlands located on FSA inventory property which were not considered cropland on the date the property was acquired and were not used for farming at any time during the period beginning on the date 5 years before the property was acquired and ending on the date the property was acquired will receive a wetland conservation easement.

(2) All wetlands or converted wetlands located on FSA inventory property that were considered cropland on the date the property was acquired or were used for farming at any time during the period beginning on the date 5 years before the property was acquired and ending on the date the property was acquired will not receive a wetland conservation easement.

(3) The following steps should be taken in determining if conservation

easements are necessary for the protection of wetlands or converted wetland on inventory property:

(i) NRCS will be contacted first to identify the wetlands or converted wetlands and wetland boundaries of each wetland or converted wetland on inventory property.

(ii) After receiving the wetland determination from NRCS, FSA will review the determination for each inventory property and determine if any of the wetlands or converted wetlands identified by NRCS were considered cropland on the date the property was acquired or were used for farming at any time during the period beginning on the date 5 years before the property was acquired and ending on the date the property was acquired. Property will be considered to have been used for farming if it was primarily used for agricultural purposes including but not limited to such uses as cropland, pasture, hayland, orchards, vineyards and tree farming.

(iii) After FSA has completed the determination of whether the wetlands or converted wetlands located on an inventory property were used for cropland or farming, the U.S. Fish and Wildlife Service (FWS) will be contacted. Based on the technical considerations of the potential functions and values of the wetlands on the property, FWS will identify those wetlands or converted wetlands that require protection with a wetland conservation easement along with the boundaries of the required wetland conservation easement. FWS may also make other recommendations if needed for the protection of important resources such as threatened or endangered species during this review.

(4) The wetland conservation easement will provide for access to other portions of the property as necessary for farming and other uses.

(5) The appraisal of the property must be updated to reflect the value of the land due to the conservation easement on the property.

(6) Easement areas shall be described in accordance with State or local laws. If State or local law does not require a survey, the easement area can be described by rectangular survey, plat map, or other recordable methods.

(7) In most cases the FWS shall be responsible for easement management and administration responsibilities for such areas unless the wetland easement area is an inholding in Federal or State property and that entity agrees to assume such responsibility, or a State fish and wildlife agency having counterpart responsibilities to the FWS is willing to assume easement management and administration responsibilities. The costs associated with such easement management responsibilities shall be the responsibility of the agency that assumes easement management and administration.

(8) County officials are encouraged to begin the easement process before the property is taken into inventory, if possible, in order to have the program completed before the statutory time requirement for sale.

(c) *Historic preservation.* (1) Pursuant to the requirements of the National Historic Preservation Act and Executive Order 11593, "Protection and Enhancement of the Cultural Environment," the Agency official responsible for the conveyance must determine if the property is listed on or eligible for listing on the National Register of Historic Places. (See subpart F of part 1901 of this chapter for additional guidance.) The State Historic Preservation Officer (SHPO) must be consulted whenever one of the following criteria are met:

(i) The property includes a structure that is more than 50 years old.

(ii) Regardless of age, the property is known to be of historical or archaeological importance; has apparent significant architectural features; or is similar to other Agency properties that have been determined to be eligible.

(iii) An environmental assessment is required prior to a decision on the conveyance.

(2) If the result of the consultations with the SHPO is that a property may be eligible or that it is questionable, an official determination must be obtained from the Secretary of the Interior.

(3) If a property is listed on the National Register or is determined eligible for listing by the Secretary of Interior, the Agency official responsible for the conveyance must consult with the

SHPO in order to develop any necessary restrictions on the use of the property so that the future use will be compatible with preservation objectives and which does not result in an unreasonable economic burden to public or private interest. The Advisory Council on Historic Preservation must be consulted by the State Director or State Executive Director after the discussions with the SHPO are concluded regardless of whether or not an agreement is reached.

(4) Any restrictions that are developed on the use of the property as a result of the above consultations must be made known to a potential bidder or purchaser through a notice procedure similar to that in § 1955.13(a)(2) of this subpart.

(d) *Highly erodible farmland.* (1) The FSA county official will determine if any inventory property contains highly erodible land as defined by the NRCS and, if so, what specific conservation practices will be made a condition of a sale of the property.

(2) If the county official does not concur in the need for a conservation practice recommended by NRCS, any differences shall be discussed with the recommending NRCS office. Failure to reach an agreement at that level shall require the State Executive Director to make a final decision after consultation with the NRCS State Conservationist.

(3) Whenever NRCS technical assistance is requested in implementing these requirements and NRCS responds that it cannot provide such assistance within a time frame compatible with the proposed sale, the sale arrangements will go forward. The sale will proceed, conditioned on the requirement that a purchaser will immediately contact (NRCS) have a conservation plan developed and comply with this plan. The county official will monitor the borrower's compliance with the recommendations in the conservation plan. If problems occur in obtaining NRCS assistance, the State Executive Director should consult with the NRCS State Conservationist.

(e) *Notification to purchasers of inventory property with reportable underground storage tanks.* If the Agency is selling inventory property containing a storage tank which was reported to the Environmental Protection Agency (EPA) pursuant to the provisions of § 1955.57 of subpart B of this part, the potential purchaser will be informed of the reporting requirement and provided a copy of the report filed by the Agency.

(f) *Real property that is unsafe.* If the Agency has in inventory, real property, exclusive of any improvements, that is unsafe, that is it does not meet the definition of "safe" as contained in § 1955.103 of this subpart and which cannot be feasibly made safe, the State Director or State Executive Director will submit the case file, together with documentation of the hazard and a recommended course of action to the National Office, ATTN: appropriate Deputy Administrator, for review and guidance.

(g) *Real property containing hazardous waste contamination.* All inventory property must be inspected for hazardous waste contamination either through the use of a preliminary hazardous waste site survey or Transaction Screen Questionnaire. If possible contamination is noted, a Phase I or II environmental assessment will be completed per the advice of the State Environmental Coordinator.

[62 FR 44401, Aug. 21, 1997, as amended at 68 FR 7700, Feb. 18, 2003; 81 FR 11032, Mar. 2, 2016; 82 FR 19320, Apr. 27, 2017]

§ 1955.138   **Property subject to redemption rights.**

If, under State law, Rural Development's interest may be sold subject to redemption rights, the property may be sold provided there is no apparent likelihood of its being redeemed.

(a) A credit sale of a program or suitable property subject to redemption rights may be made to a program applicant when the property meets the standards for the respective loan program. In areas where State law does not provide for full recovery of the cost of repairs during the redemption period, a program sale is generally precluded unless the property already meets program standards.

(b) Each purchaser will sign a statement acknowledging that:

(1) The property is subject to redemption rights according to State law, and

(2) If the property is redeemed, ownership and possession of the property would revert to the previous owner and likely result in loss of any additional investment in the property not recoverable under the State's provisions of redemption.

(c) The signed original statement will be filed in the purchaser's County or District Office case file.

(d) If real estate brokers or auctioneers are engaged to sell the property, the County Supervisor or District Director will inform them of the redemption rights of the borrower and the conditions under which the property may be sold.

(e) The State Director, with prior approval of OGC, will issue a State supplement incorporating the requirements of this section and providing additional guidance appropriate for the State.

[50 FR 23904, June 7, 1985, as amended at 53 FR 27837, July 25, 1988]

§ 1955.139 Disposition of real property rights and title to real property.

(a) *Easements, rights-of-way, development rights, restrictions or the equivalent thereof.* The State Director is authorized to convey these rights for conservation purposes, roads, utilities, and other purposes as follows:

(1) Except as provided in paragraph (a)(3) of this section, easements or rights-of-way may be conveyed to public bodies or utilities if the conveyance is in the public interest and will not adversely affect the value of the real estate. The consideration must be adequate for the inventory property being released or for a purpose which will enhance the value of the real estate. If there is to be an assessment as a result of the conveyance, relative values must be considered, including any appropriate adjustment to the property's market value, and adequate consideration must be received for any reduction in value.

(2) Except as provided in paragraph (a)(3) of this section easements or rights-of-way may be sold by negotiation for market value to any purchaser for cash without giving public notice if the conveyance would not change the classification from program/suitable to

NP or surplus, nor decrease the value by more than the price received.

(3) For FSA properties only, easements, restrictions, development rights or similar legal rights may be granted or sold separately from the underlying fee or sum of all other rights possessed by the Government if such conveyances are for conservation purposes and are transferred to a State, a political subdivision of a State, or a private nonprofit organization. Easements may be granted or sold to a Federal agency for conservation purposes as long as the requirements of § 1955.139(c)(2) of this subpart are followed. If FSA has an affirmative responsibility such as protecting an endangered species as provided for in paragraph (a)(3)(v) of this section, the requirements in § 1955.139(c) of this subpart do not apply.

(i) Conservation purposes include but are not limited to protecting or conserving the following environmental resources or land uses:

(A) Fish and wildlife habitats of local, regional, State, or Federal importance,

(B) Floodplain and wetland areas as defined in Executive Orders 11988 and 11990,

(C) Highly erodible land as defined by SCS,

(D) Important farmland, prime forest land, or prime rangeland as defined in Departmental Regulation 9500–3, Land Use Policy,

(E) Aquifer recharge areas of local, regional or State importance,

(F) Areas of high water quality or scenic value, and

(G) Historic and cultural properties.

(ii) Development rights may be sold for conservation purposes for their market value directly to a unit of local or State governmental or a private nonprofit organization by negotiation.

(iii) An easement, restriction or the equivalent thereof may be granted or sold for less than market value to a unit of local, State, Federal government or a private nonprofit organization for conservation purposes. If such a conveyance will adversely affect the Rural Development's financial interest, the State Director will submit the proposal to the Administrator for approval unless the State Director has been delegated approval authority in writing from the Administrator to approve such transactions based upon demonstrated capability and experience in processing such conveyances. Factors to be addressed in formulating such a request include the intended conservation purpose(s) and the environmental importance of the affected property, the impact to the Government's financial interest, the financial resources of the potential purchaser or grantee and its normal method of acquiring similar property rights, the likely impact to environment should the property interest not be sold or granted and any other relevant factors or concerns prompting the State Director's request.

(iv) Property interests under this paragraph may be conveyed by negotiation with any eligible recipient without giving public notice if the conveyance would not change program/suitable property to NP or surplus. Conveyances shall include terms and conditions which clearly specify the property interest(s) being conveyed as well as all appropriate restrictions and allowable uses. The conveyances shall also require the owner of such interest to permit the Rural Development, and any person or government entity designated by the Rural Development, to have access to the affected property for the purpose of monitoring compliance with terms and conditions of the conveyance. To the maximum extent possible, the conveyance should designate an organization or government entity for monitoring purposes. In developing the conveyance, the approval official shall consult with any State or Federal agency having special expertise regarding the environmental resource(s) or land uses to be protected.

(v) For FP cases except when Rural Development has an affirmative responsibility to place a conservation easement upon a farm property, easements under the authority of this paragraph will not be established unless either the rights of all prior owner(s) have been met or the prior owner(s) consents to the easement. Examples of instances where an affirmative responsibility exists to place an easement on a farm property include wetland and floodplain conservation easements re-

quired by §1955.137 of this subpart or easements designed as environmental mitigation measures for the purpose of protecting federally designated important environmental resources. These resources include: Listed or proposed endangered or threatened species, listed or proposed critical habitats, designated or proposed wilderness areas, designated or proposed wild or scenic rivers, historic or archaeological sites listed or eligible for listing on the National Register of Historic Places, coastal barriers included in Coastal Barrier Resource Systems, natural landmarks listed on national Registry of Natural Landmarks, and sole source aquifer recharge as designated by the Environmental Protection Agency.

(vi) For FP cases whenever a request is made for an easement under the authority of this paragraph and such request overlaps an area upon which Rural Development has an affirmative responsibility to place an easement, that required portion of the easement, either in terms of geographical extent or content, will not be considered to adversely impact the value of the farm property.

(4) A copy of the conveyance instrument will be retained in the County or District Office inventory file. The grantee is responsible for recording the instrument.

(b) *Mineral and water rights, mineral lease interests, air rights, and agricultural or other leases.* (1) Mineral and water rights, mineral lease interests, mineral royalty interests, air rights, and agricultural and other lease interests will be sold with the surface land and will not be sold separately, except as provided in paragrah (a) of this section and in §1955.66(a)(2)(iii) of Subpart B of Part 1955 of this chapter. If the land is to be sold in separate parcels, any rights or interests that apply to each parcel will be included with the sale.

(2) Lease or royalty interests not passing by deed will be assigned to the purchaser when property is sold. The County Supervisor or District Director, as applicable, will notify the lessee or payor of the assignment. A copy of this notice will be furnished to the purchaser.

(3) The value of such rights, interests or leases will be considered when the property is appraised.

(c) *Transfer of FSA inventory property for conservation purposes.* (1) In accordance with the provisions of this paragraph, FSA may transfer, to a Federal or State agency for conservation purposes (as defined in paragraph (a)(3)(i) of this section), inventory property, or an interest therein, meeting any one of the following three criteria and subject only to the homestead protection rights of all previous owners having been met.

(i) A predominance of the land being transferred has marginal value for agricultural production. This is land that NRCS has determined to be either highly erodible or generally not used for cultivation, such as soils in classes IV, V, VII or VIII of NRCS's Land Capability Classification, or

(ii) A predominance of land is environmentally sensitive. This is land that meets any of the following criteria:

(A) Wetlands, as defined in Executive Order 11990 and USDA Regulation 9500.

(B) Riparian zones and floodplains as they pertain to Executive Order 11988.

(C) Coastal barriers and zones as they pertain to the Coastal Barrier Resources Act or Coastal Zone Management Act.

(D) Areas supporting endangered and threatened wildlife and plants (including proposed and candidate species), critical habitat, or potential habitat for recovery pertaining to the Endangered Species Act.

(E) Fish and wildlife habitats of local, regional, State or Federal importance on lands that provide or have the potential to provide habitat value to species of Federal trust responsibility (*e.g.*, Migratory Bird Treaty Act, Anadromous Fish Conservation Act).

(F) Aquifer recharges areas of local, regional, State or Federal importance.

(G) Areas of high water quality or scenic value.

(H) Areas containing historic or cultural property; or

(iii) A predominance of land with special management importance. This is land that meets the following criteria:

(A) Lands that are in holdings, lie adjacent to, or occur in proximity to, Federally or State-owned lands or interest in lands.

(B) Lands that would contribute to the regulation of ingress or egress of persons or equipment to existing Federally or State-owned conservation lands.

(C) Lands that would provide a necessary buffer to development if such development would adversely affect the existing Federally or State-owned lands.

(D) Lands that would contribute to boundary identification and control of existing conservation lands.

(2) When a State or Federal agency requests title to inventory property, the State Executive Director will make a preliminary determination as to whether the property can be transferred.

(3) If a decision is made by the State Executive Director to deny a transfer request by a Federal or State agency, the requesting agency will be informed of the decision in writing and informed that they may request a review of the decision by the FSA Administrator.

(4) When a State or Federal agency requests title to inventory property and the State Executive Director determines that the property is suited for transfer, the following actions must be taken prior to approval of the transfer:

(i) At least two public notices must be provided. These notices will be published in a newspaper with a wide circulation in the area in which the requested property is located. The notice will provide information on the proposed use of the property by the requesting agency and request any comments concerning the negative or positive aspects of the request. A 30-day comment period should be established for the receipt of comments.

(ii) If requested, at least one public meeting must be held to discuss the request. A representative of the requesting agency should be present at the meeting in order to answer questions concerning the proposed conservation use of the property. The date and time for a public meeting should be advertised.

(iii) Written notice must be provided to the Governor of the State in which the property is located as well as at least one elected official of the county in which the property is located. The notification should provide information on the request and solicit any comments regarding the proposed transfer. All procedural requirements in paragraph (c) (3) of this section must be completed in 75 days.

(5) Determining priorities for transfer or inventory lands.

(i) A Federal entity will be selected over a State entity.

(ii) If two Federal agencies request the same land tract, priority will be given to the Federal agency that owns or controls property adjacent to the property in question or if this is not the case, to the Federal agency whose mission or expertise best matches the conservation purposes for which the transfer would be established.

(iii) In selecting between State agencies, priority will be given to the State agency that owns or controls property adjacent to the property in question or if that is not the case, to the State agency whose mission or expertise best matches the conservation purpose(s) for which the transfer would be established.

(6) In cases where land transfer is requested for conservation purposes that would contribute directly to the furtherance of International Treaties or Plans (e.g., Migratory Bird Treaty Act or North American Waterfowl Management Plan), to the recovery of a listed endangered species, or to a habitat of National importance (e.g., wetlands as addressed in the Emergency Wetlands Resources Act), priority consideration will be given to land transfer for conservation purposes, without reimbursement, over other land disposal alternatives.

(7) An individual property may be subdivided into parcels and a parcel can be transferred under the requirements of this paragraph as long as the remaining parcels to be sold make up a viable sales unit, suitable or surplus.

[50 FR 23904, June 7, 1985, as amended at 51 FR 13479, Apr. 21, 1986; 53 FR 27838, July 25, 1988; 53 FR 35781, Sept. 14, 1988; 57 FR 36592, Aug. 14, 1992; 62 FR 44403, Aug. 21, 1997; 68 FR 61332, Oct. 28, 2003; 82 FR 19320, Apr. 27, 2017]

§ 1955.140   Sale in parcels.

(a) Individual property subdivided. An individual property, other than Farm Loan Programs property, may be offered for sale as a whole or subdivided into parcels as determined by the State Director. For MFH property, guidance will be requested from the National Office for all properties other than RHS projects. When farm inventory property is larger than a family-size farm, the county official will subdivide the property into one or more tracts to be sold in accordance with § 1955.107. Division of the land or separate sales of portions of the property, such as timber, growing crops, inventory for small business enterprises, buildings, facilities, and similar items may be permitted if a better total price for the property can be obtained in this manner. Environmental review requirements must comply with 7 CFR part 1970. Any applicable State laws will be set forth in a State supplement and will be complied with in connection with the division of land. Subdivision of acquired property will be reported on Form RD 1955-3C, "Acquired Property—Subdivision," in accordance with the FMI.

(b) Grouping of individual properties. The county official for FCP cases, and the State Director for all other cases, may authorize the combining of two or more individual properties into a single parcel for sale as a suitable program property.

[62 FR 44403, Aug. 21, 1997, as amended at 81 FR 11032, Mar. 2, 2016; 82 FR 19320, Apr. 27, 2017]

§ 1955.141   Transferring title.

(a)–(c) [Reserved]

(d) Rent increases for MFH property. After approval of a credit sale for an occupied MFH project, but prior to closing, the purchaser will prepare a realistic budget for project operation (and a utility allowance, if applicable) to determine if a rent increase may be needed to continue or place project operations on a sound basis. 7 CFR part 3560, subpart E will be followed in processing the request for a rent increase. In processing the rent increase, the purchaser will have the same status as a borrower. An approved rent increase will be effective on or after the date of closing.

(e) Interest credit and rental assistance for MFH property. Interest credit and rental assistance may be granted to program applicants purchasing MFH properties in accordance with the provisions of 7 CFR part 3560, subpart F.

[53 FR 27838, July 25, 1988, as amended at 56 FR 2257, Jan. 22, 1991; 57 FR 36592, Aug. 14, 1992; 60 FR 34455, July 3, 1995; 69 FR 69106, Nov. 26, 2004]

§§ 1955.142–1955.143 [Reserved]

§ 1955.144 Disposal of NP or surplus property to, through, or acquisition from other agencies.

(a) *Property which cannot be sold.* If NP or surplus real or chattel property cannot be sold (or only token offers are received for it), the appropriate Assistant Administrator shall give consideration to disposing of the property to other Federal Agencies or State or local governmental entities through the General Services Administration (GSA). Chattel property will be reported to GSA using Standard Form 120, "Report of Excess Personal Property," with transfer documented by Standard Form 122, "Transfer Order Excess Personal Property." Real property will be reported to GSA using Standard Form 118, "Report of Excess Real Property," Standard Form 118A, "Buildings, Structures, Utilities and Miscellaneous Facilities (Schedule A)," Standard Form 118B, "Land (Schedule B)" and Standard Form 118C, "Related Personal Property (Schedule B), " with final disposition documented by a "Receiving Report," executed by the recipient with original forwarded to the Finance Office and a copy retained in the inventory file. Forms and preparation instructions will be obtained from the appropriate GSA Regional Office by the State Office.

(b) Urban Homesteading Program (UH). Section 810 of the Housing and Community Development Act of 1979, as amended, authorizes the Secretary of Housing and Urban Development (HUD) to pay for acquired Rural Development single family residential properties sold through the HUD-UH Program. Local governmental units may make application through HUD to participate in the UH Program. State Directors will be notified by the Assistant Administrator for Housing, when local governmental units in their States have obtained funding for the UH Program. The notification will provide specific guidance in accordance with the "Memorandum of Agreement between the Rural Development and the Secretary of Housing and Urban Development" dated October 2, 1981. (See Exhibit C of this subpart.) A Local Urban Homesteading Agency (LUHA) is authorized a 10 percent discount of the listed price on any SFH nonprogram property for the UH Program. No discount is authorized on program property.

[50 FR 23904, June 7, 1985, as amended at 53 FR 27839, July 25, 1988; 55 FR 3943, Feb. 6, 1990]

EDITORIAL NOTE: At 60 FR 34455, July 3, 1995, § 1955.144 was amended by removing the second through the fourth sentences. However, there are no undesignated paragraphs in the 1995 edition of this volume.

§ 1955.145 Land acquisition to effect sale.

The State Director is authorized to acquire land which is necessary to effect sale of inventory real property. This action must be considered only on a case-by-case basis and may not be undertaken primarily to increase the financial return to the Government through speculation. The State Director's authority under this section may *not* be redelegated. For MFH and other organization-type loans, prior approval must be obtained from the appropriate Assistant Administrator prior to land acquisition.

(a) *Alternate site.* Where real property has been determined to be NP due to location and where it is economically feasible to relocate the structure thereby making it a program property, the State Director may authorize the acquisition of a suitable parcel of land to relocate the structure if economically feasible. The remaining NP parcel of land will be sold for its market value.

(b) *Additional land.* Where real property has been determined NP for reasons that may be cured by the acquisition of adjacent land or an alternate site, in order to cure title defects or encroachments or where structures have been built on the wrong land *and* where it is economically feasible, the State Director may authorize the acquisition of additional land at a price not in excess of its market value.

(c) *Easements or rights-of-way.* The State Director may authorize the acquisition of easements, rights-of-way or other interests in land to cure title defects, encroachments or in order to make NP property a program property, if economically feasible.

[53 FR 27839, July 25, 1988]

§ 1955.146 Advertising.

(a) *General.* When property is being sold by Rural Development or through real estate brokers, it is the servicing official's responsibility to ensure adequate advertising of property to achieve a timely sale. The primary means of advertisements are newspaper advertisements in accordance with RD Instruction 2024–F (available in any Rural Development office), public notice using Form RD 1955–41, "Notice of Sale," and notification of known interested parties. Other innovative means are encouraged, such as the use of a bulletin board to display photographs of inventory properties for sale with a brief synopsis of the property attached; posting Forms 1955–40 or 1955–43, as appropriate, in the reception area to attract applicant and broker interest; posting notices of sale at employment centers; door-to-door distribution of sales notices at apartment complexes; radio and/or television spots; group meetings with potential applicants/investors/real estate brokers; and advertisements in magazines and other periodicals. If Rural DevelopmentFmHA or its successor agency under Public Law 103–354 personnel are not available to perform these services, Rural Development may contract for such services in accordance with Rural Development Instruction 2024–A (available in any Rural Development office).

(b) *Large-value and complex properties.* Advertising for MFH, B&I and other large-value or complex properties should also be placed in appropriate newspapers and publications designed to reach the type of particular purchasers most likely to be interested in the inventory property. The State Director will assist the District Director in determining the scope of advertising necessary to adequately market these properties. Advertising for MFH and other complex properties must also include appropriate language stressing the need to obtain and submit complete application materials for the type program involved.

(c) *MFH restrictive-use provisions.* Advertisements for multi-family housing projects will advise prospective purchasers of any restrictive-use requirements that will be attached to the project and added to the title of the property.

(d) *Racial and socio-economic considerations.* In accordance with the policies set forth in § 1901.203(c) of subpart E of part 1901 of this chapter, the approval official will make a special effort to insure that those prospective purchasers in the marketing area who traditionally would not be expected to apply for housing assistance because of existing racial or socio-economic patterns are reached.

(e) *Rejected application for SFH loan.* If an application for a SFH loan is being rejected because income is too high, a statement should be included in the rejection letter that inventory properties may be available for which they may apply.

[50 FR 23904, June 7, 1985, as amended at 53 FR 27839, July 25, 1988; 58 FR 38928, July 21, 1993]

§ 1955.147 Sealed bid sales.

This section provides guidance on the sale of all Rural Development inventory property, except suitable FP real property which will not be sold by sealed bid. Before a sealed bid sale, the State Director will determine and document the minimum sale price acceptable. In determining a minimum sale price, the State Director will consider the length of time the property has been in inventory, previous marketing efforts, the type property involved, and potential purchasers. Program financing will be offered on sales of program and suitable property. For NP or surplus property, credit may be extended to facilitate the sale. When a group of properties is to be sold at one time, advertising may indicate that Rural Development will consider bids on an individual property or a group of properties and Rural Development will accept the bid or bids which are in the best financial interest of the Government. Credit, however, may not exceed the market value of the property nor may the term exceed the period for which the property will serve as adequate security. Sealed bids will be made on Form RD 1955–46 with any accompanying deposit in the form of cashier's check, certified check, postal or bank money order or bank draft payable to Rural Development. For program and suitable property, the minimum deposit will be the same as outlined in §1955.130(e)(1) of this subpart. For NP or surplus property, the minimum deposit will be ten percent (10%). The bid will be considered delivered when actually received at the Rural Development office. All bids will be date and time stamped. Advertisements and notices will request bidders to submit their bid in a sealed envelope marked as follows:

SEALED BID OFFER _____*_____."
(*Insert "PROPERTY IDENTIFICATION NUMBER _____).

(a) *Opening bids.* Sealed bids will be held in a secured file before bid opening which will be at the place and time specified in the notice. The bid opening will be public and usually held at the Rural Development office. The County Supervisor, District Director, or State Director or his/her designee will open the bids with at least one other Rural Development employee present. Each bid received will be tabulated showing the name and address of the bidder, the amount of the bid, the amount and form of the deposit, and any conditions of the bid. The tabulation will be signed by the County Supervisor, District Director or State Director or his/her designee and retained in the inventory file.

(b) *Successful bids.* The highest complying bid meeting the minimum established price will be accepted by the approval official; however, it will be subject to loan approval by the appropriate official when a credit sale is involved. For SFH and FP (surplus property) sales, preference will be given to a cash offer on NP or surplus property sales which is at least ___*___ percent of the highest offer requiring credit [*Refer to Exhibit B of RD Instruction 440.1 (available in any Rural Development office) for the current percent-

age.] Otherwise, equal bids will be accepted by public lot drawing. For program or suitable property sales, no preference will be given to program purchasers unless two identical high bids are received, in which case the bid from the program purchaser will receive preference. If a bid is received from any purchaser with a request for credit that (considering any deposit) exceeds the market value of the property or requests a term which exceeds the period for which the property will serve as adequate security, the bidder will be given the opportunity to reduce the credit request and/or term with *no* accompanying change in the offered price.

(c) *Unsuccessful bids.* Deposits of unsuccessful bidders will be returned by certified mail with letter of explanation, return receipt requested. If there were no acceptable bids, the letter will advise each bidder of any anticipated negotiations for the sale of the property and deposits will be returned.

(d) *Disqualified bids.* Any bid that does not comply with the terms of the offer will be disqualified. Minor deviations and defects in bid submission may be waived by the Rural Development official approving the sale.

(e) *Failure to close.* If a successful bidder fails to perform under the terms of the offer, the bid deposit will be retained as full liquidated damages. However, if a credit sale complying with the Rural Development notice is an element of the offer and Rural Development disapproves the credit application, then the bid deposit will be returned to the otherwise successful bidder. Upon determination that the successful bidder will not close, the State Director may authorize either another sealed bid or auction sale of direct negotiations with the next highest bidder, all available unsuccessful bidders, or other interested parties.

(f) *No acceptable bid.* Where no acceptable bid is received although adequate competition is evident, the State Director may authorize a negotiated sale in accordance with § 1955.108(d) of this subpart.

[50 FR 23904, June 7, 1985, as amended at 53 FR 27839, July 25, 1988; 54 FR 6875, Feb. 15, 1989; 55 FR 3943, Feb. 6, 1990; 68 FR 61332, Oct. 28, 2003]

§ 1955.148   Auction sales.

This section provides guidance on the sale of all inventory property by auction, except FSA real property. Before an auction, the State Director, with the advice of the National Office for organizational property, will determine and document the minimum sale price acceptable. In determining a minimum sale price, the State Director will consider the length of time the property has been in inventory, previous marketing efforts, the type property involved, and potential purchasers. Program financing will be offered on sales of program and property. For NP property, credit may be offered to facilitate the sale. Credit, however, may not exceed the market value of the property nor may the term exceed the period for which the property will serve as adequate security. For program property sales, no preference will be given to program purchasers. The State Director will also consider whether an Agency employee will conduct an auction or whether the services of a professional auctioneer are necessary due to the complexity of the sale. When the services of a professional auctioneer are advisable, the services will be procured by contract in accordance with RD Instruction 2024-A (available in any Agency Office). Chattel property may be sold at public auction that is widely advertised and held on a regularly scheduled basis without solicitation. Form RD 1955-46 will be used for auction sales. At the auction, successful bidders will be required to make a bid deposit. For program and suitable property, the bid deposit will be the same as outlined in § 1955.130(e)(1) of this subpart. For NP property sales, a bid deposit of 10 percent is required. Deposits will be in the form of cashier's check, certified check, postal or bank money order or bank draft payable to the Agency, cash or personal checks may be accepted when deemed necessary for a successful auction by the person conducting the auction. Where

credit sales are authorized, all notices and publicity should provide for a method of prior approval of credit and the credit limit for potential purchasers. This may include submission of letters of credit or financial statements prior to the auction. The auctioneer should not accept a bid which requests credit in excess of the market value. When the highest bid is lower than the minimum amount acceptable to the Agency, negotiations should be conducted with the highest bidder or in turn, the next highest bidder or other persons to obtain an executed bid at the predetermined minimum.

[62 FR 44404, Aug. 21, 1997, as amended at 68 FR 61332, Oct. 28, 2003]

§ 1955.149   Exception authority.

(a) The Administrator may, in individual cases, make an exception to any requirement or provision of this subpart or address any omission of this subpart which is not inconsistent with the authorizing statute or other applicable law if the Administrator determines that the Government's interest would be adversely affected or the immediate health and/or safety of tenants or the community are endangered if there is no adverse effect on the Government's interest. The Administrator will exercise this authority upon request of the State Director with recommendation of the appropriate program Assistant Administrator or upon request initiated by the appropriate program Assistant Administrator. Requests for exceptions must be made in writing and supported with documentation to explain the adverse effect, propose alternative courses of action, and show how the adverse effect will be eliminated or minimized if the exception is granted.

(b) The Administrator may authorize withholding sale of surplus farm inventory property temporarily upon making a determination that sales would likely depress real estate market and preclude obtaining at that time the best price for such land.

§ 1955.150   State supplements.

State Supplements will be prepared with the assistance of OGC as necessary to comply with State laws or only as specifically authorized in this

Instruction to provide guidance to Rural Development officials. State Supplements applicable to MFH, B&I, and CP must have prior approval of the National Office. Request for approval for those affecting MFH must include complete justification, citations of State law, and an opinion from OGC.

EXHIBIT A TO SUBPART C OF PART 1955—NOTICE OF FLOOD, MUDSLIDE HAZARD OR WETLAND AREA

TO:_____
DATE:_____

This is to notify you that the real property located at _____ is in a floodplain, wetland or area identified by the Federal Insurance Administration of the Federal Emergency Management Agency as having special flood or mudslide hazards. This identification means that the area has at least one percent chance of being flooded or affected by mudslide in any given year. For floodplains and wetlands on the property, restrictions are being imposed. Specific designation(s) of this property is(are) (special flood) (mudslide hazard) (wetland)*. The following restriction(s) on the use of the property will be included in the conveyance and shall apply to the purchasers, purchaser's heirs, assigns and successors and shall be construed as both a covenant running with the property and as equitable servitude subject to release by the Farmers Home Administration or its successor agency under Public Law 103–354 (FmHA or its successor agency under Public Law 103–354) when/if no longer applicable:

(INSERT RESTRICTIONS)

The FmHA or its successor agency under Public Law 103–354 will increase the number of acres placed under easement, if requested in writing, provided that the request is supported by a technical recommendation of the U.S. Fish and Wildlife Service. Where additional acreage is accepted by FmHA or its successor agency under Public Law 103–354 for conservation easement, the purchase price of the inventory farm will be adjusted accordingly.

_____

(County Supervisor, District Director or Real Estate Broker)
*ACKNOWLEDGEMENT*_____
*DATE:*_____

I hereby acknowledge receipt of the notice that the above stated real property is in a (special flood) (mudslide hazard) (wetland) * area and is subject to use restrictions as above cited. [Also, if I purchase the property through a credit sale, I agree to insure the property against loss from (floods) (mudslide) * in accordance with require-ments of the FmHA or its successor agency under Public Law 103–354.]

_____

(Prospective Purchaser)
* Delete the hazard that does not apply.
[57 FR 31644, July 17, 1992]

PART 1956—DEBT SETTLEMENT

Subpart A [Reserved]

Subpart B—Debt Settlement—Farm Loan Programs and Multi-Family Housing

Subpart C—Debt Settlement—Community and Business Programs

AUTHORITY: 5 U.S.C. 301; and 7 U.S.C. 1989.

SOURCE: 51 FR 45434, Dec. 18, 1986, unless otherwise noted.

EDITORIAL NOTE: Nomenclature changes to part 1956 appear at 80 FR 9901, Feb. 24, 2015.

Subpart A [Reserved]

Subpart B—Debt Settlement—Farm Loan Programs and Multi-Family Housing

SOURCE: 56 FR 10147, Mar. 11, 1991, unless otherwise noted.

§ 1956.51 Purpose.

This subpart delegates authority and prescribes policy and procedures for settlement of debts owed to the United States under the Farm Credit loan programs of the Farm Service Agency (FSA) and the Multi-Family Housing (MFH) program of the Rural Housing Service (RHS). It also applies to Nonprogram (NP) loans secured by MFH property of the RHS. Settlement of claims against recipients of grant funds for reasons such as the use of funds for improper purposes is also covered by this subpart. Settlement of claims against third party converters, and Economic Opportunity (EO) loans is authorized under the Federal Claims Collection Standards, 4 CFR parts 101-105. This subpart does not apply to RHS direct Single Family Housing (SFH) loans, RHS NP loans secured by SFH property, or to the Rural Rental Housing, Rural Cooperative Housing, and Farm Labor Housing programs.

[61 FR 59779, Nov. 22, 1996, as amended at 69 FR 69106, Nov. 26, 2004]

§§ 1956.52-1956.53 [Reserved]

§ 1956.54 Definitions.

Adjustment. The reduction of a debt or claim conditioned upon completion of payment of the adjusted amount at a specific future time or times, with or without the payment of any consideration when the adjustment offer is approved. An adjustment is not a final settlement until all payments under the adjustment agreement(s) have been made.

Amount of debt. The outstanding balance of the amount loaned including principal and interest plus any outstanding advances, including interest, and subsidy to be recaptured made by the Government on behalf of the borrower.

Cancellation. The final discharge of a debt without any payment on it.

Chargeoff. The writing off of a debt and termination of collection activity without release of personal liability.

Compromise. The satisfaction of a debt or claim by the acceptance of a lump-sum payment of less than the total amount owed on the debt or claim.

Debt forgiveness. For the purposes of servicing Farm Loan Programs loans, debt forgiveness is defined as a reduction or termination of a direct FLP loan in a manner that results in a loss to the Government. Included, but not limited to, are losses from a writedown or writeoff under 7 CFR part 766, debt settlement, after discharge under the provisions of the bankruptcy code, and associated with release of liability. Debt cancellation through conservation easements or contracts is not considered debt forgiveness for loan servicing purposes.

Debtor. The borrower of funds under any of the FmHA or its successor agency under Public Law 103-354 programs. This includes co-signors, guarantors and persons or entities that initially obtained or assumed a loan. Debtor also includes grant recipients.

Farm Loan Programs (FLP) loans. Farm Ownership (FO), Operating (OL), Soil and Water (SW), Economic Emergency (EE), Emergency (EM), Recreation (RL), Special Livestock (SL), Softwood Timber (ST) loans, and/or

Rural Housing Loans for farm services buildings (RHF).

Housing programs. All programs and claims arising under programs administered by FmHA or its successor agency under Public Law 103–354 under title V of the Housing Act of 1949.

Servicing office. The FmHA or its successor agency under Public Law 103–354 office that is responsible for the account.

Settlement. The compromise, adjustment, cancellation, or chargeoff of a debt owed to FmHA or its successor agency under Public Law 103–354. The term "Settlement" is used for convenience in referring to compromise, adjustment, cancellation, or chargeoff actions, individually or collectively.

United States Attorney. An attorney for the United States Department of Justice.

[56 FR 10147, Mar. 11, 1991, as amended at 58 FR 21344, Apr. 21, 1993; 62 FR 10157, Mar. 5, 1997; 72 FR 64123, Nov. 15, 2007]

§§ 1956.55–1956.56 [Reserved]

§ 1956.57 General provisions.

(a) *Application of policies.* All debtors are entitled to impartial treatment and uniform consideration under this subpart. Accordingly. FmHA or its successor agency under Public Law 103–354 personnel charged with any responsibility in connection with debt settlement will adhere strictly to the authorizations, requirements, and limitations in this subpart, and will not substitute individual feelings or sympathies in connection with any settlement.

(b) *Information needed for debt settlement.* A debtor requesting debt settlement must submit complete and accurate information from which a full determination of his/her financial condition can be made. This should include, where applicable, but is not limited to, obtaining verification of employment, providing expense verification, verifying farm program benefits (e.g., Farm Service Agency/Commodity Credit Corporation payments), and examining county records to determine what other assets the debtor has or recently disposed of. When a FLP debtor is continuing to farm, a farm operating plan must be obtained. Also, where a

spouse is not a co-debtor the spouse's income will be considered in meeting family living expenses. If it appears that a debtor will not be able to pay in full and the indebtedness is eligible for settlement under this subpart, action should be taken, if possible, to avoid unnecessary litigation to enforce collection. If the debt is eligible for settlement, the debt settlement authorities of FmHA or its successor agency under Public Law 103–354 should be explained and the privileges thereof extended to the debtor. The information obtained from the debtor should be documented on a debt settlement form.

(c) *Negotiating a settlement.* County Supervisors may approve or reject compromises, adjustments, cancellations, or chargeoffs of SFH debts (to include recapture receivables), regardless of the amount. District Directors and County Supervisors cannot approve other debt settlement actions; therefore, other than SFH debt settlements, they will make no statements to a debtor concerning the action that may be taken upon a debtor's application. In negotiating a settlement, all of the factors which are pertinent to determining ability to pay will be discussed to assist the debtor in arriving at the proper type and terms of a settlement. The present and future repayment ability of a debtor, the factors mentioned in this subpart, and any other pertinent information will be the basis of determining whether the debt should be collected in full, compromised, adjusted, canceled, or charged off. It is impossible in cases eligible for debt settlement to forecast accurately the debtor's future repayment ability over a long period of time; consequently, the period of time during which payments on settlement offers are to be made should not exceed five years. Debtors have the right to make voluntary settlement offers in any amount should they elect to do so. Adjustment offers will not be approved in any case unless there is reasonable assurance that the debtor will be able to make the payments as they become due.

(d) *Disposition of property.* Security may be retained by the debtor only under the conditions specified in § 1956.66 of this subpart.

(e) *Proceeds from the disposal of security prior to approval of a debt settlement offer.* A debtor is not required to have disposed of the security prior to application for debt settlement for a loan to be settled. However, if a debtor has disposed of security prior to applying for debt settlement, proceeds from the disposed security must first be applied on the debtor's account, irrespective of an application for debt settlement unless the conditions specified in § 1956.66 of this subpart are met.

(f) [Reserved]

(g) *Settlement when legal or investigative action has been taken, recommended, or is contemplated.* (1) Debts cannot be settled:

(i) If the matter has been referred either to the Office of the Inspector General (OIG) under § 1962.49(a) of subpart A of part 1962 of this chapter or to Office of the General Counsel (OGC) because of suspected criminal violation, or criminal prosecution is pending because of an illegal act(s) committed by the debtor in connection with the debt or the security for that debt, the procedure outlined in paragraph (g)(3) of this section will be followed, unless, the OIG has declined to investigate the matter or, OGC has advised otherwise, or the case is in the hands of the United States Attorney.

(ii) If a request for referral to the United States Attorney to institute a civil action to protect the interest of the Government has been made by FmHA or its successor agency under Public Law 103-354.

(iii) Except as provided in paragraph (g)(3) of this section, if the case has been referred to the United States Attorney and is not closed.

(2) If a debtor's account is involved in a fiscal irregularity investigation in which final action has not been taken or the account shows evidence that a shortage may exist and an investigation will be requested, the account will not be approved for settlement.

(3) When a claim has been referred to, or a judgment has been obtained by the United States Attorney, and the debtor requests settlement, the employee in charge of the account will explain to the debtor that the United States Attorney has exclusive jurisdiction over the claim or judgment, that FmHA or

its successor agency under Public Law 103-354 has no authority to agree to a settlement offer when the United States Attorney's file is not closed, and that if the debtor wishes to make a compromise or adjustment offer when the United States Attorney's file is not closed, if will be submitted with any related payment directly to the United States Attorney for a decision on the settlement offer.

(h) *Advice from OGC.* State Directors will obtain, when necessary, advice from the OGC in handling proposed debt settlement actions which involve legal problems.

(i) *Settlement of claims against estates.* Settlement of a claim against an estate under the provisions of this subpart will be based on the recovery that may reasonably be expected, taking into consideration such items as the security, costs of administration, allowances of minor children and surviving spouse, allowable funeral expenses, and dower and courtesy rights, and specific encumbrances on the property having priority over claims of the Government.

(j) *Joint debtors.* Settlement may not be approved for one joint debtor unless approved for all debtors. "Joint debtors" includes all parties (individuals, partnerships, joint operators, cooperatives, corporations, estates) who are legally liable for payment of the debt.

(1) Separate and individual adjustment offers from joint debtors must be accepted and processed only as a joint offer. Joint debtors must be advised that all debtors will remain liable for the balance of the debt until all payments due under the joint offer have been made.

(2) A separate Form FmHA or its successor agency under Public Law 103-354 1956-1 will be completed by each debtor, unless the debtors are members of the same family and all necessary financial information on each debtor can be shown clearly on a single application. Separate applications will be sent to the State Office as a unit.

(3) If one debtor applies for compromise, adjustment, or cancellation, or if the debt is to be charged off, and the other debtor(s) is deceased or has received a discharge of the debt in bankruptcy, or the whereabouts of the

other debtor(s) is unknown, or it is impossible or impracticable to obtain the signature of the other debtor(s), Form FmHA or its successor agency under Public Law 103–354 1956–1 or Form FmHA or its successor agency under Public Law 103–354 1956–2 (for housing loans) "Cancellation or Charge-off of FmHA or its successor agency under Public Law 103–354 Indebtedness," will be prepared by showing at the top of the form the name of the debtor requesting settlement, following by the name of the other debtor.

For example, "John Doe, joint debtor with Bill Doe, deceased," "John Doe, joint debtor with Sam Doe, discharged in bankruptcy," "John Doe, joint debtor with Mary Doe, impossible or impracticable to obtain signature," as appropriate. In addition to the information concerning settlement of the debt by the applicant, information which justifies settlement of the debt as to the debtor(s) not joining in the application will be shown on Form FmHA or its successor agency under Public Law 103–354 1956–1, or 1956–2 for housing loans.

(k) *Settlement where debtor owes more than one type of Agency loan.* It is not the policy to settle any loan indebtedness of a debtor who is also indebted on another agency loan and who will continue as an active borrower. In such case, the facts will be fully documented in part VIII of Form RD 1956–1.

(l) *No previous debt forgiveness.* Debt settlement may not be approved for any direct Farm Loan Programs loan if the borrower has received debt forgiveness on any other direct loan as defined in §1956.54 of this subpart.

[56 FR 10147, Mar. 11, 1991, as amended at 58 FR 21344, Apr. 21, 1993; 62 FR 10157, Mar. 5, 1997; 68 FR 7700, Feb. 18, 2003]

§§ 1956.58–1956.65 [Reserved]

§ 1956.66 Compromise and adjustment of nonjudgment debts.

Nonjudgment debts which the debtor is unable to pay may be compromised or adjusted in accordance with applicable provisions of this section, and the debtor may retain the security property, if any. Application will be made on Form RD 1956–1 by the debtor; or if the debtor is unable to act, by another party having legal authority to act for the debtor. Collection of a lump sum offer may be deferred until the debtor is advised that the offer is approved. Upon full payment of the approved compromise or adjustment amount, the Agency will release the debtor from liability by delivering the note(s) to the debtor stamped "Satisfied by compromise or adjustment."

(a) *FLP debts.* The debt or any extension thereof on which compromise or adjustment is requested does not have to be due and payable under the terms of the note or other instrument, or because of acceleration by written notice prior to the date of application. Nonjudgment secured FLP debts may be compromised or adjusted in accordance with the following conditions:

(1) Security may be retained by the debtor if the debtor offers an amount at least equal to the current fair market value (including any crop security) less any prior lien amounts. Any remaining unsecured debt may be debt settled.

(2) Where the debtor is able to pay an amount in excess of the lump sum compromise offer, an adjustment offer must call for a lump sum payment as set out in paragraph (a)(1) of this section, plus any additional amounts the Agency determines the debtor is able to pay over a period of time not to exceed 5 years.

(3) The acceptability of a compromise or adjustment offer will be arrived at by determining and evaluating:

(i) Statement of indebtedness owed on any prior liens. Statements will be retained in the debtor's file.

(ii) Value of existing security as determined by a current appraisal made or obtained by the Agency. The appraisal will be retained in the debtor's file.

(iii) Debtor's total present income and probable sources, amount and stability of income over the next 5 years. Old age pensions, other public assistance, and veteran's disability pensions will not be considered as sources of funds for making compromise and adjustment offers.

(iv) Amount of debtor's other debts.

147

(v) Amount of debtor's essential family living expenses, and farm or business operation expenses necessary to continue the operation, if applicable.

(vi) Age and health when the debtor is largely depending on income from an occupation where manual labor is required.

(vii) Size of debtor's family, their ages and health.

(viii) Value of debtor's assets in relation to debts and liens of third parties. Reasonable equity in a modest non-security homestead occupied by the debtor will not be considered as available for settlement. Nonsecurity property in excess of minimum family living needs which is not exempt from levy and execution should be considered in determining the debtor's ability to pay.

(b) *Housing debts (both Single-family and Multi-family).* Nonjudgment secured debts may be compromised or adjusted as follows:

(1) The debt is fully matured under the terms of the note or other instrument; or has been accelerated by written notice prior to the date of the settlement application.

(2) A compromise offer must at least equal the value of the security as determined by FmHA or its successor agency under Public Law 103–354 (less any prior liens) plus any additional amount FmHA or its successor agency under Public Law 103–354 determines the debtor is able to pay based on a current financial statement.

(3) An adjustment offer must meet the requirements of paragraph (b)(2) of this section, except the debt (or the amount offered) is to be scheduled for payment over the shortest period FmHA or its successor agency under Public Law 103–354 determines is feasible based on the debtor's financial resources, but not to exceed 5 years.

(c) *Unsecured debts.* Unsecured debts considered under this paragraph (c) are most frequently account balances remaining after the debtor has sold security property to another party/entity, the security has been liquidated through foreclosure, or FmHA or its successor agency under Public Law 103–354 has accepted a deed in lieu of foreclosure and the borrower was not released from liability. An offer to compromise or adjust an unsecured debt must represent the maximum amount FmHA or its successor agency under Public Law 103–354 determines the debtor can pay based on a current financial statement and other information available to FmHA or its successor agency under Public Law 103–354. An adjustment offer is to be scheduled for payment over the shortest period FmHA or its successor agency under Public Law 103–354 determines is feasible, but not to exceed 5 years.

[56 FR 10147, Mar. 11, 1991, as amended at 58 FR 21345, Apr. 21, 1993; 62 FR 10157, Mar. 5, 1997]

§ 1956.67 **Debts which the debtor is able to pay in full but refuses to do so.**

Debts which the debtor may have the ability to pay in full but has refused to do so may be compromised or adjusted in the following situations on Form FmHA or its successor agency under Public Law 103–354 1956–1:

(a) When the full amount cannot be collected because of the refusal of the debtor to pay the debt in full and the OGC advises that the Government is unable to enforce collection in full within a reasonable time by enforced collection proceedings, the debt may be compromised. In determining inability to collect, the following factors will be considered:

(1) Availability of assets or income which may be realized by enforced collection proceedings, considering the applicable exemptions available to the debtor under State and Federal law.

(2) Inheritance prospects within 5 years.

(3) Likelihood of debtor obtaining nonexempt property or income within 5 years, out of which there could be collected a substantially larger sum than the amount of the present offer.

(4) Uncertainty as to price the security or other property will bring at forced sale.

(b) The debt may be compromised or adjusted when the OGC has advised in writing that:

(1) There is a real doubt concerning the Government's ability to prove its case in court for the full amount of the debt, and

(2) The amount offered represents a reasonable settlement considering:

(i) The probability of prevailing on the legal issues involved.

(ii) The probability of proving facts to establish full or partial recovery, with due regard to the availability of witnesses and other pertinent factors.

(iii) The probable amount of court costs and attorney's fees which may be assessed against the Government if it is unsuccessful in litigation.

(c) When the cost of collecting the debt does not justify enforced collection of the full amount, the amount accepted in compromise or adjustment may reflect an appropriate discount for administrative and litigation costs of collection. Such discount will not exceed $2,000 unless the OGC advises that in the particular case a larger discount is appropriate. The cost of collecting may be a substantial factor in settling small debts but normally will not carry great weight in settling large debts.

§ 1956.68 Compromise or adjustment without debtor's signature.

Debts of a living debtor may be compromised or adjusted if it is impossible or impracticable to obtain a signed application and all other requirements of this section applicable to compromise or adjustment with a signed application have been met. Form FmHA or its successor agency under Public Law 103–354 1956–1 will show:

(a) The sources from which the information was obtained.

(b) That a current effort was made to obtain the debtor's signature and the date(s) of such effort.

(c) The specific reasons why it was impossible or impracticable to obtain the signature of the debtor and, if the debtor refused to sign, the reason(s) given.

§ 1956.69 [Reserved]

§ 1956.70 Cancellation.

Nonjudgment debts may be canceled in the following instances:

(a) *With application.* The debt or any extension thereof on Farmer Programs debts *do not* have to be due and payable under the terms of the note or other instrument, or because of acceleration by written notice prior to the date of application. Debts due the FmHA or its successor agency under Public Law 103–354 may be canceled upon application of the debtor, or if a debtor is unable to act, upon application of a guardian, executor, or administrator, subject to the following conditions:

(1) The FmHA or its successor agency under Public Law 103–354 employee in charge of the account furnishes a report and favorable recommendation concerning the cancellation.

(2) There is no known security for the debt and the debtor has no other assets from which the debt could be collected.

(3) The debtor is unable to pay any part of the debt and has no reasonable prospect of being able to do so.

(b) *Without application.* Debts due the FmHA or its successor agency under Public Law 103–354 may be canceled upon a report and the favorable recommendation of the employee in charge of the account in the following instances:

(1) *Deceased debtors.* The following conditions must exist:

(i) There is no known security; and

(ii) An administrator or executor has not been appointed to settle the debtor's estate and the financial condition of the estate has been investigated and it has been established that there is no reasonable prospect of recovery; or

(iii) An administrator or executor has been appointed to settle the estate of the debtor; and

(A) A final settlement has been made and confirmed by the probate court and the Government's claim was recognized properly and the Government has received all funds it was entitled to, or

(B) A final settlement has not been made and confirmed by the probate court but there are no assets in the estate from which there is any reasonable prospect of recovery, or

(C) Regardless of whether a final settlement has been made, there were assets in the estate from which recovery might have been affected but such assets have been disposed of or lost in a manner which OGC advises will preclude any reasonable prospect of recovery by the Government.

(2) *Disappeared debtors.* The debt may be canceled without application where

the debtor has no known assets or future debt-paying ability, has disappeared and cannot be found without undue expense, and there is no existing security for the debt. Reasonable efforts will be made to locate the debtor. These efforts will generally include contacts, either in person or in writing, with postmasters, motor vehicle licensing and title authorities, telephone directories, city directories, utility companies, State and local governmental agencies, other Federal agencies, employees, friends, and credit agency skip locate reports, known relatives, neighbors and County Committee members. Also, the debtor's loan file should be reviewed carefully for possible leads that may be of assistance in locating the debtor. The efforts made to locate the debtor, including the names and dates of contacts, and the information furnished by each person, will be fully documented in the appropriate space on Form FmHA or its successor agency under Public Law 103–354 1956–1 or Form FmHA or its successor agency under Public Law 103–354 1956–2 for housing loans.

(3) *Debtors discharged in bankruptcy.* If there is no security for the debt, debts discharged in bankruptcy shall be cancelled by use of the appropriate Agency form with the attachments noted below. No attempt will be made to obtain the debtor's signature. If the debtor has executed a new promise to pay prior to discharge and has otherwise accomplished a valid reaffirmation of the debt in accordance with advice from OGC, the debt is not discharged.

(i) Chapter 7 Bankruptcy cases will be documented with a copy of the "Discharge of Debtor" order(s) by the court for all obligors.

(ii) For debts identified as being part of an unsecured claim under Chapter 11, the cancellation will be documented with a copy of the organization plan, copy of the order by the court confirming the plan, a copy of the order completing the plan (a similar order), and an opinion by OGC that the confirming order has discharged the obligor(s) of liability to that part of the debt.

(iii) For debts identified as being part of an unsecured claim under chapters 12 or 13, the cancellation will be documented with a copy of the reorganization plan and confirmation order, as above, a copy of the order completing the plan and closing the case, and an opinion by OGC that the completion order has discharged the obligor(s) of liability to that portion of the debt.

(c) *Signature of debtor cannot be obtained.* Debts of a living debtor may be canceled if it is impossible or impracticable to obtain a signed application and the requirements in paragraph (a) of this section concerning cancellation with application have been met or if the debt has been discharged in bankruptcy and there is no security. Form FmHA or its successor agency under Public Law 103–354 1956–1 will state:

(1) The sources of information obtained.

(2) That a current effort was made to obtain the debtor's application and the date of such effort.

(3) The specific reasons why it was impossible or impracticable to obtain the signature of the debtor and, if the debtor refused to sign, the reason(s) given.

[56 FR 10147, Mar. 11, 1991, as amended at 68 FR 7700, Feb. 18, 2003]

§ 1956.71 Settling uncollectible recapture receivables.

The settlement of uncollectible recapture receivables will be fully documented on a debt settlement form and retained in the case file.

[58 FR 21345, Apr. 21, 1993]

§§ 1956.72–1956.74 [Reserved]

§ 1956.75 Chargeoff.

(a) *Judgment debts.* Subject to the provisions of § 1956.57(g)(3), judgment debts may be charged off by use of Form FmHA or its successor agency under Public Law 103–354 1956–1 or Form FmHA or its successor agency under Public Law 103–354 1956–2 for housing upon a report and favorable recommendation of the employee in charge of the account provided:

(1) The United States Attorney's file is closed, and

(2) The requirements of § 1956.70(b)(2) have been met, or two years have elapsed since any collections were made on the judgment and the debtor(s) has no equity in property on

which the judgment is a lien or on which it can presently be made a lien.

(b) *Nonjudgment debts.* Debts which cannot be settled under other sections of this subpart may be charged off using Form FmHA or its successor agency under Public Law 103–354 1956–1 or Form FmHA or its successor agency under Public Law 103–354 1956–2 for housing loans without the debtor's signature subject to the following provisions:

(1) When the principal balance is $2,000 or less and efforts to collect have been unsuccessful or it is apparent that further collection efforts would be ineffectual or uneconomical,

(2) When the OGC advises in writing that the claim is legally without merit.

(3) Even though FmHA or its successor agency under Public Law 103–354 considers the claim to be valid, when efforts to induce voluntary payments are unsuccessful and the OGC advises in writing that evidence necessary to prove the claim in court cannot be produced, or

(4) When the employee in charge of the account recommends the chargeoff and has made the following determinations on the basis of information in FmHA or its successor agency under Public Law 103–354's official files or from other informed reliable sources:

(i) That the debtor is:

(A) Unable to pay any part of the debt and has no apparent future debt repayment ability as specified in § 1956.66(a); or

(B) Able to pay part or all of the debt but is unwilling to do so, it is clear that the Government cannot enforce collection of a significant amount from assets or income, and an opinion is received from OGC to that effect; and

(ii) There is no security for the debt.

(c) For debts identified as being part of an unsecured claim under a confirmed Chapter 11 plan, the chargeoff will be documented with a copy of the organization plan, a copy of the court order confirming the plan, an opinion by OGC that the order confirming the plan has discharged the debtor(s) of liability on the unsecured part of the debt.

§§ 1956.76–1956.83 [Reserved]

§ 1956.84 Approval or rejection.

(a)–(d) [Reserved]

(e) *Appeal rights.* A debtor whose debt settlement offer is rejected will be notified of appeal rights pursuant to 7 CFR part 11.

[58 FR 21345, Apr. 21, 1993, as amended at 68 FR 7700, Feb. 18, 2003]

§ 1956.85 Payments and receipts.

(a) *Servicing office handling.* (1) An application with which the debtor offers a lump-sum payment in compromise, or with which the debtor offers an initial payment on an adjustment offer, will be accompanied by the payments required at the time such application is filed in the servicing office.

(2) [Reserved]

(3) Checks or check transmittal letter containing restrictive notations such as "Settlement in full" or "Payment in full," or in those exceptional instances when the debtor refuses to sign the Form FmHA or its successor agency under Public Law 103–354 1956–1 in connection with a compromise offer, will be forwarded to the State Office where they will be retained until approval or rejection of the offer. The use of restrictive notations will be discouraged to the fullest extent possible.

(b) *Finance Office handling.* (1) All payments evidenced by Form FmHA or its successor agency under Public Law 103–354 451–2, "Schedule of Remittances," bearing the legend "Compromise Offer—FmHA or its successor agency under Public Law 103–354" or "Adjustment Offer—FmHA or its successor agency under Public Law 103–354," will be held in the Deposits Fund Account by the Finance Office until notification is received from the State Office of the approval or rejection of the offer. In cases of approved offers, remittances will be applied in accordance with established policies, beginning with the oldest loan included in the settlement, except that when the request for settlement includes loans made from different revolving funds the Finance Office will prorate the amount received, on the basis of the total principal balance due the respective revolving funds. Upon notification

151

of a rejection of a debtor's offer and receipt of a request from the State Director for a refund, the Finance Office will refund to the debtor, in care of the employee in charge of the account, the amount held in the Deposits Fund Account representing a rejected compromise or adjustment offer.

(2) When a debtor's adjustment offer is approved, the accounts involved will not be adjusted in the records of the Finance Office until all payments have been made. Form FmHA or its successor agency under Public Law 103–354 1956–1 will be held in a suspense file pending payment of the full amount of the approved offer. The original Form FmHA or its successor agency under Public Law 103–354 1956–1 in approved cases will be retained in the Finance Office.

[56 FR 10147, Mar. 11, 1991, as amended at 58 FR 21345, Apr. 21, 1993; 68 FR 61332, Oct. 28, 2003; 69 FR 69106, Nov. 26, 2004]

§§ 1956.86–1956.95 [Reserved]

§ 1956.96 Delinquent adjustment agreements.

A 90-day extension for making the payments may be given by the Agency when the circumstances of the case justify an extension. A decision not to extend the time for making payments is not appealable. If the debtor is delinquent under the terms of the adjustment agreement and is likely to be financially unable to meet the terms of the agreement, the Agency may cancel the existing agreement and process a different type of settlement more consistent with the debtor's repayment ability, provided the facts in the case justify such action. The cancellation of an adjustment agreement is appealable. If an agreement is cancelled, any payments received shall be retained as payments on the debt owed at the time of the adjustment agreement.

[68 FR 7700, Feb. 18, 2003]

§ 1956.97 Disposition of promissory notes.

(a) Notes evidencing debts settled by completed adjustments, completed compromise with or without signature, or canceled with signature will be returned to the debtor or to the debtor's legal representative. The original and copies of notes will be stamped "Satisfied by Approved Compromise," "Satisfied by Approved Cancellation," or "Satisfied by Completed Adjustment Offer." In such cases, the security instrument(s) will be released of record according to State law.

(b) Notes evidencing debts canceled without application will be placed in the debtor's case folder and disposed of pursuant to FmHA or its successor agency under Public Law 103–354 Instruction 2033–A (available in any FmHA or its successor agency under Public Law 103–354 office). However, if the debtor requests the notes, they may be stamped "Satisfied By Approved Cancellation" and returned.

(c) Notes evidencing charged off debts will be retained in the servicing office and will not be stamped or returned to the debtor. They will be destroyed six years after charged off pursuant to FmHA or its successor agency under Public Law 103–354 Instruction 2033–A (available in any FmHA or its successor agency under Public Law 103–354 office).

(d) In case of a transfer of security with assumption for less than the debt, the promissory note will be attached to the assumption agreement covered by the note and kept in the transferee's file.

[56 FR 10147, Mar. 11, 1991. Redesignated and amended at 58 FR 21346, Apr. 21, 1993]

§ 1956.98 [Reserved]

§ 1956.99 Exception authority.

The Administrator may, in individual cases, make an exception to any requirement or provision of this subpart which is not inconsistent with the authorizing statute or other applicable law if the Administrator determines that application of the requirement or provision would adversely affect the Government's interest. The Administrator will exercise this authority only at the request of the State Director and on the recommendation of the appropriate program Assistant Administrator. Requests for exceptions must be made in writing by the State Director and supported with documentation to explain the adverse affect on the Government's interest, propose alternative courses of action, and show how the adverse affect will be eliminated or minimized if the exception is granted. Any settlement actions approved by the Administrator under this section will be documented on Form FmHA or its successor agency under Public Law 103–354 1956–1 and returned to the State Office for submission to the Finance Office.

§ 1956.100 OMB control number.

The collection of information requirements in this regulation have been approved by the Office of Management and Budget and assigned OMB control number 0575–0118. Public reporting burden for this collection of information is estimated to vary from *15* to *20* minutes per response, with an average of *20* minutes per response including time for reviewing instructions, searching existing data sources, gathering and maintaining the data needed, and completing and reviewing the collection of information. Send comments regarding this estimate or any other aspect of this collection of information, including suggestions for reducing this burden, to Department of Agriculture, Clearance Officer, OIRM, Room 404–W, Washington, DC 20250; and to the Office of Information and Regulatory Affairs, Office of Management and Budget, Washington, DC 20503.

Subpart C—Debt Settlement—Community and Business Programs

SOURCE: 53 FR 13100, Apr. 21, 1988, unless otherwise noted.

§ 1956.101 Purpose.

This subpart delegates authority and prescribes policies and procedures for debt settlement of Community Facility loans; Association Recreation loans; Rural Renewal loans; direct Business and Industry loans; Rural Development Loan Fund loans; Intermediary Relending Program loans; and the Rural Microentrepreneur Assistance Program (RMAP) loans and repayable portions of RMAP grants; and Shift-in-land-use loans. Settlement of Economic Opportunity Cooperative loans, Claims Against Third Party Converters, Nonprogram loans, Rural Business Enterprise/Television Demonstration Grants, Nonprofit National Corporations Loans and Grants, and 601 Energy Impact Assistance Grants, is not authorized under independent statutory authority, and settlement under these programs is handled pursuant to the Federal Claims Collection Joint Standards, 31 CFR parts 900 through 904, inclusive. In addition, this subpart does not apply to Water and Waste Programs of the Rural Utilities Service, Watershed loans, and Resource Conservation and Development loans, which are serviced under part 1782 of this title.

[80 FR 13201, Mar. 13, 2015]

§ 1956.102 Application of policies.

(a) *General.* If a debt is eligible for settlement, the debt settlement authorities of the government should be explained and the privileges thereof extended to the debtor. All debtors are entitled to impartial treatment and uniform consideration under this subpart. Accordingly, Rural Development personnel charged with any responsibility in connection with debt settlement will adhere strictly to the authorizations, requirements, and limitations in this subpart.

(b) *For hospitals and health care facilities only.* Loan servicing and debt restructuring options according to § 1956.143 of this subpart must be exhausted before the other settlement authorities of this subpart are applicable.

[53 FR 13100, Apr. 21, 1988, as amended at 59 FR 46160, Sept. 7, 1994]

§§ 1956.103–1956.104 [Reserved]

§ 1956.105 Definitions.

(a) *Settlement.* The compromise, adjustment, cancellation, or chargeoff of a debt owed to Rural Development. The term "settlement" is used for convenience in referring to compromise, adjustment, cancellation, or chargeoff actions, individually or collectively.

(b) *Compromise.* The satisfaction of a debt, including a release of liability, by the acceptance of a lump-sum payment of less than the total amount owed on the debt.

(c) *Adjustment.* The satisfaction of a debt, including a release of liability, when acceptance is conditioned upon completion of payment of the adjusted amount at a specific future time or times, with or without the payment of any consideration when the adjustment offer is approved. An adjustment is not a final settlement until all payments under the adjustment agreement have been made.

(d) *Cancellation.* The final discharge of a debt with a release of liability.

(e) *Chargeoff.* To write off a debt and terminate all servicing activity *without* a release of liability. This is not a final discharge of the debt, but rather a decision upon the part of the agency to remove the debt from agency receivables.

(f) *Debtor.* The borrower of loan funds under any of Rural Development programs specified in § 1956.101.

(g) *Security.* All that serves as collateral for Rural Development loan(s), including, but not limited to, revenues, tax levies, municipal bonds, and real and chattel property.

(h) *Servicing official.* The Rural Development official who is primarily responsible for servicing the account.

(i) *United States Attorney.* An attorney for the United States Department of Justice.

(j) *Independent Qualified Fee Appraiser.* An individual who is a designated member of the American Institute of Real Estate Appraisers, Society of Real Estate Appraisers, or an equivalent organization, requiring appraisal education, testing, and experience.

[53 FR 13100, Apr. 21, 1988, as amended at 54 FR 47510, Nov. 15, 1989; 66 FR 1569, Jan. 9, 2001; 80 FR 9901, Feb. 24, 2015]

§§ 1956.106–1956.108 [Reserved]

§ 1956.109 General requirements for debt settlement.

(a) *Debt due and payable.* The debt or any extension thereof on which settlement is requested must be due and payable under the terms of the note or other instrument, or because of acceleration by written notice prior to the date of application for settlement, unless the debt is to be cancelled without application under § 1956.130(b) or charged off under § 1956.136 of this subpart.

(b) *Disposition of security.* Ordinarily, all security will be disposed of prior to

the date of application for settlement. There are exceptions:

(1) It may be necessary to abandon security through the debt settlement process. For example, a community may be rendered uninhabitable by a toxic or hazardous substance. In such cases, debt settlement may proceed provided the servicing official determines:

(i) That further collection efforts with respect to the security in question would be ineffective or uneconomical,

(ii) That it is in the best interests of the Government to proceed with debt settlement,

(iii) That the proposal otherwise meets the requirements appropriate to the type of settlement under consideration, and

(iv) The approval of the Administrator is obtained.

(2) A servicing action may have been carried out which resulted in a less than complete disposition of security. For example, the Government may have consented to a voluntary sale of a debtor's real and chattel property without reference to other security, which might include, but is not limited to: an additional lien on revenue, a third party pledge of security, or a pledge of personal liability. In such cases, debt settlement may proceed provided the requirements of § 1956.109(b)(1) of this subpart are met.

(3) Security can be retained under the compromise and adjustment offers as specified in § 1956.124 of this subpart.

(4) Settlement of a claim against an estate will be based on the recovery that may reasonably be expected, taking into consideration such items as the security, costs of administration, allowances of minor children and surviving spouse, allowable funeral expenses, dower and curtesy rights, and specific encumbrances having priority over claims of the Government.

(c) *Proceeds from the sale of security.* Proceeds from the sale of security must be applied on the debtor's account, taking into consideration the disposition requirements of any grant agreement, prior to the date of application for settlement, except when security is retained as provided for in § 1956.109(b) of this subpart. Debtors will not be allowed to sell security and use the proceeds as part or all of the debt settlement offer.

(d) *County Committee review.* Proposed settlement actions will be reviewed by the County Committee except for the cancellation of debts discharged in bankruptcy under §1956.130(b)(1) of this subpart or when a claim has been referred to a United States Attorney under §1956.112(d) of this subpart. No settlement shall be approved if it is more favorable to the debtor than recommended by the County Committee.

(e) *Assistance from Office of General Counsel (OGC).* When necessary, State Directors will obtain advice from OGC in handling proposed debt settlement actions.

(f) *Format.* Form RD 1956–1, "Application for Settlement of Indebtedness," will be utilized for all settlement actions under this subpart.

§1956.110 Joint debtors.

Settlements may not be approved for one joint debtor unless approved for all debtors. Joint debtors includes all parties, individuals, and organizations, who are legally liable for payment of the debt.

(a) Individual settlement offers from joint debtors can be accepted and processed only as a joint offer. A separate Form RD 1956–1 will be completed by each debtor unless the debtors are members of the same family and all necessary financial information on each debtor can be shown clearly on a single application.

(b) If one of the joint debtors is deceased or has received a discharge of the debt in bankruptcy, or if the whereabouts of one of the debtors is unknown, or it is otherwise impossible or impractical to obtain the signature of the debtor, the application for settlement may be accepted without that debtor's signature if it contains adequate information on each of the debtors to justify settlement of the debt as to each of the debtors. The name of the debtor requesting settlement will be shown at the top of Form RD 1956–1 followed by name and status of the other debtor. For example, "John Doe, joint debtor with Jane Doe, deceased."

(c) Joint debtors must be advised in writing that all debtors will remain liable for the balance of the debt until any payment(s) due under the joint offer have been made.

§1956.111 Debtors in bankruptcy.

Rural Development personnel will process reorganization plans of debtors filing under Chapter 9, Chapter 11, or Chapter 13 as follows:

(a) Plans submitted by debtors under Chapters 9, 11, and 13 must be sent by the servicing official to the State Director who will recommend either acceptance or rejection of the plans and refer them to the United States Attorney through OGC. When the plan calls for the adjustment of a debt to Rural Development, the State Director will obtain the advice of the Administrator before providing OGC with a recommendation on acceptance or rejection of this plan.

(b) The United States Attorney will advise the State Director, through OGC, as to approval or rejection of the debtor's reorganization plan. The State Director will then notify the Finance Office by memorandum of the terms and conditions of the bankruptcy reorganization plan, including any adjustment of the debt.

§1956.112 Debts ineligible for settlement.

Debts will not be settled:

(a) If referral to the Office of Inspector General (OIG) and/or to the OGC is contemplated or pending because of suspected criminal violation, or

(b) If civil action to protect the interests of the Government is contemplated or pending, or

(c) If an investigation for suspected fiscal irregularity is contemplated or pending, or

(d) When a claim has been referred to or a judgment has been obtained by the United States Attorney and the debtor requests settlement, the servicing official will explain to the debtor that the United States Attorney has exclusive jurisdiction over the claim or judgment, and therefore, Rural Devlopment has no authority to agree to a settlement offer. If the debtor wishes to make a settlement offer, it must be submitted with any related payment directly to the United States Attorney for consideration.

§§ 1956.113–1956.117 [Reserved]

§ 1956.118 Approval authority.

District Directors cannot approve debt settlement actions. Therefore, they will make no statements to a debtor concerning the action that may be taken upon a debtor's application. Subject to this subpart, the compromise, adjustment, cancellation, or chargeoff of debts will be approved or rejected:

(a) By the State Director when the outstanding balance of the indebtedness involved in the settlement is less then $50,000, including principal, interest, and other charges.

(b) By the Administrator or his designee when the outstanding balance of the indebtedness involved in the settlement is $50,000 or more, including principal, interest, and other charges.

§§ 1956.119–1956.123 [Reserved]

§ 1956.124 Compromise and adjustment.

Nonjudgment debts may be compromised or adjusted upon application of the debtor(s), or if the debtor is an individual and unable to act, upon application of the guardian, executor, or administrator of the debtor's estate.

(a) *General provisions.* Debts, regardless of the amount, may be compromised or adjusted subject to the following:

(1) The debt or any extension thereof on which compromise or adjustment is requested is due and payable under the terms of the note or other instrument, or because of acceleration by written notice, prior to the date of application for settlement.

(2) The period of time during which payments on adjustment offers are to be made cannot exceed five years without the approval of the Administrator.

(3) Efforts will be made to avoid applications for settlement in which debtors offer a specified amount payable upon notice of approval of the proposed settlement.

(b) *Debtor's ability to pay.* In evaluating the debtor's settlement application, it is essential that reliable information be obtained in sufficient detail to assure that the offer accurately reflects the debtor's ability to pay. The

debtor's income, expenses, and nonsecurity assets are critical factors in determining the type of settlement and the amount which the debtor can reasonably be expected to offer. Critical information should include the following:

(1) The debtor's total present income from all sources will be determined. In addition, careful consideration will be given to the probable sources, amount, and stability of income to be received over a reasonable period of years. For individuals, public welfare assistance and pensions, including old age pensions and pensions received by veterans for pensionable disabilities will not be considered as sources of funds with which to make compromise and adjustment offers.

(2) The debtor's operation and maintenance expenses, and, in the case of individuals, probable living expenses.

(3) The priority of payments on debts to third parties.

(4) When the debtor is largely dependent on income from an occupation in which manual labor is required, age and health of the individual are vital factors in determining the ability to pay. The number in the debtor's family, their ages and condition of health, will also be weighed in determining the ability to pay. However, when the debtor's income is from investments, business enterprises, or management efforts, age and health of both individual and family are of less importance.

(5) The value of the debtor's assets in relation to debts and liens of third parties is important in determining the debtor's ability to pay. It is recognized that debtors must retain a reasonable equity in essential nonsecurity property in order to continue normal operations and, in the case of an individual, to meet family living expenses over a period of years. Under this policy a reasonable equity in a modest nonsecurity homestead occupied by the debtor, whether or not exempt from levy and execution will not be considered as available for offer in settlement. Nonsecurity property which is in excess of minimum business and/or family living needs and which is not exempt from levy and execution should be considered when determining the debtor's ability to pay.

(c) *Debtor unable to pay in full.* Debts may be compromised or adjusted and security property retained by the debtor, provided:

(1) The debtor is unable to pay the indebtedness in full, and

(2) The debtor has offered an amount equal to the present fair market value of all security or facility financed, and

(3) The debtor has offered any additional amount which the debtor is able to pay, and

(4) The total amount offered represents a reasonable determination of the debtor's ability to pay.

(d) *Debtor able to pay in full but refuses to do so.* If the debtor has the ability to pay in full but refuses to do so, debts may be compromised or adjusted and security property retained by the debtor under certain conditions:

(1) The OGC advises that the Government is unable to enforce collection in full within a reasonable time by enforced collection proceedings, and the amount offered represents a reasonable settlement considering:

(i) Availability of assets or income which may be realized by enforced collection proceedings, considering the applicable exemptions available to the debtor under State and Federal law, and

(ii) Inheritance prospects within 5 years, and

(iii) Likelihood of debtor obtaining nonexempt property or income within 5 years out of which there could be collected a substantially larger sum than the amount of the present offer, and

(iv) Uncertainty as to the price that the security or other property will bring at forced sale, *or*

(2) The OGC advises that there is a real doubt concerning the Government's ability to prove its case in court for the full amount of the debt, and the amount offered represents a reasonable settlement considering:

(i) The probability of prevailing on the legal issues involved, and

(ii) The probability of proving facts to establish full or partial recovery, with due regard to the availability of witnesses and other pertinent factors, and

(iii) The probable amount of court costs and attorney's fees which may be

assessed against the Government if it is unsuccessful in litigation, *or*

(3) When the cost of collecting the debt does not justify enforced collection of the full amount. In such cases, the amount accepted in compromise or adjustment may reflect an appropriate discount for administrative and litigious costs of collection. Such discount will not exceed $600 unless the OGC advises that in the particular case a larger discount is appropriate. The cost of collecting may be a substantial factor in settling small debts but normally will not carry great weight in settling large debts.

§§ 1956.125–1956.129 [Reserved]

§ 1956.130 Cancellation.

Nonjudgment debts, regardless of the amount, may be cancelled with or without application by the debtor.

(a) *With application by debtor.* Debts may be cancelled upon application of the debtor(s), or if the debtor is an individual and unable to act, upon application of the guardian, executor, or administrator of the debtor's estate. The following conditions apply:

(1) The servicing official furnishes a favorable recommendation concerning the cancellation, and

(2) There is no known security for the debt and the debtor has no other assets from which the debt could be collected, and

(3) The debtor is unable to pay any part of the debt and has no reasonable prospect of being able to do so, and

(4) The debt or any extension thereof is due and payable under the terms of the note or other instrument, or because of acceleration by written notice prior to the date of application.

(b) *Without application by debtor.* Debts may be cancelled upon a favorable recommendation of the servicing official in the following instances:

(1) *Debtors discharged in bankruptcy.* If there is no security for the debt, debts discharged in bankruptcy shall be cancelled by the use of Form RD 1956–1 with a copy of the Bankruptcy Court's Discharge Order attached. No attempt will be made to obtain the debtor's signature and County Committee review is unnecessary. If the debtor has executed a new promise to pay prior to

discharge and has otherwise accomplished a valid reaffirmation of the debt in accordance with advice from OGC, the debt is not discharged.

(2) *Impossible or impractical to obtain a debtor's signature.* Debts may be cancelled if it is impossible or impractical to obtain a signed application and the requirements of § 1956.130(a) (1), (2), and (3) *only* of this subpart are met. Form RD 1956–1 will document:

(i) The sources of information obtained.

(ii) That a current effort was made to obtain the debtor's application and the date of such effort.

(iii) The specific reasons why it was impossible or impracticable to obtain the signature of the debtor and, if the debtor refused to sign, the reason(s) given.

(3) *Deceased debtors (individuals only).* The following conditions must exist:

(i) There is no known security,

(ii) An administrator or executor has not been appointed to settle the debtor's estate but the financial condition of the estate has been investigated and it has been established that there is no reasonable prospect of recovery, *or*

(iii) An administrator or executor has been appointed to settle the estate of the debtor, and

(A) A final settlement has been made and confirmed by the probate court and the Government's claim was recognized properly and the Government has received all funds it was entitled to, or

(B) A final settlement has not been made and confirmed by the probate court, but there are no assets in the estate from which there is any reasonable prospect of recovery, or

(C) Regardless of whether a final settlement has been made, there were assets in the estate from which recovery might have been effected but such assets have been disposed of or lost in a manner which the OGC advises will preclude any reasonable prospect of recovery by the Government.

(4) *Disappeared debtor (individuals only).* The following conditions must exist:

(i) The debtor has disappeared and cannot be found without undue expense. Reasonable efforts either in person or in writing will be made to locate the debtor. These efforts, including the

names and dates of contacts, and the information furnished by each person, will be fully documented on Form RD 1956–1,

(ii) There is no known security for the debt and the debtor has no other assets from which the debt could be collected, and

(iii) The debtor is unable to pay any part of the debt and has no reasonable prospect of being able to do so.

§§ 1956.131–1956.135 [Reserved]

§ 1956.136 Chargeoff.

(a) *Judgment debts.* Subject to the provisions of § 1956.112(d) of this subpart, judgment debts, regardless of the amount, may be charged off without the debtor's signature upon a favorable recommendation of the servicing official provided:

(1) The United States Attorney's file is closed, and

(2) The requirements of § 1956.130(b)(1), (2), (3), or (4) of this subpart have been met, as appropriate, or two years have elapsed since any collections were made on the judgment and the debtor(s) has no equity in property on which the judgment is a lien or on which it can presently be made a lien.

(b) *Nonjudgment debts.* Debts which cannot be settled under other sections of this subpart may be charged off without the debtor's signature upon a favorable recommendation of the servicing official in the following instances:

(1) When the OGC advises in writing that the claim is legally without merit, or that evidence necessary to prove the claim in court cannot be produced.

(2) When there is no known security for the debt, the debtor has no other assets from which the debt could be collected, and the debtor:

(i) Is unable to pay any party of the debt and has no reasonable prospect of being able to do so, or

(ii) Is able to pay part or all of the debt but refuses to do so, and an opinion is received from OGC to the effect that the Government cannot enforce collection of a significant amount from assets or income.

(3) When the debtor is deceased (individuals only), disappeared (individuals

only), or when it is impossible or impractical to obtain the debtor's signature, and the conditions of § 1956.136(b)(2) of this subpart are met.

§ 1956.137 [Reserved]

§ 1956.138 Processing.

(a) *Approval.* When a debt settlement application is approved, the State Director will:

(1) Send the original approved Form RD 1956–1 to the Finance Office.

(2) Notify debtors in writing of settlement approval, including the specific amount and terms of the offer that were accepted, for compromise and adjustment offers under § 1956.124 and cancellations with application under § 1956.130(a) of this subpart.

(3) Not be required to notify debtors of settlement approval when debts are cancelled without application under § 1956.130(b) or charged off under § 1956.136 of this subpart.

(b) *Requesting additional information.* When rejection appears to be necessary either because of lack of information or because the amount of a compromise or adjustment offer is inadequate, the State Director may request the servicing official to obtain the additional information or make an effort to obtain a more acceptable offer, as the circumstances justify. Notice of rejection of an offer will be withheld in such cases until sufficient time has elapsed to enable the debtor to present further information or a new offer.

(c) *Rejection.* When a debt settlement application is rejected, the State Director will:

(1) Insert the reasons for rejection on the Form FmHA or its successor agency under Public Law 103–354 1956–1.

(2) Retain the original Form RD 1956–1 in the State Office and return case files and copies of Form RD 1956–1 to the servicing official.

(3) Request the Finance Office to return any adjustment or compromise payment held by the Finance Office to the borrower, in care of the servicing official.

(4) Return any adjustment or compromise payment held by the State Office to the borrower, in care of the servicing official.

(5) Notify the debtor in writing of the reasons for the rejection for compromise and adjustment offers under § 1956.124 and cancellations with application under § 1956.130(a) of this subpart.

(d) *Appeal rights.* In accordance with subpart B of part 1900 of this chapter, the debtor will be given the right to appeal the rejection of any debt settlement offer made by the debtor under this subpart.

§ 1956.139 Collections.

(a) When the debtor offers a lump-sum payment in compromise or an initial payment on an adjustment offer, that payment will accompany the settlement application at the time the application is filed with the servicing official.

(b) [Reserved]

(c) Checks or check transmittal letters containing restrictive notations such as "Settlement in full" or "Payment in full," will be forwarded to the State Office where they will be retained until approval or rejection of the offer. The use of restrictive notations will be discouraged to the fullest extent possible.

(d) All payments evidenced by Form RD 451–2, "Schedule of Remittances," bearing the legend "Compromise Offer—Rural Development" or "Adjustment Offer—Rural Development," will be held in the Deposits Fund Account by the Finance Office until notification is received from the State Office of the approval or rejection of the offer.

(1) Upon receipt of an approved Form RD 1956–1, remittances will be applied in accordance with established policies, beginning with the oldest loan included in the settlement, except that when the request for settlement includes loans made from different revolving funds, the Finance Office will prorate the amount received on the basis of the total principal balance due the respective revolving funds.

(2) Upon notification of a rejection of a debtor's offer and receipt of a request from the State Director for a refund, the Finance Office will refund to the debtor, in care of the servicing official, the amount held in the Deposits Fund Account.

(e) When a debtor's adjustment offer is approved, the accounts involved will not be adjusted in the records of the Finance Office until all payments have been made. Form RD 1956–1 will be held in a suspense file pending payment of the full amount of the approved offer.

(f) If an approved debt settlement agreement is later voided by the State Director in accordance with § 1956.142(e) of this subpart, any payments which have been received shall be retained as payments on the debt owed at the time the compromise or adjustment offer was approved.

[53 FR 13100, Apr. 21, 1988, as amended at 68 FR 61332, Oct. 28, 2003]

§§ 1956.140–1956.141 [Reserved]

§ 1956.142 Delinquent adjustment agreements.

(a) The servicing official is responsible for notifying debtors in advance of the due dates of payments on debt settlement agreements and for monitoring compliance with the terms of settlement agreements. If a payment is delinquent, the servicing official should contact the debtor promptly to determine the reason for the delinquency and the debtor's plan for completing the agreement.

(b) Delinquencies of 30 days or more will be reported to the State Director along with other pertinent information and the recommendation of the servicing official regarding further handling of the case.

(c) The State Director may extend, for ninety days, the time for making the payments when the circumstances of the case justify an extension. Extensions for a greater period of time may be made by the State Director upon the recommendation of the County Committee and the servicing official.

(d) When the debtor is financially unable to meet the terms of the debt settlement agreement, the State Director may void the existing agreement and process a new settlement more consistent with the debtor's repayment ability, provided the facts in the case justify such action.

(e) If the State Director determines that the debtor cannot or will not meet the terms of the settlement agreement and if the facts do not justify approval

of a new settlement agreement, the State Director will void the existing agreement and direct the servicing official to take other servicing actions appropriate to the circumstances of the case.

(f) When an adjustment agreement is voided, the State Director will notify the debtor giving the reasons in writing, with a copy to the Finance Office and to the servicing official. Upon receipt, the Finance Office will return the original Form RD 1956–1 to the State Office.

§ 1956.143 Debt restructuring—hospitals and health care facilities.

This section pertains exclusively to delinquent Community Facility hospital and health care facility loans. Those facilities which are nonprogram (NP) loans as defined in § 1951.203 (f) of subpart E of part 1951 of this chapter are excluded. The purpose of debt restructuring is to keep the hospital or health care facility in operation with manageable debt.

(a) *Definitions.* As used in this section, the following definitions apply:

Consolidation. The combining of two or more debt instruments into one instrument, normally accompanied by reamortization.

Debt writedown. A one-time reduction of the debt owed to Rural Development including principal and interest. This reduction will be the minimum amount necessary to meet the level of the facility's ability to service the debt. The writedown will be applied first to interest and then principal.

Delinquency due to circumstances beyond the control of the debtor. Includes situations such as: The debtor has less money than planned due to unexpected and uncontrollable events such as unexpected loss of service area population, unforeseeable costs incurred for compliance with State or Federal regulatory requirements, or the loss of key personnel.

Delinquent debtor. For purposes of this section, delinquency is defined as being 180 days behind schedule on the Rural Development payments. That is, one full annual installment or the equivalent for monthly, quarterly, or semiannual installments.

Eligibility. Applicants must be delinquent due to circumstances beyond their control and have acted in good faith by trying to fulfill the agreements with Rural Development in connection with the delinquent loans.

Interest rate reduction. Reduction of the interest rate on the restructured loan to as low as the poverty line interest rate in effect on community and business programs loans.

Loan deferral. The temporary delay of principal and interest payments for up to 6 months. The debtor must be able to demonstrate the ability to pay the debt, as restructured, at the end of this delay period.

Net recovery value. A calculation of the net value of the collateral and other assets held by the debtor. This value would be determined by adding the fair market value of Rural Development's interest in any real property pledged as collateral for the loan, plus the value of any other assets pledged or otherwise available for the repayment of the debt, minus the anticipated administrative and legal expenses that would be incurred in connection with the liquidation of the loan. This value of the assets should be calculated based upon the facility continuing to operate as a going concern. Therefore, the facility should be valued not merely as an empty building but as a facility continuing to offer health care services which may, or may not, be similar to those offered by the current operators.

Operations review. A study of management and business operations of the facility by an independent expert. For example, a study of a hospital and nursing home would include such areas as: general and administrative, dietary, housekeeping, laundry, nursing, physical plant, social services, income potential, Federal, State, and insurance payments, and rate analysis. Also, recommendations and conclusions are to be included in the study which would indicate the creditworthiness of the facility and its ability to continue as a going concern. In analyzing a debtor's proposed restructuring plan, Rural Development may contract for the completion of an operations review. These reviews will be developed by individuals and entities who have demonstrated an expertise in the analysis

of health care facilities from an operational and administrative standpoint. Rural Development will consider the following criteria for selection: past experience in health care facility analysis, a familiarity with the problems of rural health care facilities, a knowledge of the particular area currently served by the facility in question, and a willingness to work with both Rural Development and the debtor in developing a final plan for restructuring.

Restructured loan. A revision of the debt instruments including any combination of the following: writing down of accumulated interest charges and principal, deferral, consolidation, and adjustment of the interest rates and terms, usually followed by reamortization.

(b) *Debtor notification.* All servicing actions permitted under subpart E of part 1951 of this chapter are to be exhausted prior to consideration for debt restructuring under this section. To this end, the servicing official must ensure that the casefile clearly documents that all servicing actions under subpart E of part 1951 of this chapter have been exhausted and that the debtor is at least 1 full year's debt service behind schedule for a minimum of 180 days. The debtor then should be informed of the debt restructuring available under this section by using language similar to that provided in Guide 1 of this subpart (available in any Rural Development Office) as follows:

(1) Any introductory paragraph;

(2) A paragraph concerning prior servicing attempts;

(3) A discussion of eligibility, as defined in this section, including the provision that the debtor acted in good faith in connection with their Rural Development loan and that the delinquency was caused by circumstances beyond their control;

(4) Two paragraphs that explain the goal of the debt restructuring program;

(5) A paragraph stating that debt restructuring may include a combination of servicing actions listed in paragraph (a) of this section;

(6) Information that details what the debtor must do to apply for restructuring. A response must be received within 45 days of receipt of this letter

to request consideration for debt restructuring and the request must include projected balance sheets, budgets, and cash-flow statements which include and clearly identify funding of the Rural Development reserve account for the next 3 years;

(7) A discussion of Rural Development's analysis and calculation process; and

(8) A paragraph identifying the Rural Development official who may be contacted for assistance.

(c) *State Director's restructuring determination.* Upon receipt of the delinquent debtor's request for debt restructuring consideration, the State Director will:

(1) Within 15 days of receipt of debtor's request, if an operations review is deemed necessary, send a memorandum to the Administrator asking for program authority to contract for the review in accordance with Exhibit D of Rural Development Instruction 2024-A (available in any Rural DevelopmentRural Development Office). The name of the debtor involved and the projected amount of funds anticipated to be spent for the contract should also be provided. It is anticipated that an operations review will be necessary in most cases and that the only exceptions would be for smaller health care facilities or facilities that have developed a proposed plan that is comprehensive and realistic. Upon receipt of the Administrator's program contracting approval authority, a contract is to be awarded to an organization qualified to perform an operations review as defined in paragraph (a) of this section. The operations review normally will be completed and delivered to Rural Development within 60 days of the award date.

(2) Contract for an appraisal to be performed by an independent, qualified fee appraiser. Note: To the extent possible, the appraisal should be scheduled for completion no later than the completion date of the operations review.

(3) Complete an analysis of the operations review, appraisal, and other documented information, and make an eligibility determination.

(i) *Eligibility determination.* The State Director must conclude that the debtor is eligible for debt restructuring consideration. This conclusion will be clearly documented in the casefile based on a review of the following:

(A) The debtor acted in good faith with regard to the delinquent loan. The casefile must reflect the debtor's cooperation in exploring servicing alternatives. The casefile should contain no evidence of fraud, waste, or conversion by the debtor, and no evidence that the debtor violated the loan agreement or Rural Development regulations.

(B) The delinquency was caused by circumstances beyond the control of the debtor. This determination will be based on the debtor's narrative on this issue, which is a required part of the application for debt restructuring, and a separate review of the debtor's casefile and operations.

(C) As part of the application for debt restructuring, the debtor submitted a proposed operating plan that presents feasible alternatives for addressing the delinquency.

(ii) *Debtor determined eligible.* If the debtor is determined to be eligible for debt restructuring, a determination of a net recovery value and level of debt the facility will support will be made. It is anticipated that meetings with the debtor, the contractor who performed the operations review, and others, as appropriate, could be necessary to develop these values; although it should be emphasized throughout these meetings that any calculations and conclusions reached are preliminary in nature, pending final review by the Administrator. For debt restructuring calculations and computing a feasible cash-flow projection, the following order and combinations of loan servicing actions will be followed:

(A) Loan deferral for up to 6 months.

(B) Interest rate reduction to not less than the poverty line rate as determined by Rural Development Instruction 440.1, exhibit B (available in any Rural Development Office). Interest rate reduction will be considered only in conjunction with an extension of the term of the loan to the remaining useful life of the facility or 40 years, whichever is less.

(C) Debt writedown. Other creditors of the debtor, representing a substantial portion of the total debt, are expected to participate in the development of a restructuring plan which includes debt writedown. Debt writedown participation by other creditors should be on a pro rata basis with the Rural Development writedown. However, failure of these creditors to agree to participate in the plan shall not preclude the use of principal and interest writedown by Rural Development if it is determined that this option results in the least cost to the Federal Government.

(iii) *Debtor determined ineligible.* If the State Director concludes that the debtor is not eligible for debt restructuring consideration for any of the reasons listed in paragraph (c)(3)(i) of this section, then the debtor will be notified by a letter that includes the following information:

(A) The basis for the determination;

(B) The next step in servicing the loan: possible acceleration if the delinquency is not cured; and

(C) The debtor may appeal this determination in accordance with subpart B of part 1900 of this chapter.

(iv) *State Director's recommendation.* Upon completion of the determination of net recovery value and restructured debt in accordance with paragraph (c)(3)(ii) of this section, and prior to formal presentation to the borrower, the State Director will forward a recommendation to the National Office with the following documentation:

(A) That all other servicing efforts have been exhausted as required in paragraph (b) of this section.

(B) Financial statements including balance sheets, income and expense, cash-flows for the most recent actual year, and projections for the next 3 years. The amount of Rural Development's restructured debt and reserve account requirements are to be clearly indicated on the projected statements. Also, operating statistics including number of beds, patient days of care, outpatient visits, occupancy percentage, etc., for the same periods of time must be included.

(C) Copies of the operations review, developed for the particular loan, and appraisal.

(D) Calculations of the net recovery value.

(E) Debt restructuring calculations including a listing of the various servicing combinations used in these calculations as contained in paragraph (c)(3)(ii) of this section. For example:

(1) Interest rate reduced from the applicant's current rate on all loans to the poverty line rate as determined by Rural Development instruction 440.1, exhibit B (available in any Rural Development Office); and

(2) Extension of the terms from 25 to 30 years.

(F) Information concerning discussions with the debtor and their agreement or disagreement with the calculations and recommendations.

(G) If debt restructuring is proposed:

(1) A draft of Form RD 3560–15, if applicable, and any other necessary comments or requirements that may be required by OGC and Bond Counsel in §1951.223 (c)(3) and (4) of subpart E of part 1951 of this chapter.

(2) A draft of Form RD 1956–1, if applicable. Complete only parts I, II, VI, and VIII. Part VI, "Debtor's Offer and Certification," will be in a separate attachment and contain the adjusted unpaid principal amount for which Rural Development approval is requested. In Part VI of the form, type "see attached."

(H) If the proposed restructured debt will not cash-flow or is less than the net recovery value, omit the items in paragraph (c)(3)(iv)(G) of this section.

(d) *National Office processing of State Director's request.* (1) After reviewing the recommendation to either debt restructure or liquidate for the net recovery value, the Administrator, after concurring, modifying, or not concurring in the recommendation, will return the submission for further processing.

(2) If a debt writedown is used in the restructuring process, the amount will be included in the National Office transmittal memorandum. The draft Form RD 1956–1 will not need to be finalized and returned to the Administrator for signature. The State Director's signature on the final copy will be sufficient. However, a copy of the National Office memorandum is to be attached to the form when completed.

(e) *Debtor notification of debt restructuring and net recovery value calculations.* The State Director will provide a copy of the basis for the debt restructuring or net recovery determination to the debtor.

(1) If the value of the restructured loan is equal to, or greater than, the recovery value, the debtor will be made an offer to accept the restructured debt by using language similar to that provided in Guide 2 of this subpart (available in any Rural Development Office) and including the following paragraphs:

(i) An introductory paragraph indicating that Rural Development has concluded its consideration of the debtor's request;

(ii) A paragraph indicating Rural Development's approval of the debt restructuring request and that acceptance must be received by Rural Development within 45 days from receipt of this letter; and

(iii) That the debtor's acceptance will require the execution of a Shared Appreciation Agreement similar to Guide 4 of this subpart (available in any Rural Development Office) and possible new debt instruments accompanied by Bond Counsel opinions.

(2) If the debt analysis calculations indicate that a restructured debt would be less than the net recovery value of the security, a letter using language similar to that provided in Guide 3 of this subpart (available in any Rural Development Office), will be sent to the debtor that includes the following paragraphs:

(i) An introductory paragraph indicating that Rural Development has concluded its consideration of the debtor's request;

(ii) Paragraphs indicating that:

(A) The debtor may pay Rural Development the net recovery value of the loan. The debtor will be given 30 days from receipt of this letter to inform Rural Development of its intent, 90 days to finalize the payoff, and will be notified that an election to pay off Rural Development would require the execution of a Net Recovery Buy Out Recapture Agreement, similar to that provided in Guide 5 of this subpart (available in any Rural Development Office); or

(B) If the debt is not paid off at the net recovery value, Rural Development will proceed to liquidate the loan.

(f) *Debtor responses to debt restructuring and net recovery value calculations.* Responses from the debtor will be handled as follows:

(1) *Acceptance of Rural Development's restructured debt offer.* When a debtor accepts the offer for debt restructuring, processing will be in accordance with § 1951.223 (c) of subpart E of part 1951 of this chapter using the adjusted unpaid principal and outstanding accrued interest at the Administrator's approved interest rate and terms. The debtor will be required to execute a Shared Appreciation Agreement which will provide that, should the debtor sell or transfer title to the facility within the next 10 years, Rural Development is entitled to a portion of any gain realized. This agreement will include language similar to that found in Guide 4 of this subpart (available in any Rural Development Office). The original of Form RD 1956-1, with appropriate attachments signed by the State Director, and a copy of the Shared Appreciation Agreement will be sent to the Finance Office. Note: All documents pertaining to this transaction will be sent to the Finance Office in one single complete package; and

(2) *Acceptance by debtor to pay off loan at the recovery value.* Processing of this transaction will be in accordance with § 1956.124 of this subpart. However, the account does not need to be accelerated. The debtor will be required to execute a Net Recovery Buy Out Recapture Agreement, similar to that found in Guide 5 of this subpart (available in any Rural Development Office). The original of Form RD 1956-1, with appropriate attachments signed by the State Director, and a copy of the recorded Net Recovery Buy Out Recapture Agreement will be sent to the Finance Office. The executed Net Recovery Buy Out Recapture Agreement will be recorded in the county in which the facility is located. The Finance Office will credit the accounts of debtors who entered into Net Recovery Buy Out Recapture Agreements with the amount paid by the debtor (net recovery value). Note: All documents pertaining to this

transaction will be sent to the Finance Office in one single complete package.

(g) *Collection and processing of recapture.* (1) When Rural Development becomes aware of the sale or transfer of title to the facility on which there is an effective Net Recovery Buy Out Recapture Agreement (Guide 5 of this subpart available in any Rural Development Office) or a Shared Appreciation Agreement (Guide 4 of this subpart available in any Rural Development Office) outstanding and a determination is made that a recapture is appropriate, Rural Development will notify the debtor of the following:

(i) Date and amount of recapture due; and

(ii) Rural Development action to be taken if debtor does not respond within the designated timeframe with the amount of recapture due.

(2) [Reserved]

(3) When the amount of the recapture has been paid and credited to the debtor's account, the debtor will be released from liability by using Form RD 1965–8, "Release from Personal Liability," modified as appropriate.

(h) *No recapture due.* If Rural Development determines there is no recapture due, the Net Recovery Buy Out Recapture Agreement (Guide 5 of this subpart available in any Rural Development Office) or Shared Appreciation Agreement (Guide 4 of this subpart available in any Rural Development Office) will be appropriately annotated, the Recapture Agreement released from the record, and the Agreement returned to the debtor.

[59 FR 46160, Sept. 7, 1994, as amended at 68 FR 61332, Oct. 28, 2003; 69 FR 69106, Nov. 26, 2004]

§1956.144 [Reserved]

§1956.145 Disposition of essential Rural Development records.

RD Instruction 2033–A (available in any Rural Development office) identifies an "essential Rural Development record" as the original of any document or record which provides evidence of indebtedness or obligation to RD and includes, but is not limited to: promissory notes, assumption agreements and valuable documents, such as bonds fully registered as to principal and interest.

(a) Essential Rural Development records evidencing debts settled by compromise, completed adjustment or cancelled with application will be returned to the debtor or to the debtors' legal representative. The appropriate legend, such as "Satisfied by Approved Compromise," and the date of the final action will be stamped or typed on the original document. This same information plus the date the original document is returned to the debtor will be shown on a copy to be placed in the debtor's case folder.

(b) Essential Rural Development records evidencing debts cancelled without application will be placed in the debtor's case folder and disposed of pursuant to RD Instruction 2033–A (available in any Rural Development office). However, if the debtor requests the document(s), they must be stamped "Satisfied by Approved Cancellation" and returned.

(c) Essential Rural Development records evidencing charged off debts will be retained in the servicing office and will not be stamped or returned to the debtor. They will be destroyed six years after chargeoff pursuant to RD Instruction 2033–A (available in any Rural Development office).

[80 FR 9902, Feb. 24, 2015]

§§1956.146–1956.147 [Reserved]

§1956.148 Exception authority.

The Administrator may make an exception to any requirement or provision of this subpart which is not inconsistent with the authorizing statute or other applicable law if the Administrator determines that application of the requirement or provision would adversely affect the Government's interest. Requests for exceptions must be made in writing by the State Director and supported with documentation to explain the adverse effect on the Government's interest, propose alternative courses of action, and show how the adverse effect will be eliminated or minimized if the exception is granted. Any settlement actions approved by the Administrator under this section will be documented on Form RD 1956-1 and returned to the State Office for submission to the Finance Office.

§ 1956.149 [Reserved]

§ 1956.150 OMB control number.

The reporting requirements contained in this regulation have been approved by the Office of Management and Budget and assigned OMB control number 0575-0124. Public reporting burden for this collection of information is estimated to vary from ½ hour to 30 hours per response with an average of 8.14 hours per response, including the time for reviewing instructions, searching existing data sources, gathering and maintaining the data needed, and completing and reviewing the collection of information. Send comments regarding this burden estimate or any other aspect of this collection of information, including suggestions for reducing this burden, to Department of Agriculture, Clearance Officer, OIRM, Ag Box 7630, Washington, D.C. 20250; and to the Office of Information and Regulatory Affairs, Office of Management and Budget, Washington, DC 20503.

[59 FR 46162, Sept. 7, 1994]

PART 1957—ASSET SALES

Subpart A—Rural Housing Asset Sales

AUTHORITY: Pub. L. 99–509, sec 2001(b)(1).

SOURCE: 54 FR 47958, Nov. 20, 1989, unless otherwise noted.

Subpart A—Rural Housing Asset Sales

§ 1957.1 General.

Pursuant to the Omnibus Budget Reconciliation Act of 1986, Public Law 99–509, the Rural Housing Service (RHS) sold certain of the portfolio of loans made under section 502 of the Housing Act of 1949 to the Rural Housing Trust, 1987–1. The sale was without recourse to RHS except for certain provisions providing for RHS's payment of interest credit amounts and agreement to compensate the Rural Housing Trust 1987–1 for future cash flow changes due to revised borrowers rights as set forth in RHS regulations. The sale documents to Rural Housing Trust 1987–1 recognize that the RHS loans were assigned subject to rights provided to these borrowers under documentation to recognize the rights of RHS borrowers under regulations of RHS as they may exist from time to time and to service the loans in accordance with then current RHS regulations. In addition, as provided in § 1957.6 of this subpart, RHS has retained review, but not hearing authority under the RHS Appeal Procedure, 7 CFR part 1900, Subpart B. Failure of private servicers to comply with RHS regulations in servicing loans sold to the Rural Housing Trust 1987–1 may be redressed in the review process under the Appeal Procedure.

§ 1957.2 Transfer with assumptions.

RHS regulations governing transfers and assumptions will not apply to these loans. Individuals who what to purchase property securing a loan held by the Rural Housing Trust 1987–1, and who are eligible for an RHS § 502 loan will be given the same priority by RHS as a transferee of a § 502 loan if the property is then suitable for the RHS RH program and is located in an eligible area. The Master Servicer of the Rural Housing Trust, 1987–1, may permit an assumption if it is deemed by the Master Servicer to be in the financial interest of the Trust, but in such case the transferee would not be eligible for RHS loan servicing benefits under RHS regulations.

§ 1957.3 [Reserved]

§ 1957.4 Graduation.

Borrowers will not be required to graduate to other credit.

§ 1957.5 [Reserved]

§ 1957.6 Appeal reviews.

The Master Servicer, acting through its subservicer, will have the responsibility to conduct hearings under the appeal process. Final review of an adverse decision upheld under the appeal process will remain with RHS and be conducted by the Agency's National Appeal Staff, Washington, DC, under the RHS Appeal Procedures, 7 CFR part 1900, subpart B. This review is final and will conclude the appellant's administrative appeal process.

§§ 1957.7–1957.50 [Reserved]

PART 1962—PERSONAL PROPERTY

Subpart A—Servicing and Liquidation of Chattel Security

AUTHORITY: 5 U.S.C. 301; 7 U.S.C. 1989; 42 U.S.C. 1480.

SOURCE: 50 FR 45783, Nov. 1, 1985, unless otherwise noted.

EDITORIAL NOTE: Nomenclature changes to part 1962 appear at 80 FR 9902, Feb. 24, 2015.

Subpart A—Servicing and Liquidation of Chattel Security

§ 1962.1 Purpose.

This subpart delegates authorities and gives procedures for servicing, care, and liquidation of Rural Development chattel security, Economic Opportunity (EO) loan property, and note only loans. Security servicing for Nonprogram (NP) loans on farm property will be according to subpart J of part 1951 of this chapter. This subpart is inapplicable to Farm Service Agency, Farm Loan Programs.

[50 FR 45783, Nov. 1, 1985, as amended at 58 FR 52654, Oct. 12, 1993; 72 FR 64123, Nov. 15, 2007]

§ 1962.2 Policy.

Chattel security, EO property and note only loans will be serviced to accomplish the loan objectives and protect Rural Development's financial interest. To accomplish these objectives, security will be serviced in accordance with the security instruments and related agreements, including any authorized modifications, provided the borrower has reasonable prospects of accomplishing the loan objectives, properly maintains and accounts for the security, and otherwise satisfactorily meets the loan obligations including repayment.

§ 1962.3 Authorities and responsibilities.

(a) *Redelegation of authority.* Authority will be redelegated to the maximum extent possible consistent with program requirements and available resources. The State Director, District Director and County Supervisor are authorized to redelegate, in writing, any authority delegated to them in this subpart to any employee determined by them to be qualified.

(b) *Responsibilities*—(1) *Rural Development personnel.* The State Director, District Director and County Supervisor are responsible for carrying out the policies and procedures in this subpart.

(2) *Borrower.* The borrower is responsible for repaying the loans, maintaining, protecting, and accounting to Rural Development for all chattel security, and complying with all other requirements specified in promissory notes, security instruments, and related documents.

(c) *Exception authority.* The Administrator may, in individual cases, make an exception to any requirement or provision of this subpart which is not inconsistent with the authorizing statute or other applicable law if the Administrator determines that application of the requirement or provision would adversely affect the Government's interest. The Administrator will exercise this auhority only at the request of the State Director and on the recommendation of the appropriate program Assistant Administrator. Requests for exceptions must be made in writing by the State Director and supported with documentation to explain the adverse effect on the Government's interest, propose alternative courses of action, and show how the adverse effect will be eliminated or minimized if the exception is granted.

(d) *Farms in more than one jurisdiction.* If the farm is situated in more than one State, County, or Parish, the loan will be serviced by the County Office serving the County in which the borrower's residence is located. If the borrower is a corporation, cooperative, partnership or joint operation is the borrower's residence is not on the farm, the loan will be serviced by the County Office serving the County in which the farm or a major portion of the farm is located.

[50 FR 45783, Nov. 1, 1985, as amended at 51 FR 13480, Apr. 21, 1986]

§ 1962.4 Definitions.

As used in this subpart, the following definitions apply:

Abandonment. Voluntary relinquishment by the borrower of control of security or EO property without providing for its care.

Acquired chattel property. Former security or EO property of which the government has become the owner (See § 1955.20 of *Subpart A of Part 1955* of this chapter).

Basic security. Consists of all equipment serving as security for Rural Development loans. It also consists of real estate and all foundation herds and flocks, including replacements, which serve as a basis for the farming operation outlined in the Farm and Home Plan or yearly budget which serve as security for Rural Development. With respect to livestock herds and flocks, animals that are sold as a result of the normal culling process are basic security unless the borrower has replacements that will keep numbers and production up to planned levels. However, if a borrower plans to make a significant reduction in his basic livestock herd or flocks, the animals or birds that are sold in making this reduction will be considered basic security.

Borrower. When a loan is made to an individual, the individual is the borrower. When a loan is made to an entity, the cooperative, corporation, partnership or joint operation is the borrower.

Chattel security. Chattel property which may consist of, but is not limited to, inventory; accounts; contract rights; general intangibles; crops; livestock; fish; farm, business, and recreational equipment; and supplies, and which is covered by financing statements and security agreements, chattel mortgages, and other security instruments.

Civil action. Court proceedings to protect Rural Development's financial interests such as obtaining possession of property from borrowers or third parties, judgments on indebtedness evidenced by notes or other contracts or judgments for the value of converted property, or judicial foreclosure. Bankruptcy and similar proceedings to impound and distribute the bankrupt's assets to creditors and probate and similar proceedings to settle and distribute estates of incompetents or of decendents under a will, or otherwise, and pay claims of creditors are not included.

Criminal action. Prosecution by the United States to exact punishment in the form of fines or imprisonment for alleged violations of criminal statutes. These include but are not limited to violations such as:

Unauthorized sale of security.

Purchase of security with intent to defraud and without payment of the purchase price to Rural Development;

Falsification of assets or liabilities in loan applications;

Application for a loan for an authorized purpose with intent to use and use of loan funds for an unauthorized purpose;

Decision after obtaining a loan to use and using the funds for an unauthorized purpose and then making false statements regarding their use;

By scheme, trick, or other device, covering up or concealing misuse of funds or authorized dispositions of security or EO property or other illegal action; or

Any other false statements or representations relating to Rural Development matters. To establish that a criminal act was committed by selling EO property, it is necessary to show that the borrower, at the time the loan agreement or the check on the supervised bank account was signed, intended to sell the property in violation of the loan agreement. The Federal criminal statute of limitations bars institution of criminal action 5 years after the date the act was committed. Unauthorized disposition of even minor items by the borrower will be considered criminal violations.

Default. Failure of the borrower to observe the agreements with Rural Development as contained in notes, security instruments, and similar or related instruments. Some examples of default or factors to consider in determining whether a borrower is in default are when a borrower:

Is delinquent, and the borrower's refusal or inability to pay on schedule, or as agreed upon, is due to lack of diligence, lack of sound farming or other operation, or other circumstances within the borrower's control.

Ceases to conduct farming or other operations for which the loan was made or to carry out approved changed operations.

Has disposed of security or EO property without Rural Development, has not cared properly for such property, has not accounted properly for such property or the proceeds from its sale, or taken some action which resulted in bad faith or other violations in connection with the loan.

Has progressed to the point to be able to obtain credit from other sources, and has agreed in the note or other instrument to do so but refuses to comply with that agreement.

EO property. Nonsecurity chattel property purchased, refinanced, or improved with EO loan funds.

EO property essential for minimum family living needs. Nonsecurity chattel or real property required to provide food, shelter, or other necessities for the family or to produce income without which the family would not have such necessities. This includes livestock, poultry, or other animals used as food or to produce food for the family or to produce income for minimum essential family living needs; modest amounts of real property needed for family shelter or to produce food or income for minimum essential family living needs, and items such as equipment, tools, and motor vehicles, which are of minimum value and are essential for family living needs or to produce income for that purpose. Any such item of a value in excess of the minimum need may be sold and a portion of the sale proceeds used to purchase a similar item of less value to meet such need. The remainder of the proceeds will be paid on the EO loan.

Farm income. Proceeds from the sale of chattel security which is normally sold annually during the regular course of business such as crops, feeder livestock and other farm products.

Farmer Program loans. These loans and Farm Ownership (FO), Operating (OL), Soil and Water (SW), Recreation (RL), Economic Emergency (EE), Emergency (EM), Economic Opportunity (EO) and Special Livestock (SL) loans and Rural Housing loans made for farm service buildings (RHF).

Foreclosure sale. Act of selling security either under the "Power of Sale" in the security instrument or through court proceedings.

Liquidation. The act of selling security or EO property to close the loan when no further assistance will be given; or instituting civil suit against a borrower to recover security or EO property or against third parties to recover security or its value or to recover amounts owed to Rural Development; or filing claims in bankruptcy or similar proceedings or in probate or administrative proceedings to close the loan.

Normal income security. All security not considered basic security, including crops, livestock, poultry products, Agricultural Stabilization and Conservation Service payments and Commodity Credit Corporation payments, and other property covered by Farmers Home Administration or its successor agency under Public Law 103–354 liens that is sold in conjunction with the operation of a farm or other business, but shall not include any equipment (including fixtures in States that have adopted the Uniform Commercial Code), or foundation herd or flock. that is the basis of the farming or other operation, and is the basic security for a Rural Development farmer program loan.

Office of the General Counsel (OGC). The Regional Attorneys, Attorneys-in-Charge, and National Office staff of the Office of the General Counsel of the United States Department of Agriculture.

Purchase money security interest. Special type of security interest which, if properly perfected, takes priority over an earlier-perfected security interest. A security interest is a purchase money security interest to the extent that it is taken by the seller of the collateral to secure all or part of its purchase price or by a lender who makes loans or is obligated to make loans or otherwise gives value to enable the debtor to acquire the particular collateral or obtain rights in it. Such value must be given not later than the time the debtor acquires the collateral or obtains rights in it.

Repossessed property. Security or EO property in Rural Development's custody, but still owned by the borrower.

Security. Also means "Chattel security" when appropriate.

[50 FR 45783, Nov. 1, 1985, as amended at 51 FR 13481, Apr. 21, 1986; 53 FR 35783, Sept. 14, 1988]

§ 1962.5 [Reserved]

§ 1962.6 Liens and assignments on chattel property.

(a) *Chattel property not covered by Agency lien.* (1) When additional chattel property not presently covered by an Agency lien is available and needed to protect the Government's interest, the County Supervisor will obtain one or more of the following:

(i) A lien on such property.

(ii) An assignment of the proceeds from the sale of agricultural products when such products are not covered by the lien instruments.

(iii) An assignment of other income, including FSA Farm Programs (formerly ASCS) payments.

(2) When a current loan is not being made to a borrower, a crop lien will be taken as additional security when the County Supervisor determines in individual cases that it is needed to protect the Government's interests. However, a

crop lien will not be taken as additional security for Farm Ownership (FO), Rural Housing (RH), Labor Housing (LH), and Soil and Water (SW) loans. When a new security agreement or chattel mortgage is taken, all existing security items will be described on it.

(b) [Reserved]

(c) *Assignments of upland cotton, rice, wheat and feed grain payments.* Borrowers may assign FSA Farm Programs (formerly ASCS) payments under upland cotton, rice, wheat and feed grain programs.

(1) *Obtaining assignments.* Assignments will be obtained as follows:

(i) Only when it appears necessary to collect operating-type loans.

(ii) Only for the crop year for which operating-type loans are made, and

(iii) For only the amount anticipated for payments as indicated on Form RD 1962–1, "Agreement for the Use of Proceeds/Release of Chattel Security," of the applicable upland cotton, rice, wheat and feed grain programs.

(2) *Selecting counties.* The County Supervisor then will:

(i) Determine, at the time of loan processing for indebted borrowers and new applicants, who must give assignments and obtain them no later than loan closing. Special efforts will be made to obtain the bulk of assignments before the sign-up period for enrolling in the annual Feed Grain and Wheat set aside programs.

(ii) Obtain assignments from selected borrowers on Form ASCS–36, "Assignments of Payment," which will be obtained from FSA Farm Programs.

(3) *Releasing assignments and handling checks.* (i) The County Supervisor will inform FSA Farm Programs that releasing its assignment whenever a borrower pays the amount due for the year on the operating-type loan debt or pays the debt in full.

(ii) Checks obtained as a result of an assignment will be made only to the Agency, and the proceeds used as indicated on Form RD 1962–1.

[61 FR 35929, July 9, 1996]

§ 1962.7 Securing unpaid balances on unsecured loans.

The County Supervisor will take a lien on a borrower's chattel property in accordance with § 1962.6 of this subpart if it is necessary to rely on such property for the collection of the borrower's unsecured indebtedness, or if it will assist in accomplishing loan objectives.

§ 1962.8 Liens on real estate for additional security.

The County Supervisor may take the best lien obtainable on any real estate owned by the borrower, including any real estate which already serves as security for another loan. Additional liens will be taken only when the borrower is delinquent, the existing security is not adequate to protect Rural Development interests, and the borrower has substantial equity in the real estate to be mortgaged, and taking such mortgage will not prevent making a Rural Development real estate loan, if needed, later.

(a)–(b) [Reserved]

[50 FR 45783, Nov. 1, 1985, as amended at 53 FR 35783, Sept. 14, 1988; 56 FR 15824, Apr. 18, 1991; 61 FR 35930, July 9, 1996]

§§ 1962.9–1962.12 [Reserved]

§ 1962.13 Notification to potential purchasers.

(a) In States without a Central Filing System (CFS), all Farm Credit Programs borrowers prior to loan closing or prior to any servicing actions which require taking a lien on farm products, such as crops or livestock, must provide the names and addresses of potential purchasers. A written notice will be sent by the Agency, certified mail, return receipt requested, to these potential purchasers to protect the Government's security interest.

(1) The name and address of the debtor.

(2) The name and address of any secured party.

(3) The Social Security number or tax ID number of the debtor.

(4) A description of the farm products given as security by the debtor, including the amount of such products where applicable, the crop year, the county in which the products are located, and a reasonable description of the farm products.

(5) Any payment obligation imposed on the potential purchaser by the secured party as a condition for waiver or release of lien. The original or a copy of the written notice also must be sent to the purchaser within 1 year before the sale of the farm products. The written notice will lapse on either the expiration period of the Financing Statement or the transmission of a letter signed by the County Supervisor and showing that the statement has lapsed or the borrower has performed all obligations to the Agency.

(b) Lists of borrowers whose chattels or crops are subject to an Agency lien may be made available, upon request, to business firms in a trade area, such as sale barns and warehouses, that buy chattels or crops or sell them for a commission. These lists will exclude those borrowers whose only crops for sale require FSA Farm Programs (formerly ASCS) marketing cards. The list is furnished only as a convenience and may be incomplete or inaccurate as of any particular date.

(1)–(2) [Reserved]

[61 FR 35930, July 9, 1996, as amended at 62 FR 10157, Mar. 5, 1997]

§ 1962.14 Account and security information in UCC cases.

Within 2 weeks after receipt of a written request from the borrower, the Agency must inform the borrower of the security and the total unpaid balance of the Agency indebtedness covered by the Financing Statement.

(a) If the Agency fails to provide the information, it may be liable for any loss caused the borrower and, in some States, other parties, and also may lose some of its security rights. The UCC provides that the borrower is entitled to such information once every 6 months without charge, and the Agency may charge up to $10 for each additional statement. However, the Agency provides them without charge.

(b) Although the UCC only requires the Agency to give information pursuant to the borrower's written request, the Agency will also answer oral requests. Furthermore, the UCC does not prohibit giving this information to others who have a proper need for it, such as a bank or another creditor contem-

plating advancing additional credit to the borrower.

[50 FR 45783, Nov. 1, 1985, as amended at 54 FR 47960, Nov. 20, 1989; 61 FR 35930, July 9, 1996]

§ 1962.15 [Reserved]

§ 1962.16 Accounting by County Supervisor.

The Agency will maintain a current record of each borrower's security. Whenever an inspection is performed, the borrower must advise the Agency of any changes in the security and will complete and sign Form RD 1962–1 in accordance with § 1924.56 if it has not been previously completed for the year.

(a) *Agency responsibilities.* Chattel security will be inspected annually except in cases where the Agency official has justified in assessment or analysis review that no undue risk exists. An FO borrower who has been current with the Agency and who has provided chattels as additional security is an example of a case where an inspection may not be needed. All inspections will be recorded in the running record of the borrower's file. More frequent inspections should be made for delinquent borrowers or borrowers that have been indebted for less then 1 full crop year. The Agency official will discuss the provisions of §§ 1962.17 and 1962.18 and assist the borrower in completing the form. If a borrower does not plan to dispose of any chattel security, the form should be completed to show this and should be signed. When the Agency official has other contacts with the borrower, the official should also check for dispositions and acquisitions of security. Changes will be recorded on the form, dated and initialed by the borrower and the agency official. The purpose of all inspections is to:

(1) Verify that the borrower possesses all the security,

(2) Determine security is properly maintained, and

(3) Supplement security instruments.

(b) *Dispositions.* The County Supervisor will record all dispositions of chattel security on Form RD or its successor agency under Form RD 1962–1, and on the file copy of the security agreement or chattel mortgage. The original security instrument must not be altered. Additional acquired chattel security should be entered on the file copy of the security agreement or chattel mortgage and must be described on subsequent security instruments.

(c) *Unapproved dispositions.* Unapproved dispositions of security will be handled in accordance with §§1962.18 and 1962.49 of this subpart.

[50 FR 45783, Nov. 1, 1985, as amended at 58 FR 46075, Sept. 1, 1993; 61 FR 35930, July 9, 1996]

§1962.17 Disposal of chattel security, use of proceeds and release of lien.

(a) *General.* (1) The borrower must account for all security. When the borrower sells security, the property and proceeds remain subject to the Agency's lien until the lien is released. All checks, drafts, or money orders which the borrower receives for the sale of collateral listed on Form RD 1962–1 (available in any Agency office) must be payable to both the borrower and the Agency unless all Agency loan installments for the period of the form have been paid including any past-due installments. If the borrower disposes of collateral or uses the proceeds in a way not listed on Form RD 1962–1, the borrower will have violated the loan agreement, and the Government will not release its security interest in the collateral. Releases of sales proceeds will be terminated when the borrower's accounts are accelerated.

(2) Section 1924.56 requires that there must always be a current Form RD 1962–1 in the file of a borrower with a loan secured by chattels. If a borrower asks the Agency to release proceeds from the sale of chattels and there is a current Form RD 1962–1 in the file, the request will be approved or disapproved in accordance with paragraph (b) of this section. If the borrower's request for release is denied, the borrower must be given attachment 1 of exhibit A of subpart S of part 1951 of this chapter, a written explanation of the reasons for the denial, and the opportunity for an appeal in accordance with 7 CFR part 780. Immediately upon determining that the borrower does not have a current Form RD 1962–1 in the file, the County Supervisor will immediately contact the borrower to develop one.

(3) If the borrower requests a change(s) to Form RD 1962–1, and the County Supervisor can approve the change(s), the borrower and the County Supervisor will initial and date each change in accordance with item (6) in the Forms Manual Insert (FMI) for Form RD 1962–1. The form will be marked "Revised" and the borrower will be notified in writing confirming that the change(s) has been approved.

(b) *Use of Form RD 1962–1.* (1) County Supervisors are authorized to approve or disapprove dispositions of Agency chattel security in accordance with this subpart. The County Supervisor, with the assistance of the borrower, will complete Form RD 1962–1 in accordance with the FMI (available in any Agency office) to show how, when, and to whom the borrower will sell, exchange, or consume security and use sale proceeds (include milk sale proceeds). Government payments, crop insurance and insurance proceeds derived from the loss of security will also be accounted for on Form RD 1962–1. This includes, for example, sale proceeds on hand and crops in storage. Only the proceeds from the sale of normal income security can be used to pay essential family and farm operation expenses. Proceeds from the sale of basic security will not be used for essential family living and farm operating expenses. In addition to payment of prior liens, basic security can only be released for the purposes listed in paragraphs (b)(2)(iv) through (b)(2)(vii). When proceeds from the disposition of normal income security are to be used to pay essential family living or farm operating expenses, County Supervisors must approve the disposition. Any disposition of basic or normal income security must be recorded on Form RD 1962–1. However, the borrower is responsible for providing the County Supervisor with the necessary information to update the Farm and Home Plan and Form RD 1962–1.

(2) Under all circumstances, sales proceeds must be remitted to creditors with liens on the proceeds, in order of priority of those liens. Proceeds which are released by a prior lienholder or which are in excess of the amount due to prior lienholder and which come to the Agency can be used as follows:

(i) The Form RD 1962–1 must provide for releases of normal income security so that the borrower can pay essential family living and farm operating expenses. However, proceeds from the sale of basic security will not be used to pay essential family living or farm operating expenses.

(ii) Essential expenses are those which are basic, crucial or indispensable. The following items are guidelines of what normally may be considered essential family living and farm operating expenses:

Household operating
Food, including lunches
Clothing and personal care
Health and medical expenses, including medical insurance
House repair and sanitation
School, church, recreation
Personal insurance
Transportation
Furniture
Hired labor
Machinery repair
Farm building and fence repair
Interest on loans and credit or purchase agreement
Rent on equipment, land, and buildings
Feed for animals
Seed
Fertilizer
Pesticides, herbicides, and spray materials
Farm supplies not included above
Livestock expenses, including medical supplies, artificial insemination, and veterinarian bills
Machinery hire
Fuel and oil
Personal property tax
Real estate taxes
Water charges
Property and crop insurance
Auto and truck expenses
Utilities payments
Payments on contracts or loans secured by farmland, necessary farm equipment, livestock, or other chattels
Essential farm machinery. An item of essential farm machinery which is beyond repair may be replaced when the County Supervisor determines that replacement is a better .choice than alternatives such as the lease of a similar piece of machinery or the hiring of the service.

(iii) All of the items in paragraph (b)(2)(ii) of this section may not always be considered essential for every family and farming operation. County Supervisors must consider the individual borrower's operation, what is typical for that type of operation in the area administered by the County Supervisor, and what would be an efficient method of production considering the borrower's resources. County Supervisors will refer to exhibit E of this subpart for guidance in determining whether an expense will be considered essential and the amount of proceeds which should be released. When the borrower and County Supervisor cannot agree that an expense is essential, the County Supervisor will notify the borrower, in writing, of why the requested release was denied, including why it is not basic, crucial or indispensable to the family and/or the farming operation and will give the borrower an opportunity to appeal in accordance with subpart B of part 1900 of this chapter and paragraphs (a)(2) and (b)(5) of this section.

(iv) Proceeds can be applied to the Agency debt.

(v) Proceeds can be used to purchase property better suited to the borrower's need if the Agency will acquire a lien on the new property. The new property, together with any proceeds applied to the Agency indebtedness, will have a value to the Agency at least equal to the value of the lien formerly held by the Agency on the old security.

(vi) Proceeds can be used to preserve the security because of a natural disaster or other severe catastrophe, when the need for funds cannot be met by other means or with an Agency loan or an Agency loan cannot be made in time to prevent the borrower and Agency from suffering a substantial loss.

(vii) Property can be exchanged, with prior Agency approval and in accordance with paragraph (b)(5) of this section, for property which is better suited to the borrower's needs if the Agency will acquire a lien on the new property, at least equal in value to the lien held on the property exchanged.

(viii) Property can be consumed by the borrower as follows:

(A) Livestock can be used by the borrower's family for subsistence.

(B) If crops serve as security and usually would be marketed, the County Supervisor can allow such crops to be fed to livestock, provided, this is preferable to direct marketing and also provided that the Agency obtains a lien (or assignment) on the livestock and livestock products at least equal to the lien on the crops.

(3) The borrower must maintain records of dispositions of property and the actual use of proceeds and must make these records available to the Agency at the end of the period covered by the Form RD 1962–1, or when requested by the Agency. The County Supervisor will complete the "Actual" columns on that form, indicating approval or disapproval, making sure that the dispositions of property and uses of proceeds were as agreed upon. If they were not, the County Supervisor will take the actions required by §1962.18 of this subpart. On the form, the County Supervisor will note approval or disapproval of each disposition.

(4) If, for any sale, the amount of proceeds actually received is above or below the amount of proceeds planned to be received as shown on Form RD 1962–1, the borrower will immediately notify the County Supervisor. If the borrower sells security to a purchaser not listed on the Form RD 1962–1, the borrower must immediately notify the County Supervisor of what property has been sold and of the name and business address of the purchaser. Such notification may be by telephone to the County Office, by letter, by visit to the County Office, or any other method the borrower chooses.

(5) If a borrower wants to dispose of chattel security which is not listed on Form RD 1962–1 or wants to dispose of chattel security in a way not listed in the "How" section or wants to use proceeds in a way not listed in the "Use of Proceeds" section on Form RD 1962–1, the borrower must obtain the Agency consent before the disposition or before the proceeds are used. The Agency *must* give consent for the release of normal income security if the change is necessary for the borrower to meet essential family living and farm operating expenses. The Agency *must* also give consent if the conditions set out on the form and in paragraph (b)(2) of this section are met. The borrower may obtain prior consent by telephoning the county office, by letter, by visiting the county office, or by any other method the borrower chooses. When revisions

are agreed to over the telephone, the County Supervisor *must* revise the Form RD 1962–1 contained in the borrower's case file, initial and date the change, and mark the form "Revised." The County Supervisor will then either write to the borrower and send a copy of the "Revised" form to the borrower asking the borrower to date and initial the change and return the form to the county office, or the County Supervisor will ask the borrower to date and initial the change the next time the borrower is in the county office. Changes that would result in a major change (examples of major changes are: Feeder pig to sow operation, cow/calf to feeder steer operation, dairy to row crop, etc.) in a borrower's operation will always require a visit to the county office so that the County Supervisor and the borrower can complete a new farm and home plan and revise Form RD 1962–1. The County Supervisor will be responsible for determining if the requested change is major or not. If a revision cannot be agreed upon, see §1924.56 of subpart B of part 1924 of this chapter.

(c) *Release of liens.* (1) Liens will be released by the County Supervisor when security is sold, exchanged or consumed, provided the conditions set out on Form RD 1962–1 and in this subpart are met.

(2) Junior Agency liens on chattels and crops serving as security for Agency loans can be released when such property has no present or prospective security value or enforcement of the Agency lien would be ineffectual or uneconomical. The following information will be documented in the running case record:

(i) The present market value of the chattels or crops, as determined by the County Supervisor, on which the Agency has a valueless junior lien.

(ii) The names of the prior lienholders, amount secured by each prior lien, and the present market value of any property which serves as security for the amount. The value of all property which serves as security for amounts owed to prior lienholders must be considered to determine whether the junior Agency lien has any present or prospective value.

(3) Liens obtained through a mutual mistake can be released. The reasons

for the release must be documented in the running case record.

(4) Liens can be released when there is no evidence of an existing indebtedness secured by the lien in the records of the Agency, County, State, or Finance Office.

(5) Liens on separate items of chattels can be released to another creditor for any authorized Farm Credit Programs loan purpose when it has been determined by a current appraisal that the value of the remaining security is substantially greater than the remaining Agency debt.

(d) *Processing the release of chattel security.* (1) If the borrower or an interested third party requests a release of specific items which must be recorded under the UCC or chattel mortgage laws, Form RD462–12, "Statements of Continuation, Partial Release, Assignments, etc.," Form RD 460–1, "Partial Release," or other Forms approved by OGC and required by State statute will be used. Care must be used to be sure that only specific items are released; for example, if a borrower requests a release of five cows, make sure that not all the cattle are released from the Agency lien. When specific items are listed on the security agreement, the County Supervisor should record the disposition on the *work copy* of the security agreement and on Form RD 1962–1.

(2) Assignments and consent to payment of proceeds will be processed under subpart A of part 1941 of this chapter and recorded on Form RD 1962–1.

(i) When it is necessary to temporarily amend Form RD 441–18, "Consent to Payment of Proceeds From Sale of Farm Products," or Form RD 441–25, "Assignment of Proceeds From the Sale of Dairy Products and Release of Security Interest," Form RD 462–9, "Temporary Amendment of Consent to Payment of Proceeds From Sale of Farm Products," will be used. All amendments of assignment agreements will be made on forms approved by OGC. The State Director will issue a State Supplement with the advice of OGC and prior approval of the National Office on the use of other forms. The original form after completion will be forwarded directly to the person or firm making the payment against which the assignment is effective, and a copy will be kept in the borrower's case file. All amendments of assignment agreements will be approved and recorded on Form RD 1962–1. Conditions of this section must be met. The County Supervisor will see that payments are made in accordance with the original consent when the amendment period expires. Normally, a temporary amendment will not exceed a six month period.

(ii) When the Agency is not expecting payment from the proceeds of a product on which it has a lien but the purchaser of the product inquires about payment, a letter should be written to the purchaser as follows:

Rural Development has a security interest in the (name of product) being sold to you by (name and address of borrower), but at the present time is not looking to the proceeds from the sale of that product for payment on the debt owned to this agency. Therefore, until further notice, it will not be necessary for you to make payment to the Agency for such product.

(e) *Releases of liens on wool and mohair marketed by consignment*—(1) *Conditions.* Liens on wool and mohair may be released when the security is marketed by consignment, provided all the following conditions are met:

(i) The producer assigns to the Agency the proceeds of any advances made, or to be made, on the wool or mohair by the broker, less shipping, handling, processing, and marketing costs.

(ii) The producer assigns to the Agency the proceeds of the sale of the wool or mohair, less any remaining costs in shipping, handling, processing, and marketing, and less the amount of any advance (including any interest which may have accrued on the advance) made by the broker against the wool or mohair.

(iii) The producer and broker agree that the net proceeds of any advances on, or sale of, the wool or mohair will be paid by checks made payable jointly to the producer and the Agency.

(2) *Authority.* The County Supervisor may execute releases of the Government's lien on wool and mohair on Form RD 462–4, "Assignment, Acceptance, and Release." Since Form RD 462–4 is not a binding agreement until

executed by all parties in interest, including the producer, the broker and the Government, the County Supervisor may execute it before other parties sign it.

(f) *Notice of termination of security interest to purchasers of farm products* under consents or assignments upon payment in full. County Supervisors will notify purchasers of farm products as soon as the Agency has received payment in full of indebtedness for collection of which it has accepted assignments or consents to payment of proceeds from the sale of the farm products. When Form RD 441–18 is in effect under the UCC, the notice to the purchaser will be made on Form RD 460–8, "Notice of Termination of Security Interest in Farm Products." When assignments have been used, the notice to the purchaser will be by letter or by forms prescribed by State Supplements.

(g) *Release of Agency interest in insurance policies.* When an Agency lien on property covered by insurance has been released, the County Supervisor is authorized to notify the insurance company of the release.

[50 FR 45783, Nov. 1, 1985, as amended at 51 FR 13481, Apr. 21, 1986; 52 FR 32121, Aug. 26, 1987; 53 FR 35784, Sept. 14, 1988; 56 FR 15824, Apr. 18, 1991; 57 FR 18680, Apr. 30, 1992; 57 FR 60085, Dec. 18, 1992; 58 FR 46075, Sept. 1, 1993; 61 FR 35930, 35931, July 9, 1996]

§1962.18 Unapproved disposition of chattel security.

(a) *General.* When the County Supervisor learns that a borrower has made a disposition of chattel security in a manner not provided for on the applicable Agency form or becomes aware of the misuse of proceeds by a borrower, corrective action must be taken to protect the Government's interest.

(b) *Notice to borrowers.* When a borrower has not properly accounted for the use of proceeds from the sale of chattel security, the County Supervisor must request restitution by use of a letter similar to Guide Letter 1962–A–5.

(1) If the borrower makes restitution or provides sufficient information to enable the County Supervisor to post-approve the transaction on the applicable Agency form, no further action will

be taken against the borrower. Post-approval can only be given under the conditions set out in 1962.17(b) of this subpart. Only one such transgression can be allowed in any period covered by the RD 431–2, or other similar plan of operation acceptable to Rural Development, between annual security inspections, whichever is appropriate, and this must be made clear to the borrower.

(2) If the borrower does not make restitution, if the County Supervisor cannot post-approve the transaction, or if the borrower makes a second unauthorized disposition of security or a misuse of proceeds after settling the first offense as provided in paragraphs (a) and (b) of this section, the County Supervisor will proceed in accordance with §1962.49 of this subpart.

[54 FR 14791, Apr. 13, 1989]

§1962.19 Claims against Commodity Credit Corporation (CCC).

This section is based on a Memorandum of Understanding between CCC and Rural Development (see Exhibit A of this subpart). The memorandum sets forth the procedure to follow when producers sell or pledge to CCC as loan collateral under the Price Support Program, commodities on which Rural Development holds a prior lien, and when the proceeds, or an agreed amount from them, are not remitted to Rural Development to apply against the producer's indebtedness to Rural Development . In addition to the procedures outlined in Exhibit A, the following apply:

(a) *County Office action.* (1) Claims will not be filed with CCC until it is determined that the amount involved cannot be collected from the borrower. Therefore, after preliminary notice is given of this fact to CCC by the State Director, the County Supervisor will make immediate demand on the borrower for the amount of the CCC loan or the portion of it which should have been applied to the borrower's account. If payment is made, the State Director will be notified.

(i) If payment is not made, the County Supervisor will determine whether or not the case should be liquidated in accordance with §1962.40 of this subpart. Any liquidation action will be taken immediately. If the borrower has no property from which recovery can be made through liquidation or, if after liquidation, an unpaid balance remains on the indebtedness secured by the commodity pledged or sold to CCC, the County Supervisor will make a full report to the State Director on Form RD 455–1, "Request for Legal Action," with a recommendation that a claim be filed againt CCC. However, if the indebtedness is paid through liquidation action, the State Director will be notified by memorandum.

(ii) If the facts do not warrant liquidation action, the State Director will be notified, and a recommendation will be made that no claim be filed against CCC.

(2) On receiving information from the State Director that CCC has called the borrower's loan, the County Supervisor will act to protect Rural Development's interest with respect to the commodity if CCC is repaid.

(b) *State Office action.* (1) The State Director, on receipt of reports and recommendations from the County Supervisor, will:

(i) If in agreement with the County Supervisor's recommendation not to file a claim against CCC or if notice is received that the indebtedness has been paid, forward notice to CCC.

(ii) If in agreement with the County Supervisor's recommendation to file a claim against CCC, refer the case to OGC with a statement of facts.

(iii) If OGC determines that Rural Development holds a prior lien on the commodity and the amount due on its loan is not collectible from the borrower, send CCC a copy of the OGC memorandum with a complete statement of facts supporting the claim through the applicable ASCS office or notify CCC if the OGC memorandum does not support Rural Development 's claim.

(2) The State Director will notify the County Supervisor promptly on receiving information from CCC that the borrower's loan is being called.

(3) If collection cannot be made from the borrower or other party (see paragraph 5 of Exhibit A of this Subpart), the State Director will give CCC the reasons, Rural Development will then

be paid by CCC through the applicable ASCS office.

§§ 1962.20–1952.25 [Reserved]

§ 1962.26 Correcting errors in security instruments.

The County Supervisor may use Form RD 462–12, to correct minor errors in a financing statement when the errors are not serious (i.e., a slightly misspelled name). OGC will be asked to determine whether or not such errors are in fact minor. The County Supervisor may also use Form RD 462–12 to add chattel property to the financing statement (i.e., a new type or item of chattel or crops on land not previously described).

§ 1962.27 Termination or satisfaction of chattel security instruments.

(a) *Conditions.* The County Supervisor may terminate financing statements and satisfy chattel mortgages, chattel deeds of trust, assignments, severence agreements and other security instruments when:

(1) Payment in full of all debts secured by collateral covered by the security instruments has been received; or

(2) All security has been liquidated or released and the proceeds properly accounted for, including collection or settlement of all claims against third party converters of security, even though the secured debts are not paid in full. This includes collection-only and debt settlement cases; or

(3) The U.S. Attorney has accepted a compromise offer in full settlement of the indebtedness and has asked that action be taken to satisfy or terminate such instruments; or

(4) Rural Development has a financing statement or other lien instrument which describes the real estate upon which crops are located but neither the borrower non Rural Development has an interest in the crops because the borrower no longer occupies or farms the premises described in the lien instrument. Such action will only relate to the crops.

(b) *Form of payment.* (1) Security instruments may be satisfied or the financing statements may be terminated on receipt of final payment in currency, coin, U.S. Treasury check, cashier's or certified check, bank draft, postal or bank money order, or a check issued by a party known to be financially responsible.

(2) When the final payment is tendered in a form other than those mentioned above, the security instruments will not be satisfied until 15 days after the date of the final payment. However, in UCC States the termination statement will be signed and sent to the borrower within 10 days after receipt of the borrower's written request but not until the 10th day unless it previously has been ascertained that the payment check or other instrument has been paid by the bank on which it was drawn. (See subsection (c) of this section for the reason for the 10-day requirement.)

(c) *Filing or recording termination statements.* Financing statements will be terminated by use of Form RD or its successor agency under Public Law 103–354 462–12 if provided by a State supplement. (1) Under UCC provisions if Rural Development fails to give a termination statement to the borrower within 10 days after written demand, it will be liable to the borrower for $100 and, in addition, for any loss caused to the borrower by such failure unless otherwise provided by a State supplement. In the absence of demand for a termination statement by the borrower, a termination statement will be delivered to the borrower when the notes have been paid in full.

(2) However, if Rural Development has been meeting the borrower's annual operating credit needs in the past and expects to do so the next year, the financing statements need not be terminated in the absence of such demand unless a loan for the succeeding year will not be made or earlier termination is required by a State supplement.

(d) *Filing or recording satisfactions.* Satisfactions of chattel mortgages and similar instruments will be made on Form RD 460–4, "Satisfaction," or other form approved by the State Director. The original of the satisfaction form will be delivered to the borrower for recording or filing and the copy will be retained in the borrower's case file. However, if the State supplement based

on State law requires recording or filing by the mortgagee, a second copy will be prepared for the borrower and the original will be recorded or filed by the County Supervisor. When State statutes provide that satisfactions may be accomplished by marginal entry on the records of the recording office, or when Form RD 460–4 is not legally sufficient because special circumstances require some other form of satisfaction, County Supervisors are authorized to make such satisfactions according to State supplements. In such cases, Form RD 460–4 will not be prepared but a notation of the satisfaction will be made on the copy of Form RD 451–1, "Acknowledgment of Cash Payment," or Form RD 456–3, "Journal Voucher for Write-Off or Judgment," which will be retained in the borrower's case folder.

(e) *Satisfaction or termination of lien when old loans cannot be identified.* When a request is received for the satisfaction of a crop or chattel lien, or for the termination of a financing statement and the status of the account secured by the lien cannot be ascertained from County Office records, the County Supervisor will prepare a letter to the Finance Office reflecting all the pertinent information available in the County Office regarding the account. The letter will request the Finance Office to tell the County Supervisor whether the borrower is still indebted to Rural Development and, if so, the status of the account. If the Finance Office reports to the County Supervisor that the account has been paid in full or otherwise satisfied or that there is no record of an indebtedness in the name of the borrower, the County Supervisor is authorized to issue a satisfaction of the security instruments on Form RD 460–4 or other approved form or to effect the satisfaction by marginal release, or a termination on Form RD 462–12 as appropriate.

§ 1962.28 [Reserved]

§ 1962.29 Payment of fees and insurance premiums.

(a) *Fees.* (1) Security instruments. Borrowers must pay statutory fees for filing or recording financing statements or other security instruments (including Form RD 462–12, or other renewal statements) and any notary fees for executing these instruments. They also must pay costs of obtaining lien search reports needed in properly servicing security as outlined in this subpart. Whenever possible, borrowers should pay these fees directly to the officials giving the service. When cash is accepted by Rural Development employees to pay these fees, Form RD 440–12, "Acknowledgment of Payment for Recording, Lien Search and Releasing Fees," will be executed. If the borrower cannot pay the fees, or if there are fees referred to in paragraphs (a) (2) and (3) of this section that must be paid by Rural Development, the County Supervisor may pay them as a petty purchase or as the bill of a creditor of Rural Development in accordance with Rural Development Instructions 2024–E, copies of which are available in any Rural Development office.

(2) Satisfactions. The borrower must pay fees for filing or recording satisfactions or termination statements unless a State supplement based on State law requires Rural Development to pay them.

(3) Notary fees. Rural Development will pay fees for notary service for executing releases, subordinations, and related documents for and on behalf of Rural Development if the service cannot be obtained without cost.

(b) *Insurance premiums.* County Supervisors are authorized to voucher for the payment of bills for insurance premiums on chattel security, in accordance with Rural Development Instruction 2024–A (available in any Rural Development Office). Bills may be paid when:

(1) A borrower cannot pay the premiums from the borrower's own resources at the time due;

(2) Anticipated crop income does not materialize which would normally be released for the payment of crop insurance.

(3) It is not pratical to process a loan for that purpose;

(4) It is necessary to protect Rural Development's interests; and

(5) The amount advanced can be charged to the borrower under the provisions of the security instrument.

[50 FR 45783, Nov. 1, 1985, as amended at 53 FR 35785, Sept. 14, 1988; 56 FR 15825, Apr. 18, 1991; 57 FR 36592, Aug. 14, 1992]

§ 1962.30 Subordination and waiver of liens on chattel security.

(a) *Purposes.* Subject to the limitations set out in paragraph (b) of this section, the Agency chattel liens may be subordinated to a lien of another creditor in either of the following situations:

(1) The prior lien will soon mature or has matured and the prior lienholder desires to extend or renew the obligation, or the obligation can be refinanced. The relative lien position of the Agency must be maintained; and

(2) The subordination will permit another creditor to refinance other debt or lend for an authorized direct loan purpose.

(b) *Conditions.* Agency chattel liens may be subordinated to a lien of another creditor if all of the following conditions are met:

(1) If the lien is on basic chattel security, the amount of subordination is necessary to provide the lender with the security it requires to make the loan;

(2) Approval of a subordination is limited to a specific amount and the loan to be secured by the subordination is closed within a reasonable time;

(3) Only one subordination to one creditor may be outstanding at any one time in connection with the same security;

(4) The borrower has not been convicted of planting, cultivating, growing, producing, harvesting or storing a controlled substance under Federal or state law. "Borrower" for purposes of this provision, specifically includes an individual or entity borrower and any member stockholder, partner, or joint operator, of an entity borrower and any member, stockholder, partner, or joint operator of an entity borrower. "Controlled substance" is defined at 21 CFR part 1308. The borrower will be ineligible for a subordination for the crop year in which the conviction occurred

and the four succeeding crop years. Applicants must attest on the Agency application form that it and its members, if an entity, have not been convicted of such a crime;

(5) The borrower can document the ability to repay the total amount due under the subordination and pay all other debt payments scheduled for the subject operating cycle; and

(6) The Agency loan is still adequately secured after the subordination, or the value of the loan security will be increased by at least the amount of the advances to be made under the terms of the subordination.

(c) *Subordination to make a guaranteed loan.* In addition to the requirements of this section, subordinations on chattel security to make a guaranteed loan will be approved in accordance with § 1980.108 of subpart B of part 1980 of this chapter.

(d) *Forms.* Subordinations will be requested and executed on Agency forms available in any Agency office or on any other form approved by the Agency.

(e) *Rescheduling of existing Agency debts.* The Agency may consent to rescheduling of an existing Agency debt when a subordination is granted to the debt of another lender. The rescheduling will be allowed only when the borrower cannot reasonably be expected to meet all currently scheduled installments when due and the conditions of subpart S of part 1951 of this chapter are met.

(f) *Appraisal.* The Agency will prepare a chattel appraisal report when the existing appraisal report is more than 2 years old or is inadequate to make the determination in this section. The Agency may use an appraisal submitted by the borrower if it is substantially similar to Form RD 440–21, "Appraisal of Chattel Property," and prepared by a licensed appraiser.

[63 FR 20297, Apr. 24, 1998, as amended at 82 FR 19320, Apr. 27, 2017]

§§ 1962.31–1962.33 [Reserved]

§ 1962.34 Transfer of chattel security and EO property and assumption of debts.

Chattel and EO property may be transferred to eligible or ineligible transferees who agree to assume the outstanding loan, subject to the provisions set out in this section. A transfer and assumption may also be made when one or more of the borrowers or the former spouse and co-obligor of a divorced borrower withdraws from the operation or dies. The transfer of accounts secured by real estate or both real estate and chattels will be processed under Subpart A of Part 1965 of this chapter. The transferor (borrower) must be sent Attachment 1 of exhibit A of subpart S of part 1951 of this chapter as soon as the borrower contacts the County Supervisor inquiring about a transfer. In accordance with the Food Security Act of 1985 (Pub. L. 99–198) after December 23, 1985, if a loan is being transferred and assumed by an eligible or ineligible transferee, and if an individual or any member, stockholder, partner, or joint operator of an entity transferee is convicted under Federal or State law of planting, cultivating, growing, producing, harvesting or storing a controlled substance (see 21 CFR Part 1308, which is Exhibit C of Subpart A of Part 1941of this chapter and is available in any Rural Development office, for the definition of "controlled substance") prior to the approval of the transfer and assumption in any crop year, the individual or entity shall be ineligible for a transfer and assumption of a loan for the crop year in which the individual or member, stockholder, partner, or joint operator of the entity was convicted and the four succeeding crop years. Transferee applicants will attest on Form RD 410–1, "Application for RD Services," that as individuals or that its members, if an entity, have not been convicted of such crime after December 23, 1985. A decision to reject an application for transfer and assumption for this reason is not appealable.

(a) *Transfer to eligibles.* Transfers of chattel security and EO property to a transferee who is eligible for the kind of loan being assumed or who will become eligible after the transfer may be approved, provided:

(1) The transferee assumes the total outstanding balance of the Rural Development debts or that portion of the outstanding balance equal to the

present market value of the chattel security or EO property, less any prior liens, if the property is worth less than the entire debt.

(2) Generally the debts assumed will be paid in accordance with the rates and terms of the existing notes or assumption agreements. Form RD 460-9, "Assumption Agreement (Same Terms-Eligible Transferee)," will be used. Any delinquency and any deferred interest outstanding will be scheduled for payment on or before the date the transfer is closed. If the existing loan repayment period is extended, the debt being assumed may be rescheduled using Form RD 1965-13, "Assumption Agreement (Farmer Programs Loans)." The new repayment period may not exceed that for a new loan of the same type and the current interest rate for such loans will be charged. If any deferred interest is not paid by the time the transfer takes place, it must be added to the principal balance and the loan must be assumed at new rates and terms. Upon request of an applicant assuming a loan at new rates and terms and/or an applicant eligible to receive limited resource rates and terms, the interest rate charged by Rural Development will be the lower of the interest rates in effect at the time of loan approval or loan closing. If the applicant does not indicate a choice, the loan will be closed at the rate in effect at the time of loan approval. Interest rates are specified in Exhibit B of RD Instruction 440.1 (available in any RD office) for the type assistance involved.

(3) The transfer of EM actual loss loans, or EM loans made before September 12, 1975, will be made as provided under paragraph (b) of this section. However, when one or more of the borrowers or jointly obligated partners or joint operators withdraw from the operation and those remaining desire to assume the total indebtedness and continue the operation, a transfer to the remaining borrowers, partners, or joint operators may be made as an eligible transferee.

(b) *Transfer to ineligibles.* Transfer of the chattel security and EO property to a transferee who is not eligible for the kind of loan being assumed may be approved, provided:

(1) It is in the Agency's financial interest to approve the transfer of security or EO property and assumption of the debts rather than to liquidate the security or EO property immediately.

(2) The transferee assumes the total outstanding balance of the Agency debt, or an amount equal to the present market value of the security or EO property as determined by the County Supervisor, less any prior liens, if the value is less than the entire debts.

(3) Agency debts assumed will be repaid in amortized installments not to exceed 5 years using Form FmHA 1965-13. The Farm Credit Programs NP interest rate for chattel property set forth in a National Office issuance, in effect at the time of loan approval, will be charged. Any deferred interest not paid by the time the transfer takes place must be added to the principal balance. The transferred property, including EO property, will be subject to any existing Agency lien. In the absence of an existing Agency lien, new lien instruments will be executed.

(4) The transferee can repay the Agency in accordance with the assumption agreement and can legally enter into the contract.

(5) The requirements found in Exhibit M to Subpart G of Part 1940 of this chapter are met.

(6) The transferee has never been liable for a previous Farm Loan Programs (FLP) loan or loan guarantee which was reduced or terminated in a manner that resulted in a loss to the Government.

(c) *Effect of signature.* In all cases the purpose and effect of signing an assumption agreement or other evidence of indebtedness is to engage separate and individual personal liability, regardless of any State law to the contrary.

(d) *Release of transferor from liability.* The borrower and any cosigner may be released from personal liability to Agency when all the chattel security or EO property is transferred to an eligible or ineligible applicant and the total outstanding debt or that portion of the debt equal to the present market value of the security is assumed. However, no such release will be granted to any borrower who was liable for any direct FLP loan which was reduced or terminated in a manner that resulted in a loss to the Government. The appropriate official is authorized to approve releases from liability in accordance with §1962.34(h) of this subpart. When there will be no release from liability, the transferor and co-signer of a Farm Credit Programs loan must be sent a letter similar to exhibit F of subpart A of part 1955 of this chapter (available in any Agency office).

(e) *Agency actions*—(1) *Transfer to eligible applicant.* The Agency will determine the transferee's eligibility for the type of loan to be assumed.

(2) *Release from liability.* If the total outstanding debt is not assumed, the Agency must make the following determinations before it releases the transferor from personal liability:

(i) The transferor and any cosigner do not have reasonable ability to pay all or a substantial part of the balance of the debt not assumed after considering their assets and income at the time of transfer,

(ii) The transferor and any cosigner have cooperated in good faith, used due diligence to maintain the security against loss, and have otherwise fulfilled the covenants incident to the loan to the best of their ability, and

(iii) The transferee will assume a portion of the indebtedness at least equal to the present market value of the security.

[50 FR 45783, Nov. 1, 1985]

EDITORIAL NOTE: For FEDERAL REGISTER citations affecting §1962.34, see the List of CFR Sections Affected, which appears in the Finding Aids section of the printed volume and at *www.govinfo.gov.*

§§ 1962.35–1962.39 [Reserved]

§ 1962.40 Liquidation.

(a) *Voluntary liquidation*—(1) *General.* When a borrower contacts the agency and asks about voluntarily liquidating security, the borrower will be sent attachments 1 and 2 of exhibit A of subpart S of part 1951 of this chapter or attachments 1, 3 and 4, and the preliminary application forms by certified mail, or the forms will be hand delivered at the County Office. The servicing notices which provide possible alternatives to liquidation provide a maximum of 60 days for the borrower to apply for servicing. Therefore, the agency will not discuss liquidation or methods of liquidation until 60 days after the borrower receives the notices except in serious situations which are documented in detail in the case file. During the 60-day time period the County Supervisor may answer questions regarding the servicing notices. After 60 days, the borrower will be told that liquidation can be accomplished by:

(i) Selling the security under §1962.41 of this subpart,

(ii) Transferring the security under §1962.34 of this subpart,

(iii) Conveying the security to the agency under Subpart A of Part 1955 of this chapter, or

(iv) Refinancing the debt with another lender.

The provisions of these regulations will be explained to the borrower.

(2) *Lien search.* The County Supervisor will obtain a current lien search report to determine the effect that liens of other parties will have on liquidation, the record lienholders to whom notices of sale will be given, and the distribution that will be made of the sales proceeds. Normally, lien searches should be obtained from the same source as is used when making a loan. If obtaining the searches from third party sources causes undue delay which interferes with orderly liquidation, searches may be made by the County Supervisor. If the lien search is made by third parties, the borrower will pay the cost from personal funds or if the borrower refuses, the agency will pay the cost and charge it to the borrower's account in accordance with the security instrument or EO Loan Agreement. The records to be searched and the period covered by the search will be in accordance with a State supplement.

(b) *Involuntary liquidation*—(1) *General.* When a borrower makes an unapproved disposition of security, the directions in §§1962.18 and 1962.49 of this subpart will be followed. In all other cases, when the County Supervisor, with the advice of the District Director, determines that continued servicing of the loan will not accomplish the objectives of the loan, or that further servicing cannot be justified under the policy stated in § 1962.2 of this subpart, liquidation of the account(s) will be accomplished as quickly as possible under this section and subpart A of part 1955 of this chapter. When liquidation is begun, it is the agency policy to liquidate all security and EO property, except EO property that the County Supervisor determines is essential for minimum family living needs. The present market value of security that may be retained by the borrower for minimum family living needs will not exceed $600. However, only so much of the security and EO property will be liquidated as necessary to pay the indebtedness.

(2) *Farm Loan Programs loan cases.* In Farm Loan Programs loan cases, borrowers who are 90 days past due on their payments must receive exhibit A with attachments 1 and 2 or attachments 1, 3, and 4 of exhibit A of subpart S of part 1951 of this chapter in cases involving nonmonetary default. The County Supervisor will send these forms to the borrower as soon as a decision is made to liquidate. The procedures set out in subpart S of part 1951 of this chapter shall be followed and any appeal must be concluded before any liquidation action (including termination of releases of sales proceeds) is taken. If the borrower fails to return attachment 2 of exhibit A of subpart S of part 1951 of this chapter and a preliminary application within 60 days, the County Supervisor will send attachments 9 and 10 or 9–A and 10–A, as appropriate, of exhibit A of subpart S of part 1951 of this chapter. If the borrower fails to return attachments 4, 6, 6–A, 10, or 10–A of exhibit A of subpart S of part 1951 of this chapter within 60 days, the borrower's account will be accelerated in accordance with § 1955.15(d)(2) of subpart A of part 1955 of this chapter and paragraphs (b)(2) (i) and (ii) of this section. The County Supervisor will then attempt to repossess the security in accordance with § 1962.42 of this subpart. If this is not possible, the case will be referred for civil action in accordance with § 1962.49 of this subpart. Unmatured installments will be accelerated as follows:

(i) The District Director will accelerate all unmatured installments by using exhibits D, E, or E–1 of subpart A of part 1955 of this chapter except in cases referred to OGC for civil action, if the notice has previously been given.

(ii) Exhibits D, E, or E–1 of subpart A of part 1955 of this chapter will be sent to the last known address of each obligor, with a copy to the Finance Office in those cases referred to OGC for civil action. County Office and Finance Office loan records will be adjusted to mature the entire indebtedness only.

(3) *Lien search.* The County Supervisor will follow the directions set out in paragraph (a)(2) of this section.

(c) *Multiple loans and loans secured by both real estate and chattels.* Follow the provisions of § 1965.26(c) of subpart A of part 1965 of this chapter for liquidating these loans.

(d) *Assignment of direct loans.* When liquidation of a direct loan is approved, the State Director will be asked by the official who approved the liquidation to immediately obtain an assignment of the loan to if the promissory note is not held in the County Office. Pending the assignment, preliminary steps to effect liquidation should be taken, but civil or other court action will not be started and claims will not be filed in bankruptcy or similar proceedings or in probate or administration proceedings with respect to the insured loan claim, unless essential to protect Government's interests and OGC recommends such action. However, other steps need not be held up pending assignment. If any problems are encountered in obtaining the assignment, OGC may be contacted for advice.

(e) *Protective advances.* (1) After attachments 1 and 2 or 1, 3, and 4 of exhibit A of subpart S of part 1951 of this chapter have been sent and if security is in danger of loss or deterioration, the State Director will protect Government's interest and approve protective advances in payment of:

(i) Delinquent taxes or assessments that constitute prior liens which would be paid ahead of the Agency under § 1962.44(a) of this subpart.

(ii) Premiums on insurance essential to protect Rural Development's interest, and

(iii) Other costs including transportation necessary to protect or preserve the security.

(2) However, such advances may not be made unless the amount advanced becomes a part of the debt secured by the Agency's lien, or is for expenses of administration of estates or for litigation. If a case is in the hands of the U.S. Attorney, such advances may not be made without the U.S. Attorney's concurrence. Moreover, such advances may not be made in any case to pay expenses incurred by a U.S. Marshal or other similar official such as a local sheriff. However, if the official seizes the property and delivers it to the Agency for sale by the Agency, costs incurred by the Agency after delivery to the Agency will be paid.

(3) The County Supervisor will submit a report on the need for such advances to the State Director, including:

(i) Borrower's County Office case file;

(ii) Current lien search report;

(iii) Statement of the type and value of the property and of the circumstances which may result in the loss or deterioration of such property; and

(iv) A recommendation as to whether or not the advance should be approved.

(4) [Reserved]

(f) When a borrower's security property is liquidated voluntarily or involuntarily and there is an unpaid balance on the account, the County Supervisor will meet with the borrower within 30 days to assist the borrower in developing a debt settlement offer in accordance with subpart B of part 1956 of this chapter.

[50 FR 45783, Nov. 1, 1985, as amended at 51 FR 4139, Feb. 3, 1986; 53 FR 35785, Sept. 14, 1988; 56 FR 15825, Apr. 18, 1991; 57 FR 36592, Aug. 14, 1992; 57 FR 60085, Dec. 18, 1992; 61 FR 35931, July 9, 1996; 62 FR 10157, Mar. 5, 1997; 69 FR 5267, Feb. 4, 2004]

§1962.41 Sale of chattel security or EO property by borrowers.

Borrowers who are liquidating voluntarily and who have not been sent exhibit A and attachments 1 and 2 or 1, 3 and 4 of subpart S of part 1951 of this chapter will be processed in accordance with paragraph (a)(1) of §1962.40 of this subpart before any sale occurs.

(a) *Public sale.* A borrower may voluntarily liquidate chattels by selling the property at auction in the borrower's own name. RD 455-3, "Agreement for Sale by Borrower (Chattels and/or Real Estate)", will be executed by the borrower, all lienholders, and the clerk of the sale or other person who will receive the sale proceeds before execution by the County Supervisor. When EO property is involved delete from the Agency lien wherever it appears on the forms. No Agency official is authorized to bid at such sales. The County Supervisor will arrange to promptly receive the proceeds of the sale due the Agency for application on the borrower's indebtedness.

(b) *Private sale.* The borrower may sell chattel security or EO property at a private sale if:

(1)(i) The borrower has ready purchasers and can sell *all* of the property for its present market value; or

(ii) The property is perishable; or

(iii) The property is of a type customarily sold on a recognized market; or

(iv) The property consists of items of small value or a limited number of items which do not justify public sale.

(2) Form RD 1962-1 may be used to approve liquidation of such security. The County Supervisor will document in the running case record the reasons that a public sale was not justified.

(3) Form RD 455-3 is completed before the sale.

(c) *Government takes possession.* The borrower may also turn over possession of the chattels to the agency by signing Form RD 455-4, "Agreement for Voluntary Liquidation of Chattel Security." This form authorizes the agency to sell the security at either public or private sale. If the agency hires a caretaker, services should be obtained by use of Form AD-838, "Purchase Order."

(d) *Record of Sale.* The sale will be recorded on Form RD 1962-1.

(e) *Unpaid debt.* If the sale of all security results in less than full payment of the debt, the borrower may request debt settlement of the remaining debt. The servicing official will consult with the County Committee before determining if the borrower's account can be debt settled in accordance with subpart B of part 1956 of this chapter.

[50 FR 45783, Nov. 1, 1985, as amended at 51 FR 13482, Apr. 21, 1986; 53 FR 35785, Sept. 14, 1988; 56 FR 15825, Apr. 18, 1991; 57 FR 60085, Dec. 18, 1992; 62 FR 10157, Mar. 5, 1997; 68 FR 7701, Feb. 18, 2003]

§ 1962.42 Repossession, care, and sale of chattel security or EO property by the County Supervisor.

(a) *Repossession.* Except as provided in paragraph (d) of this section, prior to any repossession of agency security a borrower and all cosigners on the note must receive exhibit A and attachments 1 and 2, or 1, 3 and 4 of subpart S of part 1951 of this chapter and the application forms. The appropriate procedures of subpart S of part 1951 of this chapter must be followed and any appeal must be concluded. The County Supervisor will take possession of security or EO property when the value of the property, based on appraisal, is substantially more than the estimated sale expenses and the amount of any prior lien, and if the prior lienholder does not intend to enforce the lien. See § 1955.20 of subpart A of part 1955 of this chapter.

(1) *Conditions.* The County Supervisor will take possession under any of the following conditions:

(i) When RD 455-4 has been executed. For EO property this form will be revised by placing a period after "interest" in the first sentence beginning "The Debtor" and deleting the remainder of that clause; deleting the words "collateral covered by the security instruments" in the second part of the sentence and inserting instead "property covered by the debtor's loan agreement which is referred to as the collateral."

(ii) When the borrower has abandoned the property.

(iii) When peaceable possession can be obtained, but the borrower has not executed RD 455-4.

(iv) When the property is delivered to the agency as a result of court action.

(v) When Form RD 455-5, "Agreement of Secured Parties to Sale of SecurityProperty," is executed by all prior lienholders. If prior lienholders will not agree to liquidate the property, their liens may be paid if their notes and liens are assigned to the agency on forms prepared or approved by OGC. When prior liens are paid, the payment will be made in accordance with RD Instruction 2024-A (available in any agency office) and charged to the borrower's account.

(vi) When arrangements cannot be made with the borrower or a member of the borrower's family to sell EO property in accordance with the loan agreement.

(2) *Recording.* A list, dated and signed by the servicing official, of all security or EO property repossessed except for those items on Form RD 455-4, will be maintained in the borrower's case file. Whenever the servicing official is transferred to another position or leaves the agency or there is a change in jurisdiction, the District Director will give the succeeding servicing official in writing, the names of such borrowers and a list of the property repossessed in the custody of the servicing official and caretakers, its location, and the names and addresses of the caretakers.

(b) *Care.* The County Supervisor will arrange for the custody and care of repossessed property as follows:

(1) *Livestock.* Care and feeding of livestock will be obtained by contract pursuant to subpart B of part 1955 of this chapter. The value of animal products (such as milk) may constitute all or part of the contractor's quotation, and if this is desired, such a statement should be included in the solicitation. Possession of the livestock will be turned over to the contractor only after the contract is awarded using Form AD-838, "Purchase Order." If a contractor's services are needed for a longer period than is authorized in paragraph (c)(4)(i) of this section, the State Director may authorize the County Supervisor to continue obtaining the necessary services for the time needed.

(2) *Machinery, equipment, tools, harvested crops, and other chattels.* Property will be stored and cared for pending sale. Storage and necessary services may be obtained by contract using Form AD-838. Use of property by the contractor is not authorized.

(3) *Crops.* Form AD-838 will be used for obtaining services for the custody,

care, and disposition of growing crops and for unharvested matured crops unless the crops are to be sold in place. Where a loanlord is involved, written consent of the landlord should be obtained. If landlord consent cannot be obtained, where applicable, the circumstances should be reported to the State Director for advice.

(c) *Sale.* Repossessed property may be sold by Rural Development at public or private sale for cash under Form RD 455–4, "Agreement for Voluntary Liquidation of Chattel Security," Form RD 1955-41, "Notice of Sale," the power of sale in security agreements under the UCC, or in crop and chattel mortgages and similar instruments if authorized by a State supplement. Also, repossessed property may be sold at private sale when the borrower executes Form RD 455–11, "Bill of Sale 'B' (Sale by Private Party).''

(1) *Tests and inspections of livestock.* If required by State law as a condition of sale, livestock will be tested or inspected before sale. A State supplement will be issued for those States.

(2) *Public sales.* Such sales will be made to the highest bidder. They may be held on the borrower's farm or other premises, at public sale barns, pavilions, or at other advantageous sales locations. No Rural Development employee will bid on or acquire property at public sales except on behalf of FmHA or its successor agency under Public Law 103–354 in accordance with §1955.20 of subpart A of part 1955 of this chapter. The County Supervisor will attend all public sales of repossessed property.

(3) *Private sales.* Rural Development will sell perishable property such as fresh fruits and vegetables for the best price obtainable. Rural Development will sell staple crops such as when, rye, oats, corn, cotton, and tobacco for a price in line with current market quotations for products of similar grade, type, or other recognized classification. Chattel property sold under Form RD 455–4, other than perishable property and staple crops, will not be sold for less than the minimum price in the agreement. Rural Development will sell other property, including that sold when the borrower executes Form RD 455–11, for its present market value.

(4) *Selling period.* Repossessed property will be sold as soon as possible. However, when notice is required by paragraph (c)(5) of this section, the sale will not be held until the notice period has expired.

(i) The sale will be made within 60 days, unless a shorter period is indicated by a State supplement because of State law. Crops will be sold when the maximum return can be realized but not later than 60 days after harvesting, or the normal marketing time for such crops. The State Director may extend the sale time within State law limits.

(ii) These requirements do not apply to irrigation or other equipment and fixtures which, together with real estate, serve as security for Rural Develoment real state loans and will be sold or transferred with the real estate. However, a State Supplement will be issued for any State having a time limit within which such items must be sold along with or as a part of the real estate.

(5) *Notice.* (i) Notice of public or private sale of repossessed property when required will be given to the borrower and to any party who has filed a financing statement or who is known by the County Supervisor to have a security interest in the property, except as set forth below. The notice will be delivered or mailed so that it will reach the borrower and any lienholder at least 5 days (or longer time if specified by a State supplement) before the time of any public sale or the time after which any private sale will be held. Form RD 1955–41, "Notice of Sale," may be used for public or private sales.

(A) Notice of the borrower or lienholder is not required when the property is sold under Form RD 455–4 because the parties are placed on notice when they execute the form. When the sale involves only collateral which is perishable, will decline quickly in value, or is a type customarily sold on a recognized market, notice is not required but may be given if time permits to maintain good public relations.

(B) Notice only to lienholder is required when repossessed property is sold at private sale and the borrower executes Form RD 455–11.

(C) If the property is to be sold under a chattel mortgage, the manner of notice will be set forth in a State supplement or on an individual case basis.

(ii) *Notice of Internal Revenue Service (IRS).* If a Federal tax lien notice has been filed in the local records more than 30 days before the sale of the repossessed security, notice to the District Director of IRS must be given at least 25 days before the sale. It should be given by sending a copy of Form RD 1955–41 and a copy of the filed Notice of Federal Tax Lien (Form IRS 668). If the security is perishable, the full 25 days' notice must be given to the District Director by registered or certified mail or by personal service before the sale. Also, the sale proceeds must be held for 30 days after the sale so that they may be claimed by IRS on the basis of its tax lien priority. In such perishable property cases, the proceeds or an amount large enough to pay the IRS tax lien will be forwarded to the Finance Office with a notation "Hold in suspense 30 days because of Federal Tax Lien." OGC will advise the Finance Office about disposing of the funds.

(6) *Advertising.* (i) Private sales and sales at established public auctions will be advertised by Rural Development only if required by a State supplement based on State law.

(ii) Other public sales, whether under power of sale in the lien instrument or under Form RD 455–4, will be widely publicized to assure large attendance and a fair sale by one or more of the following methods customarily used in the area.

(A) The sale may be advertised by posting or distributing handbills, posting Form RD 1955–41, or a revision of it approved by OGC to meet State law requirements, or by a combination of these methods. The length of time and place of giving notice will be covered by a State supplement.

(B) Advertising in newspapers or spot advertisting on local radio or TV stations may be used depending on the amount of property to be sold and the cost in relation to the value of the property, the customs in the area, and State law requirements. When newspaper advertising is required, a State supplement will indicate the types of newspapers to be used, the number and times of insertions of the advertisement, and the form of notice of sale. All advertising must contain non-discrimination clauses.

(7) *Payment of costs and prior lienholders.* If expenses must be paid before the sale or if cash proceeds are not available from the sale of the property to pay costs referred to in § 1962.44(b) of this subpart or to pay lienholders, such costs or prior liens will be paid in accordance with RD Instruction 2024–A (available in any RD office). The amount of the voucher will be charged to the borrower's account, except as limited by State law in a State Supplement. No costs in the repossession and sale of security should be incurred unless they can be charged to the borrower's account, and in no event will the Government pay them. However, if costs are legally chargeable to the borrower, they may be paid as provided in this subpart, and charged to an account set up for the officials or other persons found responsible for them.

(8) *Bill of sale or transfer of title.* If a purchaser requests a written conveyance of repossessed property sold by Rural Development at public or private sale, the County Supervisor will execute and deliver to the purchaser Form RD 455–12, "Bill of Sale 'C' (Sale Through Government as Liquidating Agent)," or other necessary instruments to convey all the rights, title, and interests of the borrower and Rural Development. A State supplement will be issued as necessary for conveying title to motor vehicles and boats.

(d) *Risk of injury.* If a farmer program loan borrower has abandoned security *and* the security is in danger of being substantially harmed or damaged, the County Supervisor will attempt to repossess the security as explained in paragraph (a) of this section. Then the County Supervisor will send the borrower and all cosigners on the note attachments 1, 3 and 4 of exhibit A of subpart S of part 1951 of this chapter. The security will be cared for as explained in paragraph (b) of this section until all appeal rights have been given and any appeal has been concluded. When the appeal process is concluded, the security will be returned to the borrower or sold in accordance with paragraph (c) of this section, depending on the outcome of any appeal. The County Supervisor will document the abandonment and the danger of substantial damage in the borrower's case file. In the case of livestock, abandonment occurs if a borrower stops caring for the animals, as determined by the County Supervisor. However, an independent third party (not a Rural Developmnet employee) must determine that livestock is in danger of substantial damage. Protective advances may be made in accordance with §1962.40(e) of this subpart.

[50 FR 45783, Nov. 1, 1985, as amended at 51 FR 13482, Apr. 21, 1986; 53 FR 35786, Sept. 14, 1988; 56 FR 15825, Apr. 18, 1991; 57 FR 36592, Aug. 14, 1992; 62 FR 10158, Mar. 5, 1997]

§1962.43 [Reserved]

§1962.44 Distribution of liquidation sale proceeds.

This section applies to proceeds of nonjudicial liquidation sales conducted under the power of sale in lien instruments or under Form RD 455–4, Form RD 455–3, or Form RD 462–2.

(a) [Reserved]

(b) *Order of payment.* Sales proceeds will be distributed in the following order of priority.

(1) To pay expenses of sale including advertising, lien searches, tests and inspection of livestock, and transportation, custody, care, storage, harvesting, marketing, and other expenses chargeable to the borrower, including reimbursement of amounts already paid by the Agency and charged to the borrower's account. Bills can be paid, after liquidation has been approved, for essential repairs and parts for machinery and equipment to place it in reasonable condition for sale, provided written agreements from any holders of liens which are prior to those of the Agency state that such bills may be paid from the sales proceeds ahead of their liens.

(i) However, any such expenses incurred by the U.S. Marshal or other similar official such as a local sheriff may not be paid from sale proceeds turned over to the Agency.

(ii) On the other hand, if the U.S. Marshal or other similar official such as a local sheriff has taken possession of the property and delivered it to the Agency for sale, such costs incurred by the Agency after delivery of the property to it may be paid from the proceeds of the sale.

(2) To pay liens which are prior to the Agency liens provided that:

(i) State and local tax liens on security or EO property which are prior to the liens of the Agency will be paid only when demand is made by tax collecting officials before distributing the sale proceeds. The sale proceeds will not be used to pay real estate, income, or other taxes which are not a lien against the security, or to pay substantial amounts of personal property taxes on nonsecurity personal property.

(ii) If action is threatened or taken by the sheriff or other official to collect taxes not authorized in suparagraph (b)(2)(i) of this section to be paid out of the security or the sale proceeds, the sale will be postponed unless an arrangement can be made to deposit in escrow with a responsible, disinterested party an amount equal to the tax claim, pending determination of priority rights. When the sale is postponed, or an escrow arrangement is made, the matter will be reported promptly to the State Director for referral to OGC.

(iii) If the Agency subordinations have been approved, their intent will be recognized in the use of sale proceeds even though the creditor in whose favor the Agency lien was subordinated did not obtain a lien. If there are other third party liens on the property, however, the lien-holders must agree to the use of the sale proceeds to pay such creditor first.

(3) To pay rent for the current crop year from the sale proceeds of other than basic security or EO property. However, there must be no liens junior to the Agency other than the landlord's lien, if any, and the borrower must consent in writing to the payment.

(4) To pay debts owed the Agency which are secured by liens on the property sold.

(5) To pay liens junior to those of the Agency in accordance with their priorities on the property sold, including any landlord's liens for rent unless

such liens already have been paid. Junior liens will not be paid unless, on request, the lienholder gives proof of the existence and the amount of his or her lien.

(6) To pay on any EO unsecured debt.

(7) To pay rent for the current crop year if the borrower consents in writing to payment and if such rent has not already been paid as provided in paragraph (b) (2), (3), or (5) of this section.

(8) To pay on any other the Agency debts, either unsecured or secured by liens on property which is not being sold. However, in justifiable circumstances, the State Director may approve the use of a part or all of the remainder of such sale proceeds by the borrower for other purposes, provided the other the Agency debts are adequately secured, or the borrower arranges to pay the other debts from income or other sources and these payments can be depended upon.

(9) To pay the remainder to the borrower.

(c) [Reserved]

[50 FR 45783, Nov. 1, 1985, as amended at 61 FR 35931, July 9, 1996; 80 FR 9903, Feb. 24, 2015]

§ 1962.45   Reporting sales.

Form RD 1955-3, "Advice of Property Acquired," will be prepared and distributed according to the FMI when property is acquired by Rural Development.

[50 FR 45783, Nov. 1, 1985, as amended at 80 FR 9904, Feb. 24, 2015]

§ 1962.46   Deceased borrowers.

Immediately on learning of the death of any person liable to the Agency, the County Supervisor will prepare Form RD 455-17, "Report on Deceased Borrower," to determine whether any special servicing action is necessary unless the County Supervisor recommends settlement of the indebtedness under subpart B of part 1956 of this chapter. If a survivor will not continue with the loan, it may be necessary to make immediate arrangements with a survivor, executor, administrator, or other interested parties to complete the year's operations or to otherwise protect or preserve the security.

(a) *Reporting.* The borrower's case files including Form 455-17 will be forwarded promptly to the State Director for use in deciding the action to take if any of the following conditions exist (When it is necessary to send an incomplete Form RD 455-17, any additional information which may affect the State Director's decision will be sent as soon as available on a supplemental Form RD 455-17 or in a memorandum.):

(1) Probate or other administration proceedings have been started or are contemplated.

(2) The debts owed to the Agency are inadequately secured and the state has other assets from which collection could be made.

(3) The Agency's security has a value in excess of the indebtedness it secures and the deceased obligor owes other debts to the Agency which are unsecured or inadequately secured.

(4) The County Supervisor recommends continuation with a survivor who is not liable for the indebtedness or recommends transfer to, and assumption by, another party.

(5) The County Supervisor recommends, but does not have authority to approve liquidation.

(6) The County Supervisor wants advice on servicing the case.

(b) *Probate or administration proceedings.* Generally, probate or administration proceedings are started by relatives or heirs of the deceased or by other creditors. Ordinarily, the Agency will not start these proceedings because of the problems of designating an administrator or other similar official, posting bond, and paying costs. If probate or administration proceedings are started by other parties or at the Agency's request, and any security is to be liquidated by the Agency instead of by the administrator or executor or other similar official, it will be liquidated in accordance with the advice of OGC. The State Director may request OGC to recommend that the U.S. Attorney bring probate or administration proceedings when it appears that:

(1) Such proceedings will not be started by other parties;

(2) The Agency's interests could best be protected by filing a proof of claim in such proceedings, and

(3) Public administrators or other similar officials or private parties, including banks and trust companies, are

eligible to, and will serve as administrator or other similar official and will provide the required bond.

(c) *Filing proof of claim.* When a proof of claim is to be filed, it will be prepared on a form approved by OGC, executed by the State Director, and transmitted to OGC. It will be filed by OGC or by the Agency official as directed by OGC or it will be referred by OGC to the U.S. Attorney for filing if representation of the Agency by counsel may be required. If a judgment claim is involved, the notification to the U.S. Attorney will be the same as for judgment claims in bankruptcy. If a direct loan is involved, the proof of claim will not be prepared until the note has been assigned to the Government. A proof of claim will be filed when probate or administration proceedings are started, unless:

(1) After considering liens and priority rights of the Agency and other parties, costs of administration, and charges against the estate, the Agency cannot reach the assets in the estate except for the Agency's own security and the Agency will liquidate the security by foreclosure or otherwise if necessary to collect its claim, or

(2) Continuation with an individual or transfer to and assumption by another party is approved, and either the debt owed to the Agency is fully secured, or the amount of the debt in excess of the value of the security which could be collected by filing a claim is obtained in cash or additional security, or

(3) The debt owed to the Agency by the estate is settled under subpart B of part 1956 of this chapter, well ahead of the deadline for filing proof of claim.

(d) *Priority of claims.* (1) Each secured claim will take its relative lien priority to the extent of the value of the property serving as security for it. These claims include those secured by mortgages, deeds of trust, landlord's contractual liens, and other contractual liens or security instruments executed by the borrower or real or personal property. However, tax, judgment, attachment, garnishment, laborer's, mechanic's, materialmen's, landlord's statutory liens, and other noncontractual lien claims may or may not be secured claims. Therefore, if any noncontractual claims are allowed as secured claims and the Agency claim is not paid in full, the advice of OGC will be obtained as to whether they constitute secured claims and as to their relative priorities.

(2) Unsecured claims will be handled as follows:

(i) The remaining assets of the estate, including any value of security for more than the amount of the secured claims against it, are to be applied first to payment of administration costs and charges against the estate and second to unsecured debts of the deceased.

(ii) If the total of the remaining assets in the estate being administered is not enough to pay all administration costs, charges against the estate, and unsecured debts of the deceased, the Government's unsecured claims against the remaining assets will have priority over all other unsecured claims, except the costs of administration and charges against the estate. Under such circumstances unsecured claims are payable in the following order of priority:

(A) Costs of administration and charges against the estate unless under State law they are payable after the Government's unsecured claims. Such costs and charges include costs of administration of the estate, allowable funeral expenses, allowances of minor children and surviving spouse, and dower and curtesy rights.

(B) The Government's unsecured claims.

(3) A State supplement will be issued as needed taking into consideration 31 U.S.C. §3713 lien waivers and subordinations, and notice and other statutory provisions which affect lien priorities.

(e) *Withdrawal of claim.* It may not be necessary to withdraw a claim when it is paid in full by someone other than the estate or when compromised. However, when it is necessary to permit closing of an estate, compromise of a claim, or for other justifiable reasons, the State Director will recommend to OGC that the claim be withdrawn on receipt of cash or security, or both, of a value at least equal to the amount that could be recovered under the claim against the estate. When the

Agency keeps existing security, arrangements must be made to assure that withdrawal of the claim will not affect the Agency's rights under the existing notes or security instruments with respect to the retained security. In some cases, with OGC's advice, the claim may be properly handled without filing a formal petition for withdrawal of the claim. However, if the claim has been referred to the U.S. Attorney, or if a formal withdrawal of the claim is necessary, the matter will be referred by OGC to the U.S. Attorney.

(f) *Liquidation of security.* When the County Supervisor determines that the account of a deceased borrower is in monetary or nonmonetary default, and liquidation is necessary because no survivor or third party has applied to assume the borrower's the Agency loan, chattel security and real estate security will be liquidated promptly in accordance with this subpart and subpart A of part 1965 of this chapter. Before liquidation, the notices required by subpart S of part 1951 of this chapter will be sent to the executor of the estate and/or other appropriate person(s) or entity(ies) as advised by OGC. If a suvivor(s) or heir(s) who will continue with the borrower's operation applies for servicing, the Agency will determine whether these individuals meet the requirements of paragraph (g) of this section. If a third party who will not continue with the borrower's operation applies for servicing, the requirements of § 1962.34 of this subpart, or § 1965.47 of subpart A of part 1965 of this chapter, as applicable, must be met. To qualify for servicing, the eligibility and feasibility requirements in § 1951.909 of subpart S of part 1951 of this chapter must also be met. However, the borrower's estate is not eligible for servicing. After the provisions of subpart S of part 1951 of this chapter have been complied with, and the opportunity to appeal has expired, the State Director will request OGC to effect collection if the proceeds from the sale of security are insufficient to pay in full the indebtedness owed to the Agency and other assets are available in the estate or in the hands of heirs.

(g) *Continuation of secured debt and transfer or security.* When a surviving member of a deceased borrower's family or other person is interested in continuing the loan and taking over the security for the benefit of all or a part of the deceased borrower's family who were directly dependent on the borrower for their support at the time of the borrower's death, continuation may be approved subject to the following:

(1) Any individual who is liable for the indebtedness of the deceased borrower may continue with the loan provided that individual can comply with the obligations of the notes or other evidence of debt and chattel or real estate security instruments and so long as liquidation is not necessary to protect the interest of the Agency. When an individual who is liable for the indebtedness is to continue with the account, Form 450-10, "Advice of Borrower's Change of Address or Name," will be sent to the Finance Office to change the account to that individual's name. A new case number will be assigned or, if the continuing individual already has a case number, that number will be used regardless of whether that individual assumed all or a portion of the amount of the debt owed by the estate of the deceased.

(2) When a surviving member of a deceased borrower's family, a relative or other individual who is not liable for the indebtedness desires to continue with the farming or other operations and the loan, the State Director may approve the transfer of chattel or real estate security or both to the individual and the assumption of the debt secured by such property without regard to whether the transferee is eligible for the type of loan being assumed, subject to the following conditions:

(i) The transferee will continue the farming or other operations for the benefit of all or a part of the deceased borrower's family who were directly dependent on the borrower for their support at the time of death.

(ii) The amount to be assumed and the repayment rates and terms will be the same as provided in § 1962.34(a) of this Subpart.

(iii) The State Director determines that the continuation will not adversely affect repayment of the loan.

(iv) The transferee has never been liable for a previous Farm Loan Programs direct farm loan or loan guarantee which was reduced or terminated in a manner that resulted in a loss to the Government.

(3) In determining whether to continue with individuals, whether they are already liable or assume the indebtedness, all pertinent factors will be considered including whether:

(i) Probate or administration proceedings have been or will be started and, with OGC's advice, whether the filing of a claim on the debt owed to the Agency in such proceedings is necessary to protect the Agency's interests.

(ii) Arrangements can be made with the heirs, creditors, executors, administrators, and other interested parties to transfer title to the security to the continuing individual and to avoid liquidating the assets so that the individual can continue with the loan on a feasible basis.

(4) If continuation is approved, all reasonable and practical steps, short of foreclosure or other litigation, will be taken to vest title to the security in the joint debtor or transferee.

(5) The deceased borrower's estate may be released from liability for the Agency indebtedness if title to the security is vested in the joint debtor or transferee, and:

(i) The full amount of the debt is assumed, or

(ii) If only a portion of the debt is assumed, the amount assumed equals the amount as determined by OGC which could be collected from the assets of the estate of the deceased borrower, including the value of any security or EO property.

(h) *Special servicing of deceased EO borrower cases.* If the EO loan is secured, all paragraphs in this section will be followed. If the EO loan is unsecured, paragraphs (a), (b), (c), (d), and (e) of this section will be followed along with the following requirements.

(1) An individual who is liable for the indebtedness of the deceased borrower and wishes to continue with the EO debt and the EO property, may do so in accordance with paragraph (g)(1) of this section.

(2) A surviving member of the deceased borrower's family, a joint operator with the deceased borrower, a relative, or other individual who is not liable for the EO debt who desires to continue with the farming or other operation may do so in accordance with paragraph (g)(2) of this section. This individual must execute a loan agreement in addition to the assumption agreement and secure the EO debt with a lien on the remaining EO property when title to the property is vested in the individual and the County Supervisor determines that security is necessary to protect the interests of the deceased borrower's family or the Agency.

(3) If no individual listed in paragraphs (h) (1) and (2) of this section wishes to continue, but a member of the borrower's family turns over to the Agency the EO property in which the estate has an interest and which is not essential for minimum family living needs, the County Supervisor will take possession of EO property and sell it in accordance with §1962.42 of this Subpart. If this cannot be done, or if real property is involved, the case will be referred to OGC. If the property is sold, notice will be delivered to any of the borrower's heirs who are in possession of the property and to any administrator or executor of the borrower's estate.

[50 FR 45783, Nov. 1, 1985, as amended at 51 FR 4140, Feb. 3, 1986; 51 FR 45439, Dec. 18, 1986; 56 FR 15826, Apr. 18, 1991; 61 FR 35931, July 9, 1996; 62 FR 10158, Mar. 5, 1997; 68 FR 7701, Feb. 18, 2003; 80 FR 9903, Feb. 24, 2015]

§1962.47 **Bankruptcy and insolvency.**

(a) *Borrower files bankruptcy.* When the Agency becomes aware that a Farm Loan Programs borrower has filed for protection under Title 11 of the United States Code (bankruptcy), the borrower and the borrower's attorney, if any, will be notified in writing of the borrower's remaining servicing options.

(1) If the borrower wishes to apply for servicing options remaining, the borrower, or the borrower's attorney on behalf of the borrower, must sign and return the appropriate response form, or similar written request for servicing, and any forms or information as requested by the Agency, within 60

days from the date the borrower or the borrower's attorney received the notification, or the time remaining from a previous notification that was suspended when the borrower filed bankruptcy, whichever is greater.

(2) The Agency will consider a request for servicing options to be an acknowledgment that the Agency will not be interfering with any rights or protections under the Bankruptcy Code and its automatic stay provisions.

(3) The Agency's processing of any request for servicing may include consideration of primary and preservation loan servicing options, notification of the Agency's decision on the request or application for servicing, mediation, and holding of any meetings or appeals requested by the borrower.

(4) If court approval is required for the borrower to exercise these servicing rights, it will be the borrower or the borrower's attorney's responsibility to obtain that approval.

(5) If a plan is confirmed before servicing and any appeal is completed under 7 CFR part 11, the Agency will complete the servicing or appeals process and may consent to a post-confirmation modification of the plan if it is consistent with the Bankruptcy Code and 7 CFR part 1951, subpart S, as appropriate.

(6) In chapter 7 cases, the Agency will not provide primary loan servicing to a borrower discharged in bankruptcy unless the borrower reaffirms the entire Agency debt. If the chapter 7 debtor obtains the permission of the court and reaffirms the debt, the loan servicing application will be processed in accordance with 7 CFR part 1951, subpart S. If the borrower reaffirms the Agency debt in order to be considered for restructuring but is later denied restructuring, the borrower may revoke the reaffirmation subject to the provisions of the Bankruptcy Code. No reaffirmation is necessary for any discharged chapter 7 borrower to be eligible for preservation loan servicing in accordance with 7 CFR part 1951, subpart S.

(b) *Borrower defaults on plan or bankruptcy is dismissed*—(1) *90 days past due on a reorganization plan while still under court jurisdiction.* (i) If allowed by the Bankruptcy Code or court, the borrower and the borrower's attorney, if any, will be notified of any remaining servicing options under 7 CFR part 1951, subpart S, that were not exhausted prior to filing bankruptcy or during the bankruptcy proceedings according to paragraph (a) of this section.

(ii) No notices will be sent if the account was previously accelerated, such action is inconsistent with the provisions of the confirmed bankruptcy plan or the Bankruptcy Code, or the case has been referred to the Department of Justice.

(iii) If a borrower operating under a confirmed bankruptcy plan desires to apply for loan servicing and qualifies for servicing under 7 CFR part 1951, subpart S, the borrower must also comply with Bankruptcy Code rules and requirements concerning modification of the plan.

(2) *Bankruptcy is dismissed without a confirmed plan.* If the borrower's bankruptcy is dismissed without a confirmed plan, and the borrower is in default on Farm Loan Programs loans, the borrower's account will be liquidated after all remaining servicing options under 7 CFR part 1951, subpart S are exhausted. The borrower will be notified of any servicing options remaining according to 7 CFR part 1951, subpart S. Notwithstanding the previous sentence, no notices will be sent if the account was previously accelerated, the Agency is advised that such an act is inconsistent with the confirmed bankruptcy plan or the Bankruptcy Code, or the account has been referred to the Department of Justice.

(3) *Bankruptcy is dismissed after a confirmed reorganization plan.* If a bankruptcy is dismissed after a reorganization plan was confirmed, the account will be serviced as follows:

(i) If the borrower has substantially complied with the plan, but later defaults for reasons beyond the borrower's control, (see 7 CFR 1951.909(c)), the borrower will be notified of loan servicing in accordance with 7 CFR 1951.907. No notices will be sent if the account was previously accelerated; such action is inconsistent with the provisions of the confirmed bankruptcy plan or the Bankruptcy Code; or the case has been referred to the Department of Justice.

(ii) If the borrower failed to make one full payment under the plan, or did not comply with the plan for reasons not beyond the borrower's control, the borrower will be serviced according to paragraph (b)(2) of this section.

(c) *Servicing of bankruptcy loans after the case is closed.* In chapter 11, 12, or 13 cases after the case is closed and the discharge order is issued by the court, if the borrower becomes delinquent after performing as agreed under the plan, the borrower will be sent a notice explaining the loan servicing options available under 7 CFR part 1951, subpart S. The borrower's attorney of record will be sent a courtesy copy if the bankruptcy has not been closed for at least 2 years. No notices will be sent if the account has been accelerated, such act is inconsistent with the provisions of a confirmed bankruptcy plan or other provisions of the Bankruptcy Code, or the account has been referred to the Department of Justice.

(d) *Liquidation.* The account will be liquidated after obtaining any necessary relief, if required, from the automatic stay. In chapter 7 cases after discharge, the account can be liquidated if the debt has not been reaffirmed and the property is no longer part of the estate. Liquidation can proceed prior to discharge if allowed by the court.

(1) If the borrower or borrower's attorney was not previously notified of any remaining servicing options available under 7 CFR part 1951, subpart S before or during the course of the bankruptcy proceedings, the borrower and the borrower's attorney will be sent the notices referenced in paragraph (c) of this section prior to liquidating any security property.

(2) If the borrower or the borrower's attorney had been previously notified of loan servicing options remaining, the account will be liquidated.

[63 FR 29341, May 29, 1998]

§1962.48 [Reserved]

§1962.49 Civil and criminal cases.

All cases in which court actions to effect collection or to enforce Rural Development rights are recommended, as well as actions relating to apparent violations of Federal criminal statutes, will be handled under this section.

(a) *Criminal action.* When facts or circumstances indicate that criminal violations may have been committed by an applicant, a borrower, or third party purchaser, the State Director will refer the case to the appropriate Regional Inspector General for Investigations, Office of Inspector General (OIG), USDA, in accordance with RD Instruction 2012–B (available in any Rural Development office) for criminal investigation. Any questions as to whether a matter should be referred will be resolved through consultation with OIG for Investigations and the State Director and confirmed in writing. In order to assure protection of the financial and other interest of the government, a duplicate of the notification will be sent to the Office of General Counsel (OGC). After OIG has accepted any matter for investigation, Rural Development staff must coordinate with OIG in advance regarding any administrative action on the matter/borrower other than routine servicing actions on existing loans. Cases requiring further action by OGC will be handled in accordance with paragraph (c) of this section.

(b) *Civil action.* Court action or other judicial process will be recommended to OGC when all other reasonable and proper efforts and methods to obtain payment, to remove other defaults, and to protect Rural Development 's property/financial interests have been exhausted. However, if an emergency situation exists or criminal action is to be recommended, the case will be submitted to OGC without taking the action necessary to report the information required by Part II of Form RD 455–22, "Information for Litigation." This is because delay in submitting cases in emergency situations may affect the financial interests of Rural Development and collection efforts may adversely affect the criminal investigation and/or criminal prosecution.

(1) Civil action will be recommended when one or more of the following conditions exists:

(i) There is a need to repossess security or EO property or to foreclose a lien and such action cannot be accomplished by other means authorized in this subpart.

(ii) There is a need for filing claims against third parties because of a conversion of security or other action.

(iii) Payment due on debts are not made in accordance with the borrower's ability to pay, and the borrower has assets or income from which collection can be made.

(iv) The borrower has progressed to the point that credit can be obtained from other sources, has agreed in the note or other instrument to do so, but refuses to comply with that agreement.

(v) Rural Development or its security becomes involved in court action through foreclosure by a third-party lienholder or through some other action.

(vi) Other conditions exist which indicate that court action may be necessary to protect Rural Development 's interests.

(2) Claims of less than $600 principal will not be referred to OGC for court action unless:

(i) A statement of facts is submitted as to the exact manner in which the interest of Rural Development, other than recovery of the amount involved, would be adversely affected if suit were not filed; and

(ii) Collection of a substantial part of the claim can be made from assets and income that are not exempt under State or Federal law. A State supplement will be issued to set forth such exemptions or a summary of those exemptions with respect to property to which Rural Development normally would look for payment such as real estate, livestock, equipment, and income.

(3) When a borrower has not properly accounted for the proceeds of the sale of security, it is the general policy to look first to the borrower for restitution rather than to third-party purchasers. In line with this policy the remaining chattel security on which Rural Development holds a first lien usually will be liquidated before demand is made, or civil action to recover from third-party purchasers.

(i) When the County Supervisor determines that full collection cannot be made from the borrower and that it will be necessary to collect the full value of the security purchased by a converter, a demand (see Guide Letter 1962-A-1, a copy of which is available in any Rural Development county office) will be sent to the purchaser at the same time that exhibit D or E of subpart A of part 1955 of this chapter, is sent to the borrower.

(ii) When the County Supervisor determines that it is likely that action will have to be taken to collect from third-party pruchasers, the County Supervisor will notify such purchasers by letter (see Guide Letter 1962-A-2, a copy of which is available in any Rural Development county office) that Rural Development security has been purchased by them and that they may be called upon to return the property or pay the value thereof in the event restitution is not made by the borrower. If it later becomes necessary to make demand on such third-party purchasers, Rural Development will do so unless the case already has been referred to OGC or the U.S. Attorney, in which event the demand will be made by one of those offices.

(iii) When restitution is made by the borrower, or a determination is made, with the advice of OGC, that the facts in the case do not support the claim against the third-party purchaser, the third-party purchaser will be informed by the County Supervisor that Rural Development will take no adverse action (see Guide Letter 1962-A-3, a copy of which is available in any Rural Development county office). Ordinarily, it will not be necessary to inform the third-party purchaser of OGC's decision when OGC determines that the facts support the claim against the third-party purchaser but no substantial part of the claim can be collected. If OGC makes such a determination and the third-party purchaser asks what determination has been made, the County Supervisor will say that no further action is to be taken on the claim "at this time."

(iv) In addition, unless personal contacts with the third-party purchaser, or other efforts to collect demonstrate that further demand would be futile, and a satisfactory compromise offer has not been received, a follow-up letter (see Guide Letter 1962-A-4, a copy of which is available in any Rural Development county office) will be sent by the State Director as soon as possible after the 15-day period set forth in the demand letter has expired. Unless response to the State Director's followup letter or personal contacts or other efforts indicate that further demand would be futile, an additional follow-up letter will be sent to the third-party purchaser by OGC after the case has been referred to that office.

(c) *Handling civil and criminal cases.* All cases in which court actions to effect collection or to enforce the rights of Rural Development are recommended, will be forwarded to OGC by the State Director in accordance with paragraph (c)(3) of this section.

(1) *County Office actions.* Forms RD 455-1, "Request for Legal Action," and RD 455-22 will be prepared. Form RD 455-2, "Evidence of Conversion," will be prepared for each unauthorized disposal. The original and two copies of Forms RD 455-1 and RD 455-22 and, wh = n applicable, Rural Development 455-2 together with the borrower's case file, will be submitted to the State Office. Signed statements should be obtained, if possible, from the borrower, any third party purchasers, or others to support the information contained on Form RD 455-1. Appropriate recommendations regarding civil actions will be made on Forms RD 455-1 and RD 455-22 against the borrower or others. When a case is referred to the State Office the County Supervisor will keep that office informed of any future developments in the case. If Attachments 1, 2 and other appropriate attachments to exhibit A of subpart S of part 1951 of this chapter have not been sent, they will now be sent to the borrower and any other obligor(s) on the note. Any appeal must be concluded before a civil action can be filed.

(2) *District Office actions.* Exhibits D, E, or E-1 of subpart A of part 1955 of this chapter will be prepared and sent after any appeal is concluded.

(3) *State Office actions.* (i) upon receipt of Form RD 455-1 and, when applicable, Form RD 455-2, the State Director will analyze each form to determine if all of the necessary information is documented and, if not, whether an appropriate effort was made to obtain the information. If all the necessary information is not documented, the State Director will return the case and request the County Supervisor to obtain the information to complete Forms RD 455-1 and 455-2. The State Director may assign any qualified Rural Development employee to help a County Supervisor obtain the information necessary to complete the reports. After diligent efforts, if Rural Development employees are unable to obtain the additional information, the case will be returned to the State Office with an explanation of why the information is unavailable.

(ii) After all of the pertinent information available has been obtained, the State Director will refer the case to OGC for civil action, if referral is required under the policy expressed in this section. If such referral is not required, the State Director will set forth in Item 19 of Form RD 455-1 the basis for the determination not to refer the case and instructions for follow-up servicing action. The State Director will not recommend a third-party conversion claim to the OGC if more than one year has run from the date of the annual accounting following the disposition of security, unless the Administrator or delegate determines a longer period of time should be applied either because of compelling circumstances such as the case has been referred to and accepted by OIG for criminal or civil investigation. The period of time during which a suit may be filed is set by federal statute and is not changed by this section. Demands on third-party purchasers will be made in accordance with paragraph (b) of this section. In cases referred to OGC, the State Director will make comments and recommendations regarding the civil aspects of the case on Form RD 455-1.

(A) When cases are referred to OGC, the County Office case file, Form RD 455-1, and, when appropriate, Form RD 455-2 will be transmitted. In addition, when the institution of civil court proceedings by Rural Development is recommended, the notes, financing statements, security agreements, loan agreements, other legal instruments and copies thereof, as required by OGC, and Form RD 451–11, "Statement of Account," and Form RD 455–22 will be submitted to OGC. The State Director, with the advice of OGC, will determine the number of copies of such instruments needed and the information required on the certified statement of account. Each request for a certified statement of account will specify the type of information needed.

(B) Notes, statements of account, files, or other documents and copies thereof needed in referring cases to OGC for civil court or other action will be obtained from the Finance Office, or County Office, by the State Director. When the time required for obtaining the above material or documents may jeopardize Rural Development's interest by permitting the diversion or dissipation of assets which otherwise could be expected as a source of payment, the Finance Office, upon the request of the State Director, will forward such material or documents directly to OGC or (at the State Director's direction) to the U.S. Attorney.

(d) *Actions on cases referred to OGC.* When a civil case is referred to OGC, the State Director will notify the County Supervisor of the referral and will return the County Office case file when it is no longer needed. The State Director will also prepare and distribute Form RD 1951–6 according to the FMI. The Rural Development field office will process the descriptive code via the Rural Development field office terminal system. This will flag the borrower's account indicating court action is pending (CAP). After notice of the referral is received by the County Supervisor, no collection or servicing action will be taken except upon specific instructions from the State Director or OGC. However, when a borrower voluntarily proposes to make a payment on an account, the County Supervisor will accept the collection unless notice has been received that the case has been referred to the U.S. Attorney for civil action. The County Supervisor will immediately notify OGC directly by memorandum, with a copy sent to the State Director, of any collections received. The County Supervisor also will notify the State Director and OGC of any developments which may affect a case which has been referred to OGC.

(e) *Actions on cases referred to the U.S. Attorney and on judgement cases (including third-party judgements).* OGC will notify the State Director, the Finance Office, and the County Supervisor when a case is referred to the U.S. Attorney or is otherwise closed. When a case is referred to the U.S. Attorney, the Finance Office will discontinue mailing Form RD 1951–9, Annual "Statement of Loan Account," to such borrowers. OGC will also notify the State Director when a judgement (including third-party) is obtained.

(1) When the County Supervisor receives notice from OGC that a judgment (including third-party) has been obtained, the County Supervisor will establish a judgment account by completing Form RD 1962–20, "Notice of Judgment," in accordance with the FMI. The Rural Development field office will process the judgment or the third party judgment via the Rural Development field office terminal.

(2) After notice has been received that a case has been referred to the U.S. Attorney or a judgment has been obtained and has not been returned to Rural Development by the U.S. Attorney, no action will be taken by the County Supervisor except upon specific instructions from the State Director, OGC, or the U.S. Attorney. However, the County Supervisor will keep the State Director informed of any developments which may affect the Rural Development security interest or any pending court action to enforce collection. If information is obtained indicating that such debtors have assets or income not previously reported by the County Supervisor to the State Director from which collection of such judgment accounts can be obtained, the facts will be reported to the State Director. The State Director immediately will notify OGC of any developments which might have a bearing on cases referred to the U.S. Attorney, including such judgment cases.

(i) If the debtor proposes to make a payment, Rural Development employees will not accept such payment but will offer to assist in preparing a letter for the debtor's signature to be used in transmitting the payment to the U.S. Attorney. In such case, the debtor will be advised to make payment by check or money order payable to the Treasurer of the United States.

(ii) Collection items received through the mail from the debtor or from other sources by the County Office to be applied to such accounts will be forwarded by the County Supervisor through OGC to the appropriate U.S. Attorney. Likewise, collections received by the District Director or the State Office will be forwarded through OGC to the appropriate U.S. Attorney. Such items will be forwarded in the form received except that cash will be converted into money orders made payable to the Treasurer of the United States. The money order receipts will remain attached to the money orders. Form FmHA or its successor agency under Public Law 103–354 451–1 will not be issued in any such case. The debtor will be informed in writing by the County Supervisor of the disposition of the amount received.

(3) When the U.S. Attorney has returned a judgment case to Rural Development, the County Supervisor is responsible for servicing it as follows:

(i) When the judgment debtor has the ability to make periodic payments, action will be taken by the County Supervisor to make arrangements for the judgment debtor to do so.

(ii) [Reserved]

(iii) At the time of the annual review of collection-only or delinquent and problem cases, the County Supervisor will determine whether such judgment debtors, whose judgments have not been charged off and who are not making regular and satisfactory payments, have assets or income from which the judgment can be collected. If such debtors have either assets or income from which collection can be made and they have declined to make satisfactory arrangements for payment, the facts will be reported by the County Supervisor to the State Director. The State Director will notify OGC of developments when it appears that collections can be enforced out of income or assets.

(iv) Such judgments will not be renewed or revived unless there is a reason to believe that substantial assets have or may become subject thereto.

(v) Such judgments may be released only by the U.S. Attorney when they are paid in full or compromised.

(4) In all judgment cases, any proposed compromise or adjustment will be handled in accordance with subpart B of part 1956 of this chapter.

(5) If the debtor requests information as to the amount of outstanding indebtedness, such information, including court costs, should be obtained from the Finance Office if the County Supervisor does not have that information. If questions arise as to the payment of court costs, information as to such costs will be obtained through the State Office from OGC.

[50 FR 45783, Nov. 1, 1985, as amended at 51 FR 45439, Dec. 18, 1986; 53 FR 35787, Sept. 14, 1988; 54 FR 42799, Oct. 18, 1989; 55 FR 35296, Aug. 29, 1990; 57 FR 60085, Dec. 18, 1992; 68 FR 61332, Oct. 28, 2003; 80 FR 9904, Feb. 24, 2015]

§ 1962.50 [Reserved]

EXHIBIT A TO SUBPART A OF PART 1962— MEMORANDUM OF UNDERSTANDING BETWEEN COMMODITY CREDIT CORPORATION AND FARMERS HOME ADMINISTRATION OR ITS SUCCESSOR AGENCY UNDER PUBLIC LAW 103–354

IT IS HEREBY AGREED by and between the Farmers Home Administration or its successor agency under Public Law 103–354 (hereinafter referred to as "FHA") and the Commodity Credit Corporation (hereinafter referred to as "CCC") that the following procedure will be observed in those cases where producers sell to CCC or pledge to CCC as loan collateral under the Price Support Program, agricultural commodities such as, but not limited to, cotton, tobacco, peanuts, rice, soybeans, grains, on which FHA holds a prior lien and the proceeds from such sales or loans are not remitted to FHA for application against the loan(s) secured by such lien:

1. When an FHA County Supervisor learns that an FHA borrower has obtained a loan from CCC on a commodity or sold a commodity to CCC under such circumstances, he shall immediately notify his State Director. The State Director, immediately upon receipt of the notice, shall furnish CCC (see Appendix 1) with the name and address of such borrower, the county of his location at the time the commodity was placed under loan or sold, and the amount of the FHA loan secured by the lien.

2. When CCC receives such a notice from FHA, CCC shall take steps to prevent the making of any further loans on or purchases of the commodity of the borrower. If the CCC loan is still outstanding and CCC calls the

loan, CCC shall notify the FHA State director of the demand.

3. If the CCC loan is repaid, whether prior to or after the receipt by CCC of the notice from FHA, the FHA State Director shall be notified immediately, at which time CCC will have discharged its responsibility under this agreement.

4. FHA shall, in each case in which the CCC loan is not repaid or the commodity has been sold to CCC, endeavor to collect from the borrower the amount due on the FHA loan. Such collection efforts shall include the making of demand on the borrower and the following of FHA's normal administrative policies with respect to the collection of debts, but shall not include the making of demand for payment upon the area peanut producer cooperative marketing associations through which CCC makes price support available to producers. If collection efforts are not successful, the FHA County Supervisor shall make a complete report on the matter to his State Director. If the State Director determines that the amount due on the FHA lien is not collectible by administrative action, he shall refer the matter to the appropriate local office of the General Counsel, with a full statement of the facts, for a determination of the validity of the FHA lien. If it is determined by the General Counsel's Office that FHA holds a valid prior lien on the commodity, the State Director shall furnish CCC with a copy of such determination, together with all other pertinent information, and shall request payment to FHA of the lesser of (1) the amount due on its loan, or (2) the value of the commodity at the time the CCC loan or purchase was made (based on the market value of the commodity on the local market nearest to the place where the commodity was stored). The information to be furnished CCC shall include (a) the principal balance plus interest due FHA on the date of the request, (b) the amount due on the FHA loan at the time the CCC loan or purchase was made, and (c) the amount of the CCC loan or purchase proceeds, if any, applied by the producer against the FHA loan. FHA shall continue to make collection efforts and shall notify CCC of any amount collected from the producer or any other party.

5. Upon receipt of evidence, including a copy of the determination of the Office of the General Counsel, from the State Director of FHA that the proceeds from the CCC loan or purchase have not been received by FHA from the borrower, and that collection cannot be made by FHA, CCC will if the CCC loan has not been repaid or if CCC has purchased the commodity, pay FHA the amount specified in paragraph 4 above or deliver the commodity (or warehouse receipts representing the commodity) to FHA: *Provided,* That if CCC has any information indicating that collection may be made by FHA from the borrower or any other party, it may notify FHA and delay payment pending additional collection efforts by FHA.

6. It is the desire of both FHA and CCC that claims to be processed under this agreement receive prompt attention by both parties and be disposed of as soon as possible. Instructions for the implementation of these procedures at the field office level will be developed and issued by the Washington offices of FHA and CCC.

7. Any question with regard to the handling of any claim hereunder shall be reported by the applicable ASCS office to ASCS in Washington and by the FHA State Director to the National Office of FHA.

This Memorandum of Understanding supersedes the agreement entered into between FmHA or its successor agency under Public Law 103–354 and CCC on November 5, 1951.

Entered into as of this 29th day of May, 1973.

FARMERS HOME ADMINISTRATION OR ITS SUCCESSOR AGENCY UNDER PUBLIC LAW 103–354,

FRANK B. ELLIOTT,
Acting Administrator.

COMMODITY CREDIT CORPORATION,

KENNETH E. FRICK,
Executive-Vice President.

APPENDIX 1—FURNISHING NOTICE OR INFORMATION TO COMMODITY CREDIT CORPORATION

Commodity	Direct to
Cotton	Prairie Village, Kansas, ASCS Commodity Office.
Tobacco	Applicable tobacco association.
Peanuts	Applicable peanut association.
All other commodities	Applicable State ASCS office.

[44 FR 4437, Jan. 22, 1979]

EXHIBIT B TO SUBPART A OF PART 1962— MEMORANDUM OF UNDERSTANDING AND BLANKET COMMODITY LIEN WAIVER

The Farmers Home Administration or its successor agency under Public Law 103–354 (FmHA or its successor agency under Public Law 103–354) sometimes makes loans to farmers on the security of agricultural commodities that are eligible for price support under loan and purchase programs conducted by the Commodity Credit Corporation (CCC). FmHA or its successor agency under Public Law 103–354 and CCC desire that price support be made available to farmers without unnecessarily impairing or undermining the respective security interests of FmHA or its successor agency under Public Law 103–354 and CCC in and without undue inconvenience to producers and FmHA or its successor agency under Public Law 103–354 and CCC in securing lien waivers on such commodities.

Now, therefore, it is agreed as follows:

(1) Upon request of an official of a State ASCS office, the FmHA or its successor agency under Public Law 103–354 State Director in such State shall furnish designated county ASCS offices with the names of producers in the trade area from whom FmHA or its successor agency under Public Law 103–354 holds currently effective liens on commodities with respect to which CCC conducts price support programs. FmHA or its successor agency under Public Law 103–354 will try to furnish a complete and current list of the names of such producers; however, FmHA or its successor agency under Public Law 103–354's liens with respect to any commodity will not be affected by an error in or omission from such lists.

(2) For a loan disbursed by a county ASCS office, CCC will issue a draft in the amount (less fees and charges due under CCC program regulations) of the loan on, or purchase price of, the commodity payable jointly to FmHA or its successor agency under Public Law 103–354 and the producer if (a) his name is on the list furnished by FmHA or its successor agency under Public Law 103–354, or (b) he names FmHA or its successor agency under Public Law 103–354 as lienholder. The draft will indicate the commodity covered by the loan or purchase.

(3) On issuance of the draft, the security interest of FmHA or its successor agency under Public Law 103–354 shall be subordinated to the rights of CCC in the commodity with respect to which the loan or purchase is made. The word "subordinated" means that, in the case of a loan, CCC's security interest in the commodity shall be superior and prior in right to that of FmHA or its successor agency under Public Law 103–354 and that, on purchase of a commodity by CCC or its acquisition by CCC in satisfaction of a loan, the security interest of FmHA or its successor agency under Public Law 103–354 in such commodity shall terminate.

(4) Nothing contained in this Memorandum of Understanding shall be construed to affect the rights and obligations of the parties except as specifically provided herein.

(5) This agreement may be terminated by either party on 30 days' written notice to the other party.

Dated: July 20, 1980.

RAY V. FITTZERALD,
Executive Vice President. CCC.

Dated: July 14, 1980.

GORDON CAVANAUGH,
Administrator, FmHA or its successor agency under Public Law 103–354.

[53 FR 35787, Sept. 14, 1988]

EXHIBIT C TO SUBPART A OF PART 1962—
MEMORANDUM OF UNDERSTANDING
BETWEEN FARMERS HOME ADMINISTRATION OR ITS SUCCESSOR AGENCY
UNDER PUBLIC LAW 103–354 AND COMMODITY CREDIT CORPORATION

Rotation of Grain Crops

Under the Commodity Credit Corporation (CCC) Farmer-Owned Grain Reserve Program, a producer may request to rotate or exchange new crop grain for the original crop grain that is in the Farmer-Owned Grain Reserve Program and already encumbered by CCC. The Farmers Home Administration or its successor agency under Public Law 103–354 (FmHA or its successor agency under Public Law 103–354) may have subordinated their first lien position to CCC on the original grain placed in reserve and/or may have a first lien on the new crop. FmHA or its successor agency under Public Law 103–354 and CCC desire to devise a mechanism whereby the CCC can relinquish its first lien position on the original grain reserve crop to FmHA or its successor agency under Public Law 103–354 and in turn the FmHA or its successor agency under Public Law 103–354 can relinquish its first lien position to CCC on the replacement grain reserve crop.

Now, therefore, it is agreed as follows:

(1) Upon receipt of a memorandum from an Agricultural Stabilization and Conservation Service (ASCS) County Executive Director or other designated county office official requesting the rotation of a grain reserve crop for a producer borrower(s), the FmHA or its successor agency under Public Law 103–354 County Supervisor and the ASCS county office official will jointly indicate approval or rejection of the request on the bottom of the original and a copy of the memorandum (Approval Memorandum) as follows:

"We hereby agree to and authorize the rotation of the subject producer's grain crops in accordance with the provisions of the Memorandum of Understanding between Farmers Home Administration or its successor agency under Public Law 103–354 and Commodity Credit Corporation dated_____."

FmHA or its successor agency under Public Law 103–354 _____

ASCS _____

In the memorandum, ASCS will include the name(s) of the producer(s) desiring to rotate the grain crops, the approximate number of bushels being rotated, the type of crop, years' crop being rotated and the location of the original grain reserve crop (approximate land and facility description).

(2) Upon execution of the Approval Memorandum by both ASCS and FmHA or its successor agency under Public Law 103–354, the security interest of FmHA or its successor agency under Public Law 103–354 in the new

crop grain shall be subordinated to the security interest of CCC in such grain and the security interest of CCC in the original crop grain shall be subordinated to the security interest of FmHA or its successor agency under Public Law 103–354 in such grain. At that point in time it will be the responsibility of each agency and the borrower to account for their respective interests in the grain crops and/or proceeds from the sale of the grain. The crop rotation and subordination of liens will only involve the amount of grain that has been specifically provided for in the memorandum from ASCS.

(3) If there is an intervening third party lien and it is impossible for FmHA or its successor agency under Public Law 103–354 or CCC to have a first lien on their respective grain crops, the request of the producer to rotate crops will not be granted.

(4) Nothing contained in this Memorandum of Understanding shall be construed to affect the rights and obligations of the parties except as specifically provided herein.

(5) This agreement may be terminated by either party on 30 days written notice to the other party.

[44 FR 4437, Jan. 22, 1979]

EXHIBITS D—D–1 TO SUBPART A OF PART 1962 [RESERVED]

EXHIBIT E TO SUBPART A OF PART 1962— RELEASING SECURITY SALES PROCEEDS AND DETERMINING "ESSENTIAL" FAMILY LIVING AND FARM OPERATING EXPENSES

Family Living Expenses

Expenses for household operating, food, clothing, medical care, house repair, transportation, insurance and household appliances, i.e., stove, refrigerator, etc., are essential family living expenses. We do not expect there will be any disagreements over this. However, when proceeds are less than expenses, there might be disagreements about the amounts FmHA or its successor agency under Public Law 103–354 should release to pay for particular items within these broad categories. For example, FmHA or its successor agency under Public Law 103–354 has to release for transportation expenses, but should FmHA or its successor agency under Public Law 103–354 release so that a borrower can buy a new car? If at planning time or during the crop year it appears that there will be sales proceeds available to pay for the borrower's operating and living expenses, including the expense of a new car, the Form FmHA or its successor agency under Public Law 103–354 1962–1 can be completed to show that FmHA or its successor agency under Public Law 103–354 plans to release for a new car. On the other hand,

it would also be proper to complete the Form FmHA or its successor agency under Public Law 103–354 1962–1 to release for a used car or for gas and repairs to the borrower's present car. Since it is necessary for FmHA or its successor agency under Public Law 103–354 to release for essential family living expenses and because transportation is an essential family living expense, some proceeds must be released for transportation. However, nothing requires FmHA or its successor agency under Public Law 103–354 to release for a specific expense; usually, there will be several ways to use proceeds to provide for essential family living expenses. We must provide the borrower with a written decision and an opportunity to appeal whenever there is a disagreement over the use of proceeds or whenever we reject a request for a release.

Farm Operating Expenses

We would expect farm operating expenses to present more of a problem than family living expenses. There will probably be a few disagreements over whether an expense is an operating expense (as opposed to a capital expense), but it is more likely that there will be disagreements over the amount FmHA or its successor agency under Public Law 103–354 should release for operating expenses and whether a particular farm operating expense is "essential." As is the case with family living expenses, disagreements will most likely arise when proceeds are less than expenses.

To resolve disputes over the amount to be released, remember that we must be reasonable and release enough to pay for essential farm operating expenses. Although a borrower might not always agree that enough money is being released, if the borrower's essential farm operating expenses are being paid, we are fulfilling the requirements of the statute. We must provide the borrower with an opportunity to appeal when there is a disagreement over the use of proceeds or when we reject a request for a release.

Section 1962.17 of this subpart states that essential expenses are those which are "basic, crucial or indispensable." Whether an expense is basic, crucial or indispensable depends on the circumstances. For example, feed is a farm operating expense, but it is not always an essential expense. If adequate pasture is available to meet the needs of the borrower's animals, feed is not essential. Feed is essential if animals are confined in lots. Hiring a custom harvester is a farm operating expense, but is not an essential expense if the farmer has the equipment and labor to harvest the crop just as well as a custom harvester. Hired labor is an operating expense which might be essential in a dairy operation but not in a beef cattle operation. Payments to creditors are essential if the creditor is unable to restructure the debt or to carry the debt delinquent. Renting land is not essential if the borrower plans to use it to grow corn which can be purchased for less than the cost of production. Paying outstanding bills is essential if a supplier is refusing to provide additional credit but not if the supplier is willing to carry a balance due. Of course, the long term goal of any farming operation is to pay all of its expenses, but when this is not possible, FmHA or its successor agency under Public Law 103–354 and the borrower must work together to decide which farm operating expenses are essential and demand immediate attention and cannot be neglected. These are the essential expenses.

We absolutely must release to pay for essential family living and farm operating expenses; there are no exceptions to this. When deciding whether an expense is essential and when deciding how much to release, the choices we make must be rational, reasonable, fair and not extreme. They must be based on sound judgment, supported by facts, and explained to the borrower. Following these rules will help us avoid disagreements with borrowers.

[56 FR 15829, Apr. 18, 1991]

EXHIBIT F TO SUBPART A OF PART 1962
[RESERVED]

PART 1965—REAL PROPERTY

AUTHORITY: 5 U.S.C. 301; 7 U.S.C. 1989, 42 U.S.C. 1480.

Subparts A–E [Reserved]

PART 1970—ENVIRONMENTAL POLICIES AND PROCEDURES

Subpart A—Environmental Policies

Sec.
1970.1   Purpose, applicability, and scope.
1970.2   [Reserved]
1970.3   Authority.
1970.4   Policies.
1970.5   Responsible parties.
1970.6   Definitions and acronyms.
1970.7   [Reserved]
1970.8   Actions requiring environmental review.
1970.9   Levels of environmental review.
1970.10   Raising the level of environmental review.
1970.11   Timing of the environmental review process.
1970.12   Limitations on actions during the NEPA process.
1970.13   Consideration of alternatives.
1970.14   Public involvement.
1970.15   Interagency cooperation.
1970.16   Mitigation.

1970.17   Programmatic analysis and tiering.
1970.18   Emergencies.
1970.19–1970.50   [Reserved]

Subpart B—NEPA Categorical Exclusions

1970.51   Applying CEs.
1970.52   Extraordinary circumstances.
1970.53   CEs involving no or minimal disturbance without an environmental report.
1970.54   CEs involving small-scale development with an environmental report.
1970.55   CEs for multi-tier actions.
1970.56–1970.100   [Reserved]

Subpart C—NEPA Environmental Assessments

1970.101   General.
1970.102   Preparation of EAs.
1970.103   Supplementing EAs.
1970.104   Finding of No Significant Impact.
1970.105–1970.150   [Reserved]

Subpart D—NEPA Environmental Impact Statements

1970.151   General.
1970.152   EIS funding and professional services.
1970.153   Notice of Intent and scoping.
1970.154   Preparation of the EIS.
1970.155   Supplementing EISs.
1970.156   Record of decision.
1970.157–1970.200   [Reserved]

AUTHORITY: 7 U.S.C. 6941 *et seq.*, 42 U.S.C. 4241 *et seq.*; 40 CFR parts 1500–1508; 5 U.S.C. 301; 7 U.S.C. 1989; and 42 U.S.C. 1480.

SOURCE: 81 FR 11032, Mar. 2, 2016, unless otherwise noted.

Subpart A—Environmental Policies

§ 1970.1 **Purpose, applicability, and scope.**

(a) *Purpose.* The purpose of this part is to ensure that the Agency complies with the National Environmental Policy Act of 1969, as amended (NEPA) (42 U.S.C. 4321, *et seq.*), and other applicable environmental requirements in order to make better decisions based on an understanding of the environmental consequences of proposed actions, and take actions that protect, restore, and enhance the quality of the human environment.

(b) *Applicability.* The environmental policies and procedures contained in this part are applicable to programs administered by the Rural Business-Cooperative Service (RBS), Rural Housing

Service (RHS), and Rural Utilities Service (RUS); herein referred to as "the Agency."

(c) *Scope.* This part integrates NEPA with other planning, environmental review processes, and consultation procedures required by other Federal laws, regulations, and Executive Orders applicable to Agency programs. This part also supplements the Council on Environmental Quality (CEQ) regulations implementing the procedural provisions of NEPA, 40 CFR parts 1500 through 1508. To the extent appropriate, the Agency will take into account CEQ guidance and memoranda. This part also incorporates and complies with the procedures of Section 106 (36 CFR part 800) of the National Historic Preservation Act (NHPA) and Section 7 (50 CFR part 402) of the Endangered Species Act (ESA).

§ 1970.2 [Reserved]

§ 1970.3 Authority.

This part derives its authority from a number of statutes, Executive Orders, and regulations, including but not limited to those listed in this section. Both the Agency and the applicant, as appropriate, must comply with these statutes, Executive Orders, and regulations, as well as any future statutes, Executive Orders, and regulations that affect the Agency's implementation of this part.

(a) National Environmental Policy Act of 1969 (42 U.S.C. 4321 *et seq.*);

(b) Council on Environmental Quality Regulations Implementing the Procedural Provisions of the National Environmental Policy Act (40 CFR parts 1500 through 1508);

(c) U. S. Department of Agriculture, NEPA Policies and Procedures (7 CFR part 1b).

(d) Department of Agriculture, Enhancement, Protection and Management of the Cultural Environment (7 CFR parts 3100 through 3199);

(e) Archaeological and Historic Preservation Act of 1960, as amended, (16 U.S.C. 469 *et seq.*);

(f) Archaeological Resources Protection Act of 1979 (16 U.S.C. 470aa *et seq.*);

(g) Bald and Golden Eagle Protection Act (16 U.S.C. 668 *et seq.*);

(h) Clean Air Act (42 U.S.C. 7401 *et seq.*);

(i) Clean Water Act (Federal Water Pollution Control Act, 33 U.S.C. 1251 *et seq.*);

(j) Coastal Barrier Resources Act (16 U.S.C. 3501 *et seq.*);

(k) Coastal Barrier Improvement Act (42 U.S.C. 4028 *et seq.*);

(l) Coastal Zone Management Act (16 U.S.C. 1456);

(m) Comprehensive Environmental Response, Compensation, and Liability Act (42 U.S.C. 103) (CERCLA);

(n) Consolidated Farm and Rural Development Act, Sections 307(a)(6)(A) (7 U.S.C. 1927(a)(6)(A)) and 363 (7 U.S.C. 2006e);

(o) Endangered Species Act of 1973 (16 U.S.C. 1531 *et seq.*);

(p) Farmland Protection Policy Act (7 U.S.C. 4201 *et seq.*);

(q) Historic Sites, Buildings and Antiquities Act (16 U.S.C. 461 *et seq.*);

(r) Housing and Community Development Act of 1992 (42 U.S.C. 542(c)(9));

(s) Migratory Bird Treaty Act (16 U.S.C. 703–711);

(t) National Historic Preservation Act (16 U.S.C. 470 *et seq.*);

(u) National Trails System Act (16 U.S.C. 1241 *et seq.*);

(v) Native American Graves Protection and Repatriation Act (25 U.S.C. 3001 *et seq.*);

(w) Noise Control Act (42 U.S.C. 4901 *et seq.*);

(x) Pollution Prevention Act of 1990 (42 U.S.C. 13101 *et seq.*);

(y) Resource Conservation and Recovery Act (42 U.S.C. 6901);

(z) Safe Drinking Water Act—(42 U.S.C. 300f *et seq.*);

(aa) Wild and Scenic Rivers Act (16 U.S.C. 1271 *et seq.*);

(bb) Wilderness Act (16 U.S.C. 1131 *et seq.*);

(cc) Compact of Free Association between the United States and the Republic of the Marshall Islands and between the United States and the Federated States of Micronesia (Public Law 108–188);

(dd) Compact of Free Association between the United States and the Republic of Palau (Public Law 99–658);

(ee) Executive Order 11514, Protection and Enhancement of Environmental Quality;

(ff) Executive Order 11593, Protection and Enhancement of the Cultural Environment;

(gg) Executive Order 11988, Floodplain Management;

(hh) Executive Order 11990, Protection of Wetlands;

(ii) Executive Order 12898, Federal Actions to Address Environmental Justice in Minority Populations and Low Income Populations;

(jj) Executive Order 12372, Intergovernmental Review;

(kk) Executive Order 13112, Invasive Species;

(ll) Executive Order 13175, Consultation and Coordination with Indian Tribal Governments;

(mm) Executive Order 13287, Preserve America;

(nn) Executive Order 13016, Federal Support of Community Efforts along American Heritage Rivers;

(oo) Executive Order 13352, Facilitation of Cooperative Conservation;

(pp) Executive Order 13423, Strengthening Federal Environmental, Energy, and Transportation Management;

(qq) Executive Order 13653, Preparing the United States for the Impacts of Climate Change;

(rr) Executive Order 13690, Establishing a Federal Flood Risk Management Standard and a Process for Further Soliciting and Considering Stakeholder Input;

(ss) Executive Order 13693, Planning for Federal Sustainability in the Next Decade;

(tt) Agriculture Departmental Regulation (DR) 5600–2, Environmental Justice;

(uu) Agriculture Departmental Regulation (DR) 9500–3, Land Use Policy;

(vv) Agriculture Departmental Regulation (DR) 9500–4, Fish and Wildlife Policy;

(ww) Agriculture Departmental Regulation (DR) 1070–001, U.S. Department of Agriculture (USDA) Policy Statement on Climate Change Adaptation; and

(xx) Agriculture Departmental Manual (DM) 5600–001, Environmental Pollution Prevention, Control, and Abatement Manual.

§1970.4 Policies.

(a) Applicants' proposals must, whenever practicable, avoid or minimize adverse environmental impacts; avoid or minimize conversion of wetlands or important farmlands (as defined in the Farmland Protection Policy Act and its implementing regulations issued by the USDA Natural Resources Conservation Service) when practicable alternatives exist to meet development needs; avoid unwarranted alterations or encroachment on floodplains when practicable alternatives exist to meet developmental needs; and avoid or minimize potentially disproportionate and adverse impacts to minority or low-income populations within the proposed action's area of impact. Avoiding development in floodplains includes avoiding development in the 500-year floodplain, as shown on the Federal Emergency Management Agency's (FEMA) Flood Insurance Rate Maps, where the proposed actions and facilities are defined as critical actions in §1970.6. The Agency shall not fund the proposal unless there is a demonstrated, significant need for the proposal and no practicable alternative exists to the proposed conversion of the above resources.

(b) The Agency encourages the reuse of real property defined as brownfields per Section 101 of the Comprehensive Environmental Response, Compensation, and Liability Act (CERCLA) where the reuse of such property is complicated by the presence or potential presence of a hazardous substance, pollutant, or other contaminant, provided that the level of such presence does not threaten human health and the environment for the proposed land use. The Agency will defer to the agency with regulatory authority under the appropriate law in determining the appropriate level of contaminant for a specific proposed land use. The Agency will evaluate the risk based upon the applicable regulatory agency's review and concurrence with the proposal.

(c) The Agency and applicant will involve other Federal agencies with jurisdiction by law or special expertise, state and local governments, Indian tribes and Alaska Native organizations, Native Hawaiian organizations, and the public, early in the Agency's environmental review process to the fullest extent practicable. To accomplish this objective, the Agency and applicant will:

(1) Ensure that environmental amenities and values be given appropriate consideration in decision making along with economic and technical considerations;

(2) At the earliest possible time, advise interested parties of the Agency's environmental policies and procedures and required environmental impact analyses during early project planning and design; and

(3) Make environmental assessments (EA) and environmental impact statements (EIS) available to the public for review and comment in a timely manner.

(d) The Agency and applicant will ensure the completion of the environmental review process prior to the irreversible and irretrievable commitment of Agency resources in accordance with §1970.11. The environmental review process is concluded when the Agency approves the applicability of a Categorical Exclusion (CE), issues a Finding of No Significant Impact (FONSI), or issues a Record of Decision (ROD).

(e) If an applicant's proposal does not comply with Agency environmental policies and procedures, the Agency will defer further consideration of the application until compliance can be demonstrated, or the application may be rejected. Any applicant that is directly and adversely affected by an administrative decision made by the Agency under this part may appeal that decision, to the extent permissible under 7 CFR part 11.

(f) The Agency recognizes the worldwide and long-range character of environmental problems and, where consistent with the foreign policy of the United States, will lend appropriate support to initiatives, resolutions, and programs designed to maximize international cooperation in anticipating and preventing a decline in the quality of humankind's world environment in accordance with NEPA, 42 U.S.C. 4321 *et seq.*

(g) The Agency will use the NEPA process, to the maximum extent feasible, to identify and encourage opportunities to reduce greenhouse gas (GHG) emissions caused by proposed Federal actions that would otherwise result in the emission of substantial quantities of GHG.

§ 1970.5 **Responsible parties.**

(a) *Agency.* The following paragraphs identify the general responsibilities of the Agency.

(1) The Agency is responsible for all environmental decisions and findings related to its actions and will encourage applicants to design proposals to protect, restore, and enhance the environment.

(2) If the Agency requires an applicant to submit environmental information, the Agency will outline the types of information and analyses required in guidance documents. This guidance is available on the Agency's Web site. The Agency will independently evaluate the information submitted.

(3) The Agency will advise applicants and applicable lenders of their responsibilities to consider environmental issues during early project planning and that specific actions listed in §1970.12, such as initiation of construction, cannot occur prior to completion of the environmental review process or it could result in a denial of financial assistance.

(4) The Agency may act as either a lead agency or a cooperating agency in the preparation of an environmental review document. If the Agency acts as a cooperating agency, the Agency will fulfill the cooperating agency responsibilities outlined in 40 CFR 1501.6.

(5) Mitigation measures described in the environmental review and decision documents must be included as conditions in Agency financial commitment documents, such as a conditional commitment letter.

(6) The Agency, guaranteed lender, or multi-tier recipients will monitor and track the implementation, maintenance, and effectiveness of any required mitigation measures.

(b) *Applicants.* Applicants must comply with provisions found in paragraphs (b)(1) through (8) of this section.

(1) Consult with Agency staff to determine the appropriate level of environmental review and to obtain publicly available resources at the earliest

possible time for guidance in identifying all relevant environmental issues that must be addressed and considered during early project planning and design throughout the process.

(2) Where appropriate, contact state and Federal agencies to initiate consultation on matters affected by this part. This part authorizes applicants to coordinate with state and Federal agencies on behalf of the Agency. However, applicants are not authorized to initiate consultation in accordance with Section 106 of the National Historic Preservation Act with Indian tribes on behalf of the Agency. In those cases, applicants need the express written authority of the Agency and consent of Indian tribes in order to initiate consultation.

(3) Provide information to the Agency that the Agency deems necessary to evaluate the proposal's potential environmental impacts and alternatives.

(i) Applicants must ensure that all required materials are current, sufficiently detailed and complete, and are submitted directly to the Agency office processing the application. Incomplete materials or delayed submittals may jeopardize consideration of the applicant's proposal by the Agency and may result in no award of financial assistance.

(ii) Applicants must clearly define the purpose and need for the proposal and inform the Agency promptly if any other Federal, state, or local agencies are involved in financing, permitting, or approving the proposal, so that the Agency may coordinate and consider participation in joint environmental reviews.

(iii) As necessary, applicants must develop and document reasonable alternatives that meet their purpose and need while improving environmental outcomes.

(iv) Applicants must prepare environmental review documents according to the format and standards provided by the Agency. The Agency will independently evaluate the final documents submitted. All environmental review documents must be objective, complete, and accurate in order for them to be finally accepted by the Agency. Applicants may employ a design or environmental professional or technical

service provider to assist them in the preparation of their environmental review documents.

(A) Applicants are not generally required to prepare environmental documentation for proposals that involve Agency activities with no or minimal disturbance listed in §1970.53. However, the Agency may request additional environmental documentation from the applicant at any time, specifically if the Agency determines that extraordinary circumstances may exist.

(B) For CEs listed in §1970.54, applicants must prepare environmental documentation as required by the Agency; the environmental documentation required for CEs is referred to as an environmental report(ER).

(C) When an EA is required, the applicant must prepare an EA that meets the requirements in subpart C of this part, including, but not limited to, information and data collection and public involvement activities. When the applicant prepares the EA, the Agency will make its own independent evaluation of the environmental issues and take responsibility for the scope and content of the EA.

(D) Applicants must cooperate with and assist the Agency in all aspects of preparing an EIS that meets the requirements specified in subpart D of this part, including, but not limited to, information and data collection and public involvement activities. Once authorized by the Agency in writing, applicants are responsible for funding all third-party contractors used to prepare the EIS.

(4) Applicants must provide any additional studies, data, and document revisions requested by the Agency during the environmental review and decision-making process. The studies, data, and documents required will vary depending upon the specific project and its impacts. Examples of studies that the Agency may require an applicant to provide are biological assessments under the ESA, archeological surveys under the NHPA, wetland delineations, surveys to determine the floodplain elevation on a site, air quality conformity analysis, or other such information needed to adequately assess impacts.

(5) Applicants must ensure that no actions are taken (such as any demolition, land clearing, initiation of construction, or advance of interim construction funds from a guaranteed lender), including incurring any obligations with respect to their proposal, that may have an adverse impact on the quality of the human environment or that may limit the choice of reasonable alternatives during the environmental review process. Limitations on actions by an applicant prior to the completion of the Agency environmental review process are defined in CEQ regulations at 40 CFR 1506.1 and 7 CFR 1970.12.

(6) Applicants must promptly notify the Agency processing official when changes are made to their proposal so that the environmental review and documentation may be supplemented or otherwise revised as necessary.

(7) Applicants must incorporate any mitigation measures identified and any required monitoring in the environmental review process into the plans and specifications and construction contracts for the proposals. Applicants must provide such mitigation measures to consultants responsible for preparing design and construction documents, or provide other mitigation action plans. Applicants must maintain, as applicable, mitigation measures for the life of the loans or refund term for grants.

(8) Applicants must cooperate with the Agency on achieving environmental policy goals. If an applicant is unwilling to cooperate with the Agency on environmental compliance, the Agency will deny the requested financial assistance.

§ 1970.6 Definitions and acronyms.

(a) *Definitions.* Terms used in this part are defined in 40 CFR part 1508, 36 CFR 800.16, and this section. If a term is defined in this section and in one or both of the other referenced regulations, such term will have the meaning as defined in this subpart.

Agency. USDA Rural Development, which includes RBS, RHS, and RUS, and any successor agencies.

Applicant. An individual or entity requesting financial assistance including but not limited to loan recipients,

grantees, guaranteed lenders, or licensees.

Average megawatt. The equivalent capacity rating of a generating facility based on the gross energy output generated over a 12-month period or one year.

Construction work plan. An engineering planning study that is used in the Electric Program to determine and document a borrower's 2- to 4-year capital construction investments that are needed to provide and maintain adequate and reliable electric service to a borrower's new and existing members.

Cooperative agreement. For the purposes of this part, a cooperative agreement is a form of financial assistance in which the Agency provides funding that is authorized by public statute, not to be repaid, and for a purpose that includes substantial involvement and a mutual interest of both the Agency and the cooperator.

Critical action. Any activity for which even a slight chance of flooding would be hazardous as determined by the Agency. Critical actions include activities that create, maintain, or extend the useful life of structures or facilities that produce, use, or store highly volatile, flammable, explosive, toxic, or water-reactive materials; maintain irreplaceable records; or provide essential utility or emergency services (such as data storage centers, electric generating facilities, water treatment facilities, wastewater treatment facilities, large pump stations, emergency operations centers including fire and police stations, and roadways providing sole egress from flood-prone areas); or facilities that are likely to contain occupants who may not be sufficiently mobile to avoid death or serious injury in a flood.

Design professional. An engineer or architect providing professional design services to applicants during the planning, design, and construction phases of proposals submitted to the Agency for financial assistance.

Distributed resources. Sources of electrical power that are not directly connected to a bulk power transmission system, having an installed capacity of not more than 10 Mega volt-amperes (MVA), connected to an electric power system through a point of common coupling. Distributed resources include both generators (distributed generation) and energy storage technologies.

Emergency. A disaster or a situation that involves an immediate or imminent threat to public health or safety as determined by the Agency.

Environmental report. The environmental documentation that is required of applicants for proposed actions eligible for a CE under §1970.54.

Environmental review. Any or all of the levels of environmental analysis described under this part.

Financial assistance. A loan, grant, cooperative agreement, or loan guarantee that provides financial assistance, provided by the Agency to an applicant. In accordance with 40 CFR 1505.1(b), the Agency defines the major decision point at which NEPA must be complete, as the approval of financial assistance.

Grant. A form of financial assistance for a specified purpose without scheduled repayment.

Guaranteed lender. The organization making, servicing, or collecting the loan which is guaranteed by the Agency under applicable regulations, excluding the Federal Financing Bank.

Historic property. Any prehistoric or historic district, site, building, structure, or object included in, or eligible for inclusion in, the National Register of Historic Places maintained by the Secretary of the Interior. This term includes artifacts, records, and remains that are related to and located within such properties. The term includes properties of traditional religious and cultural importance to an Indian tribe or Native Hawaiian organization and that meet the National Register criteria. (See 36 CFR 800.16(1)).

Indian tribe. An Indian tribe, band, nation, or other organized group or community, including a native village, regional corporation or village corporation, as those terms are defined in Section 3 of the Alaska Native Claims Settlement Act (43 U.S.C. 1602), which is recognized as eligible for the special programs and services provided by the United States to Indians because of their status as Indians (see 36 CFR 800.16(m)).

Lien sharing. Agreement to pro rata payment on shared secured collateral without priority preference.

Lien subordination. The circumstance in which the Agency, as a first lien holder, provides a creditor with a priority security interest in secured collateral.

Loan. The provision of funds by the Agency directly to an applicant in exchange for repayment with interest and collateral to secure repayment.

Loan guarantee. The circumstance in which the Agency guarantees all or a portion of payment of a debt obligation to a lender.

Loan/System design. An engineering study, prepared to support a loan application under this part, demonstrating that a system design provides telecommunication services most efficiently to proposed subscribers in a proposed service area, in accordance with the Telecommunications Program guidance.

Multi-tier action. Financial assistance provided by specific programs administered by the Agency, that provides financial assistance to eligible recipients, including but not limited to: Intermediaries; community-based organizations, such as housing or community development non-profit organizations; rural electric cooperatives; or other organizations with similar financial arrangements who, in turn, provide financial assistance to eligible recipients. The entities or organizations receiving the initial Agency financial assistance are considered "primary recipients." As the direct recipient of this financial assistance, "primary recipients" provide the financial assistance to other parties, referred to as "secondary recipients" or "ultimate recipients." The multi-tier action programs include Housing Preservation Grants (42 U.S.C. 1490m), Multi-Family Housing Preservation Revolving Loan Fund (7 CFR part 3560), Intermediary Relending Program (7 U.S.C. 1932 note and 42 U.S.C. 9812), Rural Business Development Grant Program (7 U.S.C. 940c and 7 U.S.C. 1932(c)), Rural Economic Development Loan and Grant Program (7 U.S.C. 940c), Rural Microentrepreneur Assistance Program (7

U.S.C. 1989(a), 7 U.S.C. 2008s), Household Water Well System Grant Program (7 U.S.C. 1926e), Revolving Funds for Financing Water and Wastewater Projects (Revolving Fund Program) (7 U.S.C. 1926(a)(2)(B)), Energy Efficiency and Conservation Loan Program (7 U.S.C. 901), Section 313A, Guarantees for Bonds and Notes Issued for Electrification or Telephone Purposes (7 U.S.C. 940c–1), Rural Energy Savings Program (7 U.S.C. 8107a), and any other such programs or similar financial assistance actions to primary recipients as described above.

No action alternative. An alternative that describes the reasonably foreseeable future environment in the event a proposed Federal action is not taken. This forms the baseline condition against which the impacts of the proposed action and other alternatives are compared and evaluated.

Preliminary Architectural/Engineering Report. Documents prepared by the applicant's design professional in accordance with applicable Agency guidance for Preliminary Architectural Reports for housing, business, and community facilities proposals and for Preliminary Engineering Reports for water and wastewater proposals.

Previously disturbed or developed land. Land that has been changed such that its functioning ecological processes have been and remain altered by human activity. The phrase encompasses areas that have been transformed from natural cover to non-native species or a managed state, including, but not limited to, utility and electric power transmission corridors and rights-of-way, and other areas where active utilities and currently used roads are readily available.

Servicing actions. All routine, ministerial, or administrative actions for Agency-provided financial assistance that do not involve new financial assistance, including, but not limited to:

(1) Advancing of funds, billing, processing payments, transfers, assumptions, refinancing involving only a change in an interest rate, and accepting prepayments;

(2) Monitoring collateral; foreclosure; compromising, adjusting, reducing, or charging off debts or claims; and modifying or releasing the terms of security instruments, leases, contracts, and agreements; and

(3) Consents or approvals provided pursuant to loan contracts, agreements, and security instruments.

Substantial improvement. Any repair, reconstruction or other improvement of a structure or facility, which has been damaged in excess of, or the cost of which equals or exceeds, 50% of the market value of the structure or replacement cost of the facility (including all "public facilities" as defined in the Disaster Relief Act of 1974) before the repair or improvement is started, or, if the structure or facility has been damaged and is proposed to be restored, before the damage occurred. If a facility is an essential link in a larger system, the percentage of damage will be based on the relative cost of repairing the damaged facility to the replacement cost of the portion of the system which is operationally dependent on the facility. The term "substantial improvement" does not include any alteration of a structure or facility listed on the National Register of Historic Places or a State Inventory of Historic Places. (*See* 44 CFR 59.1.)

Third-party contractor. Contractors for the preparation of EISs, under the Agency's direction, and paid by the applicant. Under the Agency's direction and in compliance with 40 CFR 1506.5(c), the applicant may undertake the necessary paperwork for the solicitation of a field of candidates. Federal procurement requirements do not apply to the Agency because it incurs no obligations or costs under the contract, nor does the Agency procure anything under the contract.

(b) *Acronyms.*

aMW—Average megawatt
CE—Categorical Exclusion
CERCLA—Comprehensive Environmental Response, Compensation, and Liability Act
CEQ—Council on Environmental Quality
EA—Environmental Assessment
ER—Environmental Report
EIS—Environmental Impact Statement
EPA—United States Environmental Protection Agency
ESA—Endangered Species Act

FEMA—Federal Emergency Management Agency

FONSI—Finding of No Significant Impact

GHG—Greenhouse Gas

kV—kilovolt (kV)

kW—kilowatt (kW)

MW—megawatt

MVA—Mega volt-amperes

NEPA—National Environmental Policy Act

NHPA—National Historic Preservation Act

NOI—Notice of Intent

RBIC—Rural Business Investment Company

RBS—Rural Business-Cooperative Service

RHS—Rural Housing Service

RUS—Rural Utilities Service

ROD—Record of Decision

SEPA—State Environmental Policy Act

USDA—United States Department of Agriculture

USGS—United States Geological Survey

§1970.7 [Reserved]

§1970.8 Actions requiring environmental review.

(a) The Agency must comply with the requirements of NEPA for all Federal actions within the:

(1) United States borders and any other commonwealth, territory or possession of the United States such as Guam, American Samoa, U.S. Virgin Islands, the Commonwealth of the Northern Mariana Islands, and the Commonwealth of Puerto Rico; and

(2) Republic of the Marshall Islands, the Federated States of Micronesia and the Republic of Palau, subject to applicable Compacts of Free Association.

(b) Except as provided in paragraphs (c), (d), and (e) of this section, the provisions of this part apply to administrative actions by the Agency with regard to the following to be Federal actions:

(1) Providing financial assistance;

(2) Certain post-financial assistance actions with the potential to have an effect on the environment, including:

(i) The sale or lease of Agency-owned real property;

(ii) Lien subordination; and

(iii) Approval of a substantial change in the scope of a project receiving financial assistance not previously considered.

(3) Promulgation of procedures or regulations for new or significantly revised programs; and

(4) Legislative proposals (see 40 CFR 1506.8).

(c) For environmental review purposes, the Agency has identified and established categories of proposed actions (§§1970.53 through 1970.55, 1970.101, and 1970.151). An applicant may propose to participate with other parties in the ownership of a project. In such a case, the Agency will determine whether the applicant participants have sufficient control and responsibility to alter the development of the proposed project prior to determining its classification. Only if there is such control and responsibility as described below will the Agency consider its action with regard to the project to be a Federal action for purposes of this part. Where the applicant proposes to participate with other parties in the ownership of a proposed project and all applicants cumulatively own:

(1) Five percent (5%) or less, the project is not considered a Federal action subject to this part;

(2) Thirty-three and one-third percent (33⅓%) or more, the project shall be considered a Federal action subject to this part;

(3) More than five percent (5%) but less than thirty-three and one-third percent (33⅓%), the Agency will determine whether the applicant participants have sufficient control and responsibility to alter the development of the proposal such that the Agency's action will be considered a Federal action subject to this part. In making this determination, the Agency will consider such factors as:

(i) Whether construction would be completed regardless of the Agency's financial assistance or approval;

(ii) The stage of planning and construction;

(iii) Total participation of the applicant;

(iv) Participation percentage of each participant; and

(v) Managerial arrangements and contractual provisions.

(d) Lien sharing is not an action for the purposes of this part.

(e) Servicing actions are directly related to financial assistance already provided, do not require separate NEPA review, and are not actions for the purposes of this part.

§ 1970.9 Levels of environmental review.

(a) The Agency has identified classes of actions and the level of environmental review required for applicant proposals and Agency actions in subparts B (CEs), C (EAs), and D (EISs) of this part. An applicant seeking financial assistance from the Agency must sufficiently describe its proposal so that the Agency can properly classify the proposal for the purposes of this part.

(b) If an action is not identified in the classes of actions listed in subparts B, C, or D of this part, the Agency will determine what level of environmental review is appropriate.

(c) A single environmental document will evaluate an applicant's proposal and any other activities that are connected, interdependent, or likely to have significant cumulative effects. When a proposal represents one segment of a larger interdependent proposal being funded jointly by various entities, the level of environmental review will normally include the entire proposal.

(d) Upon submission of multi-year planning documents, such as Telecommunications Program Loan/System Designs or multi-year Electric Program Construction Work Plans, the Agency will identify the appropriate classification for all proposals listed in the applicable design or work plan and may request any additional environmental information prior to the time of loan approval.

§ 1970.10 Raising the level of environmental review.

Environmental conditions, scientific controversy, or other characteristics unique to a specific proposal can trigger the need for a higher level of environmental review than described in subparts B or C of this part. As appropriate, the Agency will determine whether extraordinary circumstances

(see § 1970.52) or the potential for significant environmental impacts warrant a higher level of review. The Agency is solely responsible for determining the level of environmental review to be conducted and the adequacy of environmental review that has been performed.

§ 1970.11 Timing of the environmental review process.

(a) Once an applicant decides to request Agency financial assistance, the applicant must initiate the environmental review process at the earliest possible time to ensure that planning, design, and other decisions reflect environmental policies and values, avoid delays, and minimize potential conflicts. This includes early coordination with the Agency, all funding partners, and regulatory agencies, in order to minimize duplication of effort.

(b) The environmental review process must be concluded before completion of the obligation of funds.

(c) The environmental review process is formally concluded when all of the following have occurred:

(1) The Agency has reviewed the appropriate environmental review document for completeness;

(2) All required public notices have been published and public comment periods have elapsed;

(3) All comments received during any established comment period have been considered and addressed, as appropriate by the Agency;

(4) The environmental review documents have been approved by the Agency; and

(5) The appropriate environmental decision document has been executed by the Agency after paragraphs (c)(1) through (4) of this section have been concluded.

(d) For proposed actions listed in § 1970.151 and to ensure Agency compliance with the conflict of interest provisions in 40 CFR 1506.5(c), the Agency is responsible for selecting any third-party EIS contractor and participating in the EIS preparation. For more information regarding acquisition of professional services and funding of a third-party contractor, refer to § 1970.152.

Effective Date Note: At 83 FR 59271, Nov. 23, 2018, § 1970.11 was amended by revising paragraph (b), effective Jan. 7, 2019. For the convenience of the user, the revised text is set forth as follows:

§ 1970.11 Timing of the environmental review process.

*   *   *   *   *

(b) The environmental review process must be concluded before the obligation of funds; except for infrastructure projects where the assurance that funds will be available for community health, safety, or economic development has been determined as necessary by the Agency Administrator. At the discretion of the Agency Administrator, funds may be obligated contingent upon the conclusion of the environmental review process prior to any action that would have an adverse effect on the environment or limit the choices of any reasonable alternatives. Funds so obligated shall be rescinded if the Agency cannot conclude the environmental review process before the end of the fiscal year after the year in which the funds were obligated, or if the Agency determines that it cannot proceed with approval based on findings in the environmental review process. For the purposes of this section, infrastructure projects shall include projects such as broadband, telecommunications, electric, energy efficiency, smart grid, water, sewer, transportation, and energy capital investments in physical plant and equipment, but not investments authorized in the Housing Act of 1949.

*   *   *   *   *

§ 1970.12 Limitations on actions during the NEPA process.

(a) *Limitations on actions.* Applicants must not take actions concerning a proposal that may potentially have an environmental impact or would otherwise limit or affect the Agency's decision until the Agency's environmental review process is concluded. If such actions are taken prior to the conclusion of the environmental review process, the Agency may deny the request for financial assistance.

(b) *Anticipatory demolition.* If the Agency determines that an applicant has intentionally significantly adversely affected a historic property with the intent to avoid the requirements of Section 106 of the NHPA (such as demolition or removal of all or part of the property) the Agency may deny the request for financial assistance in accordance with section 110(k) of the NHPA.

(c) *Recent construction.* When construction is in progress or has recently been completed by applicants who can demonstrate no prior intent to seek Agency assistance at the time of application submittal to the Agency, the following requirements apply:

(1) In cases where construction commenced within 6 months prior to the date of application, the Agency will determine and document whether the applicant initiated construction to avoid environmental compliance requirements. If any evidence to that effect exists, the Agency may deny the request for financial assistance.

(2) If there is no evidence that an applicant is attempting to avoid environmental compliance requirements, the application is subject to the following additional requirements:

(i) The Agency will promptly provide written notice to the applicant that the applicant must halt construction if it is ongoing and fulfill all environmental compliance responsibilities before the requested financing will be provided;

(ii) The applicant must take immediate steps to identify any environmental resources affected by the construction and protect the affected resources; and

(iii) With assistance from the applicant and to the extent practicable, the Agency will determine whether environmental resources have been adversely affected by any construction and this information will be included in the environmental document.

(d) *Minimal expenditures.* In accordance with 40 CFR 1506.1(d), the Agency will not be precluded from approving minimal expenditures by the applicant not affecting the environment (*e.g.*, long lead-time equipment, purchase options, or environmental or technical documentation needed for Agency environmental review). To be minimal, the expenditure must not exceed the amount of loss which the applicant could absorb without jeopardizing the Government's security interest in the event the proposed action is not approved by the Agency, and must not compromise the objectivity of the

Agency's environmental review process.

§ 1970.13 Consideration of alternatives.

The purpose of considering alternatives to a proposed action is to explore and evaluate whether there may be reasonable alternatives to that action that may have fewer or less significant negative environmental impacts. When considering whether the alternatives are reasonable, the Agency will take into account factors such as economic and technical feasibility. The extent of the analysis on each alternative will depend on the nature and complexity of the proposal. Environmental review documents must discuss the consideration of alternatives as follows:

(a) For proposals subject to subpart C of this part, the environmental effects of the "No Action" alternative must be evaluated. All EAs must evaluate other reasonable alternatives whenever the proposal involves potential adverse effects to environmental resources.

(b) For proposals subject to subpart D of this part, the Agency will follow the requirements in 40 CFR part 1502.

§ 1970.14 Public involvement.

(a) *Goal.* The goal of public involvement is to engage affected or interested parties and share information and solicit input regarding environmental impacts of proposals. This helps the Agency to better identify potential environmental impacts and mitigation measures and allows the public to review and comment on proposals under consideration by the Agency. The nature and extent of public involvement will depend upon the public interest and the complexity, sensitivity, and potential for significant environmental impacts of the proposal.

(b) *Responsibility to involve the public.* The Agency will require applicant assistance throughout the environmental review process, as appropriate, to involve the public as required under 40 CFR 1506.6. These activities may include, but are not limited to:

(1) Coordination with Federal, state, and local agencies; Federally recognized American Indian tribes; Alaska Native organizations; Native Hawaiian organizations; and the public;

(2) Providing meaningful opportunities for involvement of affected minority or low-income populations, which may include special outreach efforts, so that potential disproportionate effects on minority or low-income populations are reduced to the maximum extent practicable;

(3) Publication of notices;

(4) Organizing and conducting meetings; and

(5) Providing translators, posting information on electronic media, or any other additional means needed that will successfully inform the public.

(c) *Scoping.* In accordance with 40 CFR 1501.7, scoping is an early and open process to identify significant environmental issues deserving of study, de-emphasize insignificant issues, and determine the scope of the environmental review process.

(1) Public scoping meetings allow the public to obtain information about a proposal and to express their concerns directly to the parties involved and help determine what issues are to be addressed and what kinds of expertise, analysis, and consultation are needed. For proposals classified in §§ 1970.101 and 1970.151, scoping meetings may be required at the Agency's discretion. The Agency may require a scoping meeting whenever the proposal has substantial controversy, scale, or complexity.

(2) If required, scoping meetings will be held at reasonable times, in accessible locations, and in the geographical area of the proposal at a location the Agency determines would best afford an opportunity for public involvement.

(3) When held, applicants must attend and participate in all scoping meetings. When requested by the Agency, the applicant must organize and arrange meeting locations, publish public notices, provide translation, provide for any equipment needs such as those needed to allow for remote participation, present information on their proposal, and fulfill any related activities.

(d) *Public notices.* (1) The Agency is responsible for meeting the public notice requirements in 40 CFR 1506.6, but will require the applicant to provide public notices of the availability of environmental documents and of public meetings so as to inform those persons and agencies who may be interested in or affected by an applicant's proposal. The Agency will provide applicants with guidance as to specific notice content, publication frequencies, and distribution requirements. Public notices issued by the Agency or the applicant must describe the nature, location, and extent of the applicant's proposal and the Agency's proposed action; notices must also indicate the availability and location of pertinent information.

(2) Notices generally must be published in a newspaper(s) of general circulation (both in print and online) within the proposal's affected areas and other places as determined by the Agency. The notice must be published in the non-classified section of the newspaper. If the affected area is largely non-English speaking or bilingual, the notice must be published in both English and non-English language newspapers serving the affected area, if both are available. The Agency will determine the use of other distribution methods for communicating information to affected individuals and communities if those are more likely to be effective. The applicant must obtain an "affidavit of publication" or other such evidence from all publications (or equivalent verification if other distribution methods were used) and must submit such evidence to the Agency to be made a part of the Agency's Administrative Record.

(3) The number of times notices regarding EAs must be published is specified in §1970.102(b)(6)(ii). Other distribution methods may be used in special circumstances when a newspaper notice is not available or is not adequate. Additional distribution methods may include, but are not limited to, direct public notices to adjacent property owners or occupants, mass mailings, radio broadcasts, internet postings, posters, or some other combination of public announcements.

(4) Formal notices required for EIS-level proposals pursuant to 40 CFR part 1500 will be published by the Agency in the FEDERAL REGISTER.

(e) *Public availability.* Documents associated with the environmental review process will be made available to the public at convenient locations specified in public notices and, where appropriate, on the Agency's internet site. Environmental documents that are voluminous or contain hard-to-reproduce graphics or maps should be made available for viewing at one or more locations, such as an Agency field office, public library, or the applicant's place of business. Upon request, the Agency will promptly provide interested parties copies of environmental review documents without charge to the extent practicable, or at a fee not to exceed the cost of reproducing and shipping the copies.

(f) *Public comments.* All comments should be directed to the Agency. Comments received by applicants must be forwarded to the Agency in a timely manner. The Agency will assess and consider all comments received.

§1970.15 Interagency cooperation.

In order to reduce delay and paperwork, the Agency will, when practicable, eliminate duplication of Federal, state, and local procedures by participating in joint environmental document preparation, adopting appropriate environmental documents prepared for or by other Federal agencies, and incorporating by reference other environmental documents in accordance with 40 CFR 1506.2 and 1506.3.

(a) *Coordination with other Federal agencies.* When other Federal agencies are involved in an Agency action listed in §1970.101 or §1970.151, the Agency will coordinate with these agencies to determine cooperating agency relationships as appropriate in the preparation of a joint environmental review document. The criteria for making this determination can be found at 40 CFR 1501.5.

(b) *Adoption of documents prepared for or by other Federal agencies.* The Agency may adopt EAs or EISs prepared for or by other Federal agencies if the proposed actions and site conditions addressed in the environmental document are substantially the same as those associated with the proposal being considered by the Agency. The Agency will consider age, location, and other reasonable factors in determining the usefulness of the other Federal documents. The Agency will complete an independent evaluation of the environmental document to ensure it meets the requirements of this part. If any environmental document does not meet all Agency requirements, it will be supplemented prior to adoption. Where there is a conflict in the two agencies' classes of action, the Agency may adopt the document provided that it meets the Agency's requirements.

(c) *Cooperation with state and local governments.* In accordance with 40 CFR 1500.5 and 1506.2, the Agency will cooperate with state and local agencies to the fullest extent possible to reduce delay and duplication between NEPA and comparable state and local requirements.

(1) Joint environmental documents. To the extent practicable, the Agency will participate in the preparation of a joint document to ensure that all of the requirements of this part are met. Applicants that request Agency assistance for specific proposals must contact the Agency at the earliest possible date to determine if joint environmental documents can be effectively prepared. In order to prepare joint documents the following conditions must be met:

(i) Applicants must also be seeking financial, technical, or other assistance such as permitting or approvals from a state or local agency that has responsibility to complete an environmental review for the applicant's proposal; and

(ii) The Agency and the state or local agency may agree to be joint lead agencies where practicable. When state laws or local ordinances have environmental requirements in addition to, but not in conflict with those of the Agency, the Agency will cooperate in fulfilling these requirements.

(2) Incorporating other documents. The Agency cannot adopt a non-Federal environmental document under NEPA. However, if an environmental document is not jointly prepared as described in paragraph (c)(1) of this section (*e.g.*, prepared in accordance with a state environmental policy act [SEPA]), the Agency will evaluate the document as reference or supporting material for the Agency's environmental document.

§ 1970.16 Mitigation.

(a) The goal of mitigation is to avoid, minimize, rectify, reduce, or compensate for the adverse environmental impacts of an action. The Agency will seek to mitigate potential adverse environmental impacts resulting from Agency actions. All mitigation measures will be included in Agency commitment or decision documents.

(b) Mitigation measures, where necessary for a FONSI or a ROD, will be discussed with the applicant and with any other relevant agency and, to the extent practicable, incorporated into Agency commitment documents, plans and specifications, and construction contracts so as to be legally binding.

(c) The Agency, applicable lenders, or any intermediaries will monitor implementation of all mitigation measures during development of design, final plans, inspections during the construction phase of projects, as well as in future servicing visits. The Agency will direct applicants to take necessary measures to bring the project into compliance. If the applicant fails to achieve compliance, all advancement of funds and the approval of cost reimbursements will be suspended. Other measures may be taken by the Agency to redress the failed mitigation as appropriate.

§ 1970.17 Programmatic analyses and tiering.

In accordance with 40 CFR 1502.20 and to foster better decision making, the Agency may consider preparing programmatic-level NEPA analyses and tiering to eliminate repetitive discussions of the same issues and to focus on the actual issues ripe for decision at each level of environmental review.

§ 1970.18 Emergencies.

When an emergency exists and the Agency determines that it is necessary to take emergency action before preparing a NEPA analysis and any required documentation, the provisions of this section apply.

(a) *Urgent response.* The Agency and the applicant, as appropriate, may take actions necessary to control the immediate impacts of an emergency (see § 1970.53(e)). Emergency actions include those that are urgently needed to re-store services and to mitigate harm to life, property, or important natural or cultural resources. When taking such actions, the Agency and the applicant, when applicable, will take into account the probable environmental consequences of the emergency action and mitigate foreseeable adverse environmental effects to the extent practicable.

(b) *CE- and EA-level actions.* If the Agency proposes longer-term emergency actions other than those actions described in paragraph (a) of this section, and such actions are not likely to have significant environmental impacts, the Agency will document that determination in a finding for a CE or in a FONSI for an EA prepared in accordance with this part. If the Agency finds that the nature and scope of proposed emergency actions are such that they must be undertaken prior to preparing any NEPA analysis and documentation associated with a CE or EA, the Agency will identify alternative arrangements for compliance with this part with the appropriate agencies.

(1) Alternative arrangements for environmental compliance are limited to actions necessary to control the immediate impacts of the emergency.

(2) Alternative arrangements will, to the extent practicable, attempt to achieve the substantive requirements of this part.

(c) *EIS-level actions.* If the Agency proposes emergency actions other than those actions described in paragraphs (a) or (b) of this section and such actions are likely to have significant environmental impacts, then the Agency will consult with the CEQ about alternative arrangements in accordance with CEQ regulations at 40 CFR 1506.11 as soon as possible.

§§ 1970.19–1970.50 [Reserved]

Subpart B—NEPA Categorical Exclusions

§ 1970.51 Applying CEs.

(a) The actions listed in §§ 1970.53 through 1970.55 are classes of actions that the Agency has determined do not individually or cumulatively have a significant effect on the human envi-ronment (referred to as "categorical exclusions" or CEs).

(1) Actions listed in § 1970.53 do not normally require applicants to submit environmental documentation with their applications. However, these applicants may be required to provide environmental information at the Agency's request.

(2) Actions listed in § 1970.54 normally require the submission of an environmental report (ER) by an applicant to allow the Agency to determine whether extraordinary circumstances (as defined in § 1970.52(a)) exist. When the Agency determines that extraordinary circumstances exist, an EA or EIS, as appropriate, will be required and, in such instances, applicants may be required to provide additional environmental information later at the Agency's request.

(3) Actions listed in § 1970.55 relate to financial assistance whereby the applicant is a primary recipient of a multitier program providing financial assistance to secondary or ultimate recipients without specifying the use of such funds for eligible actions at the time of initial application and approval. The decision to approve or fund such initial proposals has no discernible environmental effects and is therefore categorically excluded provided the primary recipient enters into an agreement with the Agency for future reviews. The primary recipient is limited to making the Agency's financial assistance available to secondary recipients for the types of projects specified in the primary recipient's application. Second-tier funding of proposals to secondary or ultimate recipients will be screened for extraordinary circumstances by the primary recipient and monitored by the Agency. If the primary recipient determines that extraordinary circumstances exist on any second-tier proposal, it must be referred to the Agency for the appropriate level of review under this part in accordance with subparts C and D.

(b) To find that a proposal is categorically excluded, the Agency must determine the following:

(1) The proposal fits within a class of actions that is listed in §§ 1970.53 through 1970.55;

(2) There are no extraordinary circumstances related to the proposal (see § 1970.52); and

(3) The proposal is not "connected" to other actions with potentially significant impacts (see 40 CFR 1508.25(a)(1)) or is not considered a "cumulative action" (see 40 CFR 1508.25(a)(2)), and is not precluded by 40 CFR 1506.1.

(c) A proposal that consists of more than one action may be categorically excluded only if all components of the proposed action are eligible for a CE.

(d) If, at any time during the environmental review process, the Agency determines that the proposal does not meet the criteria listed in §§ 1970.53 through 1970.55, an EA or EIS, as appropriate, will be required.

(e) Failure to achieve compliance with this part will postpone further consideration of an applicant's proposal until such compliance is achieved or the applicant withdraws the proposal. If compliance is not achieved, the Agency will deny the request for financial assistance.

§ 1970.52 Extraordinary circumstances.

(a) Extraordinary circumstances are unique situations presented by specific proposals, such as characteristics of the geographic area affected by the proposal, scientific controversy about the environmental effects of the proposal, uncertain effects or effects involving unique or unknown risks, and unresolved conflicts concerning alternate uses of available resources within the meaning of section 102(2)(E) of NEPA. In the event of extraordinary circumstances, a normally excluded action will be the subject of an additional environmental review by the Agency to determine the potential of the Agency action to cause any significant adverse environmental effect, and could, at the Agency's sole discretion, require an EA or an EIS, prepared in accordance with subparts C or D of this part, respectively.

(b) Significant adverse environmental effects that the Agency considers to be extraordinary circumstances include, but are not limited to:

(1) Any violation of applicable Federal, state, or local statutory, regulatory, or permit requirements for environment, safety, and health.

(2) Siting, construction, or major expansion of Resource Conservation and Recovery Act permitted waste storage, disposal, recovery, or treatment facilities (including incinerators), even if the proposal includes categorically excluded waste storage, disposal, recovery, or treatment actions.

(3) Any proposal that is likely to cause uncontrolled or unpermitted releases of hazardous substances, pollutants, contaminants, or petroleum and natural gas products.

(4) An adverse effect on the following environmental resources:

(i) Historic properties;

(ii) Federally listed threatened or endangered species, critical habitat, Federally proposed or candidate species;

(iii) Wetlands (Those actions that propose to convert or propose new construction in wetlands will require consideration of alternatives to avoid adverse effects and unwarranted conversions of wetlands. For actions involving linear utility infrastructure where utilities are proposed to be installed in existing, previously disturbed rights-of-way or that are authorized under applicable Clean Water Act, Section 404 nationwide permits will not require the consideration of alternatives. Those actions that require Section 404 individual permits would create an extraordinary circumstance);

(iv) Floodplains (those actions that introduce fill or structures into a floodplain or propose substantial improvements to structures within a floodplain will require consideration of alternatives to avoid adverse effects and incompatible development in floodplains. Actions that do not adversely affect the hydrologic character of a floodplain, such as buried utility lines or subsurface pump stations, would not create an extraordinary circumstance; or purchase of existing structures within the floodplain will not create an extraordinary circumstance but may require consideration of alternatives to avoid adverse effects and incompatible development in floodplains when determined appropriate by the Agency);

(v) Areas having formal Federal or state designations such as wilderness areas, parks, or wildlife refuges; wild and scenic rivers; or marine sanctuaries;

(vi) Special sources of water (such as sole source aquifers, wellhead protection areas, and other water sources that are vital in a region);

(vii) Coastal barrier resources or, unless exempt, coastal zone management areas; and

(viii) Coral reefs.

(5) The existence of controversy based on effects to the human environment brought to the Agency's attention by a Federal, tribal, state, or local government agency.

§1970.53 CEs involving no or minimal disturbance without an environmental report.

The CEs in this section are for proposals for financial assistance that involve no or minimal alterations in the physical environment and typically occur on previously disturbed land. These actions normally do not require an applicant to submit environmental documentation with the application. However, based on the review of the project description, the Agency may request additional environmental documentation from the applicant at any time, specifically if the Agency determines that extraordinary circumstances may exist. In accordance with Section 106 of the National Historic Preservation Act (54 U.S.C. 300101 *et seq.*) and its implementing regulations under 36 CFR 800.3(a), the Agency has determined that the actions in this section are undertakings, and in accordance with 36 CFR 800.3(a)(1) has identified those undertakings for which no further review under 36 CFR part 800 is required because they have no potential to cause effects to historic properties. In accordance with section 7 of the Endangered Species Act (16 U.S.C. 1531 *et seq.*) and its implementing regulations at 50 CFR part 402, the Agency has determined that the actions in this section are actions for purposes of the Endangered Species Act, and in accordance with 50 CFR 402.06 has identified those actions for which no further review under 50 CFR part 402 is required

because they will have no effect to listed threatened and endangered species.

(a) *Routine financial actions.* The following are routine financial actions and, as such, are classified as categorical exclusions identified in paragraphs (a)(1) through (7) of this section.

(1) Financial assistance for the purchase, transfer, lease, or other acquisition of real property when no or minimal change in use is reasonably foreseeable.

(i) Real property includes land and any existing permanent or affixed structures.

(ii) "No or minimal change in use is reasonably foreseeable" means no or only a small change in use, capacity, purpose, operation, or design is expected where the foreseeable type and magnitude of impacts would remain essentially the same.

(2) Financial assistance for the purchase, transfer, or lease of personal property or fixtures where no or minimal change in operations is reasonably foreseeable. These include:

(i) Approval of minimal expenditures not affecting the environment such as contracts for long lead-time equipment and purchase options by applicants under the terms of 40 CFR 1506.1(d) and 7 CFR 1970.12;

(ii) Acquisition of end-user equipment and programming for telecommunication distance learning;

(iii) Purchase, replacement, or installation of equipment necessary for the operation of an existing facility (such as Supervisory Control and Data Acquisition Systems (SCADA), energy management or efficiency improvement systems (including heat rate efficiency), replacement or conversion to enable use of renewable fuels, standby internal combustion electric generators, battery energy storage systems, and associated facilities for the primary purpose of providing emergency power);

(iv) Purchase of vehicles (such as those used in business, utility, community, or emergency services operations);

(v) Purchase of existing water rights where no associated construction is involved;

(vi) Purchase of livestock and essential farm equipment, including crop storing and drying equipment; and

(vii) Purchase of stock in an existing enterprise to obtain an ownership interest in that enterprise.

(3) Financial assistance for operating (working) capital for an existing operation to support day-to-day expenses.

(4) Sale or lease of Agency-owned real property, if the sale or lease of Agency-owned real property will have no or minimal construction or change in current operations in the foreseeable future.

(5) The provision of additional financial assistance for cost overruns where the purpose, operation, location, and design of the proposal as originally approved has not been substantially changed.

(6) Rural Business Investment Program (7 U.S.C. 1989 and 2009cc *et seq.*) actions as follows:

(i) Non-leveraged program actions that include licensing by USDA of Rural Business Investment Companies (RBIC); or

(ii) Leveraged program actions that include licensing by USDA of RBIC and Federal financial assistance in the form of technical grants or guarantees of debentures of an RBIC, unless such Federal assistance is used to finance construction or development of land.

(7) A guarantee provided to a guaranteed lender for the sole purpose of refinancing outstanding bonds or notes or a guarantee provided to the Federal Financing Bank pursuant to Section 313A(a) of the Rural Electrification Act of 1936 for the purpose of:

(i) Refinancing existing debt instruments of a lender organized on a not-for-profit basis; or

(ii) Prepaying outstanding notes or bonds made to or guaranteed by the Agency.

(b) *Information gathering and technical assistance.* The following are CEs for financial assistance, identified in paragraphs (b)(1) through (3) of this section.

(1) Information gathering, data analysis, document preparation, real estate appraisals, environmental site assessments, and information dissemination. Examples of these actions are:

(i) Information gathering such as research, literature surveys, inventories, and audits;

(ii) Data analysis such as computer modeling;

(iii) Document preparation such as strategic plans; conceptual designs; management, economic, planning, or feasibility studies; energy audits or assessments; environmental analyses; and survey and analyses of accounts and business practices; and

(iv) Information dissemination such as document mailings, publication, and distribution; and classroom training and informational programs.

(2) Technical advice, training, planning assistance, and capacity building. Examples of these actions are:

(i) Technical advice, training, planning assistance such as guidance for cooperatives and self-help housing group planning; and

(ii) Capacity building such as leadership training, strategic planning, and community development training.

(3) Site characterization, environmental testing, and monitoring where no significant alteration of existing ambient conditions would occur. This includes, but is not limited to, air, surface water, groundwater, wind, soil, or rock core sampling; installation of monitoring wells; and installation of small-scale air, water, or weather monitoring equipment.

(c) *Minor construction proposals.* The following are CEs that apply to financial assistance for minor construction proposals:

(1) Minor amendments or revisions to previously approved projects provided such activities do not alter the purpose, operation, geographic scope, or design of the project as originally approved;

(2) Repair, upgrade, or replacement of equipment in existing structures for such purposes as improving habitability, energy efficiency (including heat rate efficiency), replacement or conversion to enable use of renewable fuels, pollution prevention, or pollution control;

(3) Any internal modification or minimal external modification, restoration, renovation, maintenance, and replacement in-kind to an existing facility or structure;

(4) Construction of or substantial improvement to a single-family dwelling, or a Rural Housing Site Loan project or multi-family housing project serving up to four families and affecting less than 10 acres of land;

(5) Siting, construction, and operation of new or additional water supply wells for residential, farm, or livestock use;

(6) Replacement of existing water and sewer lines within the existing right-of-way and as long as the size of pipe is either no larger than the inner diameter of the existing pipe or is an increased diameter as required by Federal or state requirements. If a larger pipe size is required, applicants must provide a copy of written administrative requirements mandating a minimum pipe diameter from the regulatory agency with jurisdiction;

(7) Modifications of an existing water supply well to restore production in existing commercial well fields, if there would be no drawdown other than in the immediate vicinity of the pumping well, no resulting long-term decline of the water table, and no degradation of the aquifer from the replacement well;

(8) New utility service connections to individual users or construction of utility lines or associated components where the applicant has no control over the placement of the utility facilities; and

(9) Conversion of land in agricultural production to pastureland or forests, or conversion of pastureland to forest.

(d) *Energy or telecommunication proposals.* The following are CEs that apply to financial assistance for energy or telecommunication proposals:

(1) Upgrading or rebuilding existing telecommunication facilities (both wired and wireless) or addition of aerial cables for communication purposes to electric power lines that would not affect the environment beyond the previously-developed, existing rights-of-way;

(2) Burying new facilities for communication purposes in previously developed, existing rights-of-way and in areas already in or committed to urbanized development or rural settlements whether incorporated or unincorporated that are characterized by high human densities and within contiguous, highly disturbed environments with human-built features. Covered actions include associated vaults and pulling and tensioning sites outside rights-of-way in nearby previously disturbed or developed land;

(3) Changes to electric transmission lines that involve pole replacement or structural components only where either the same or substantially equivalent support structures at the approximate existing support structure locations are used;

(4) Phase or voltage conversions, reconductoring, upgrading, or rebuilding of existing electric distribution lines that would not affect the environment beyond the previously developed, existing rights-of-way. Includes pole replacements but does not include overhead-to-underground conversions;

(5) Collocation of telecommunications equipment on existing infrastructure and deployment of distributed antenna systems and small cell networks provided the latter technologies are not attached to and will not cause adverse effects to historic properties;

(6) Siting, construction, and operation of small, ground source heat pump systems that would be located on previously developed land;

(7) Siting, construction, and operation of small solar electric projects or solar thermal projects to be installed on or adjacent to an existing structure and that would not affect the environment beyond the previously developed facility area and are not attached to and will not cause adverse effects to historic properties;

(8) Siting, construction, and operation of small biomass projects, such as animal waste anaerobic digesters or gasifiers, that would use feedstock produced on site (such as a farm where the site has been previously disturbed) and supply gas or electricity for the site's own energy needs with no or only incidental export of energy;

(9) Construction of small standby electric generating facilities with a rating of one average megawatt (MW) or less, and associated facilities, for the purpose of providing emergency power for or startup of an existing facility;

(10) Additions or modifications to electric transmission facilities that would not affect the environment beyond the previously developed facility area including, but not limited to, switchyard rock, grounding upgrades, secondary containment projects, paving projects, seismic upgrading, tower modifications, changing insulators, and replacement of poles, circuit breakers, conductors, transformers, and cross-arms; and

(11) Safety, environmental, or energy efficiency (including heat rate efficiency) improvements within an existing electric generation facility, including addition, replacement, or upgrade of facility components (such as precipitator, baghouse, or scrubber installations), that do not result in a change to the design capacity or function of the facility and do not result in an increase in pollutant emissions, effluent discharges, or waste products.

(e) *Emergency situations.* Repairs made because of an emergency situation to return to service damaged facilities of an applicant's utility system or other actions necessary to preserve life and control the immediate impacts of the emergency.

(f) *Promulgation of rules or formal notices.* The promulgation of rules or formal notices for policies or programs that are administrative or financial procedures for implementing Agency assistance activities.

(g) *Agency proposals for legislation.* Agency proposals for legislation that have no potential for significant environmental impacts because they would allow for no or minimal construction or change in operations.

(h) *Administrative actions.* Agency procurement activities for goods and services; routine facility operations; personnel actions, including but not limited to, reduction in force or employee transfers resulting from workload adjustments, and reduced personnel or funding levels; and other such management actions related to the operation of the Agency.

§ 1970.54　CEs involving small-scale development with an environmental report.

The CEs in this section are for proposals for financial assistance that re-quire an applicant to submit an ER with their application to facilitate Agency determination of extraordinary circumstances. At a minimum, the ER will include a complete description of all components of the applicant's proposal and any connected actions, including its specific location on detailed site plans as well as location maps equivalent to a U.S. Geological Survey (USGS) quadrangle map; and information from authoritative sources acceptable to the Agency confirming the presence or absence of sensitive environmental resources in the area that could be affected by the applicant's proposal. The ER submitted must be accurate, complete, and capable of verification. The Agency may request additional information as needed to make an environmental determination. Failure to submit the required environmental report will postpone further consideration of the applicant's proposal until the ER is submitted, or the Agency may deny the request for financial assistance. The Agency will review the ER and determine if extraordinary circumstances exist. The Agency's review may determine that classification as an EA or an EIS is more appropriate than a CE classification.

(a) *Small-scale site-specific development.* The following CEs apply to proposals where site development activities (including construction, expansion, repair, rehabilitation, or other improvements) for rural development purposes would impact not more than 10 acres of real property and would not cause a substantial increase in traffic. These CEs are identified in paragraphs (a)(1) through (a)(9) of this section. This paragraph does not apply to new industrial proposals (such as ethanol and biodiesel production facilities) or those classes of action listed in §§ 1970.53, 1970.101, or 1970.151.

(1) Multi-family housing and Rural Housing Site Loans.

(2) Business development.

(3) Community facilities such as municipal buildings, libraries, security services, fire protection, schools, and health and recreation facilities.

(4) Infrastructure to support utility systems such as water or wastewater facilities; headquarters, maintenance,

equipment storage, or microwave facilities; and energy management systems. This does not include proposals that either create a new or relocate an existing discharge to or a withdrawal from surface or ground waters, or cause substantial increase in a withdrawal or discharge at an existing site.

(5) Installation of new, commercial-scale water supply wells and associated pipelines or water storage facilities that are required by a regulatory authority or standard engineering practice as a backup to existing production well(s) or as reserve for fire protection.

(6) Construction of telecommunications towers and associated facilities, if the towers and associated facilities are 450 feet or less in height and would not be in or visible from an area of documented scenic value.

(7) Repair, rehabilitation, or restoration of water control, flood control, or water impoundment facilities, such as dams, dikes, levees, detention reservoirs, and drainage ditches, with minimal change in use, size, capacity, purpose, operation, location, or design from the original facility.

(8) Installation or enlargement of irrigation facilities on an applicant's land, including storage reservoirs, diversion dams, wells, pumping plants, canals, pipelines, and sprinklers designed to irrigate less than 80 acres.

(9) Replacement or restoration of irrigation facilities, including storage reservoirs, diversion dams, wells, pumping plants, canals, pipelines, and sprinklers, with no or minimal change in use, size, capacity, or location from the original facility(s).

(10) Vegetative biomass harvesting operations of no more than 15 acres, provided any amount of land involved in harvesting is to be conducted managed on a sustainable basis and according to a Federal, state, or other governmental unit approved management plan.

(b) *Small-scale corridor development.* The following CEs apply to financial assistance for:

(1) Construction or repair of roads, streets, and sidewalks, including related structures such as curbs, gutters, storm drains, and bridges, in an existing right-of-way with minimal change

in use, size, capacity, purpose, or location from the original infrastructure;

(2) Improvement and expansion of existing water, waste water, and gas utility systems:

(i) Within one mile of currently served areas irrespective of the percent of increase in new capacity, or

(ii) Increasing capacity not more than 30 percent of the existing user population;

(3) Replacement of utility lines where road reconstruction undertaken by non-Agency applicants requires the relocation of lines either within or immediately adjacent to the new road easement or right-of-way; and

(4) Installation of new linear telecommunications facilities and related equipment and infrastructure.

(c) *Small-scale energy proposals.* The following CEs apply to financial assistance for:

(1) Construction of electric power substations (including switching stations and support facilities) or modification of existing substations, switchyards, and support facilities;

(2) Construction of electric power lines and associated facilities designed for or capable of operation at a nominal voltage of either:

(i) Less than 69 kilovolts (kV);

(ii) Less than 230 kV if no more than 25 miles of line are involved; or

(iii) 230 kV or greater involving no more than three miles of line, but not for the integration of major new generation resources into a bulk transmission system;

(3) Reconstruction (upgrading or rebuilding) or minor relocation of existing electric transmission lines (230 kV or less) 25 miles in length or less to enhance environmental and land use values or to improve reliability or access. Such actions include relocations to avoid right-of-way encroachments, resolve conflict with property development, accommodate road/highway construction, allow for the construction of facilities such as canals and pipelines, or reduce existing impacts to environmentally sensitive areas;

(4) Repowering or uprating modifications or expansion of an existing unit(s) up to a rating of 50 average MW at electric generating facilities in

order to maintain or improve the efficiency, capacity, or energy output of the facility. Any air emissions from such activities must be within the limits of an existing air permit;

(5) Installation of new generating units or replacement of existing generating units at an existing hydroelectric facility or dam which results in no change in the normal maximum surface area or normal maximum surface elevation of the existing impoundment. All supporting facilities and new related electric transmission lines 10 miles in length or less are included;

(6) Installation of a heat recovery steam generator and steam turbine with a rating of 200 average MW or less on an existing electric generation site for the purpose of combined cycle operations. All supporting facilities and new related electric transmission lines 10 miles in length or less are included;

(7) Construction of small electric generating facilities (except geothermal and solar electric projects), including those fueled with wind or biomass, with a rating of 10 average MW or less. All supporting facilities and new related electric transmission lines 10 miles in length or less are included;

(8) Siting, construction, and operation of small biomass projects (except small electric generating facilities projects fueled with biomass) producing not more than 3 million gallons of liquid fuel or 300,000 million british thermal units annually, developed on up 10 acres of land;

(9) Geothermal electric power projects or geothermal heating or cooling projects developed on up to 10 acres of land and including installation of one geothermal well for the production of geothermal fluids for direct use application (such as space or water heating/cooling) or for power generation. All supporting facilities and new related electric transmission lines 10 miles in length or less are included;

(10) Solar electric projects or solar thermal projects developed on up to 10 acres of land including all supporting facilities and new related electric transmission lines 10 miles in length or less;

(11) Distributed resources of any capacity located at or adjacent to an existing landfill site or wastewater treatment facility that is powered by refuse-derived fuel. All supporting facilities and new related electric transmission lines 10 miles in length or less are included;

(12) Small conduit hydroelectric facilities having a total installed capacity of not more than 5 average MW using an existing conduit such as an irrigation ditch or a pipe into which a turbine would be placed for the purpose of electric generation. All supporting facilities and new related electric transmission lines 10 miles in length or less are included; and

(13) Modifications or enhancements to existing facilities or structures that would not substantially change the footprint or function of the facility or structure and that are undertaken for the purpose of improving energy efficiency (including heat rate efficiency), promoting pollution prevention or control, safety, reliability, or security. This includes, but is not limited to, retrofitting existing facilities to produce biofuels and replacing fossil fuels used to produce heat or power in biorefineries with renewable biomass. This also includes installation of fuel blender pumps and associated changes within an existing fuel facility.

§ 1970.55 CEs for multi-tier actions.

The CEs in this section apply solely to providing financial assistance to primary recipients in multi-tier action programs.

(a) The Agency's approval of financial assistance to a primary recipient in a multi-tier action program is categorically excluded under this section only if the primary recipient agrees in writing to:

(1) Conduct a screening of all proposed uses of funds to determine whether each proposal that would be funded or financed falls within § 1970.53 or § 1970.54 as a categorical exclusion;

(2) Obtain sufficient information to make an evaluation of those proposals listed in § 1970.53 and prepare an ER for proposals under § 1970.54 to determine if extraordinary circumstances (as described in § 1970.52) are present;

(3) Document and maintain its conclusions regarding the applicability of a CE in its official records for Agency verification; and

(4) Refer all proposals that do not meet listed CEs in § 1970.53 or § 1970.54, and proposals that may have extraordinary circumstances (as described in § 1970.52) to the Agency for further review in accordance with this part.

(b) The primary recipient's compliance with this section will be monitored and verified in Agency compliance reviews and other required audits. Failure by a primary recipient to meet the requirements of this section will result in penalties that may include written warnings, withdrawal of Agency financial assistance, suspension from participation in Agency programs, or other appropriate action.

(c) Nothing in this section is intended to delegate the Agency's responsibility for compliance with this part. The Agency will continue to maintain ultimate responsibility for and control over the environmental review process in accordance with this part.

§§ 1970.56–1970.100   [Reserved]

Subpart C—NEPA Environmental Assessments

§ 1970.101   **General.**

(a) An EA is a concise public document used by the Agency to determine whether to issue a FONSI or prepare an EIS, as specified in subpart D of this part. If, at any point during the preparation of an EA, it is determined that the proposal will have a potentially significant impact on the quality of the human environment, an EIS will be prepared.

(b) Unless otherwise determined by the Agency, EAs will be prepared for all "Federal actions" as described in § 1970.8, unless such actions are categorically excluded, as determined under subpart B of this part, or require an EIS, as provided under subpart D of this part;

(c) Preparation of an EA will begin as soon as the Agency has determined the proper classification of the applicant's proposal. Applicants should consult as early as possible with the Agency to determine the environmental review requirements of their proposals. The EA must be prepared concurrently with the early planning and design phase of the proposal. The EA will not be considered complete until it is in compliance with this part.

(d) Failure to achieve compliance with this part will postpone further consideration of the applicant's proposal until such compliance is achieved or the applicant withdraws the application. If compliance is not achieved, the Agency will deny the request for financial assistance.

§ 1970.102   **Preparation of EAs.**

The EA must focus on resources that might be affected and any environmental issues that are of public concern.

(a) The amount of information and level of analysis provided in the EA should be commensurate with the magnitude of the proposal's activities and its potential to affect the quality of the human environment. At a minimum, the EA must discuss the following:

(1) The purpose and need for the proposed action;

(2) The affected environment, including baseline conditions that may be impacted by the proposed action and alternatives;

(3) The environmental impacts of the proposed action including the No Action alternative, and, if a specific project element is likely to adversely affect a resource, at least one alternative to that project element;

(4) Any applicable environmental laws and Executive Orders;

(5) Any required coordination undertaken with any Federal, state, or local agencies or Indian tribes regarding compliance with applicable laws and Executive Orders;

(6) Mitigation measures considered, including those measures that must be adopted to ensure the action will not have significant impacts;

(7) Any documents incorporated by reference, if appropriate, including information provided by the applicant for the proposed action; and

(8) A listing of persons and agencies consulted.

(b) The following describes the normal processing of an EA under this subpart:

(1) The Agency advises the applicant of its responsibilities as described in

subpart A of this part. These responsibilities include preparation of the EA as discussed in § 1970.5(b)(3)(iv)(B).

(2) The applicant provides a detailed project description including connected actions.

(3) The Agency verifies that the applicant's proposal should be the subject of an EA under § 1970.101. In addition, the Agency identifies any unique environmental requirements associated with the applicant's proposal.

(4) The Agency or the applicant, as appropriate, coordinates with Federal, state, and local agencies with jurisdiction by law or special expertise; tribes; and interested parties during EA preparation.

(5) Upon receipt of the EA from the applicant, the Agency evaluates the completeness and accuracy of the documentation. If necessary, the Agency will require the applicant to correct any deficiencies and resubmit the EA prior to its review.

(6) The Agency reviews the EA and supporting documentation to determine whether the environmental review is acceptable.

(i) If the Agency finds the EA unacceptable, the Agency will notify the applicant, as necessary, and work to resolve any outstanding issues.

(ii) If the Agency finds the EA acceptable, the Agency will prepare or review a "Notice of Availability of the EA" and direct the applicant to publish the notice in local newspapers or through other distribution methods as approved by the Agency. The notice must be published for three consecutive issues (including online) in a daily newspaper, or two consecutive weeks in a weekly newspaper. If other distribution methods are approved, the Agency will identify equivalent requirements. The public review and comment period will begin on the day of the first publication date or equivalent if other distribution methods are used. A 14- to 30-day public review and comment period, as determined by the Agency, will be provided for all Agency EAs.

(7) After reviewing and evaluating all public comments, the Agency determines whether to modify the EA, prepare a FONSI, or prepare an EIS that conforms with subpart D of this part.

(8) If the Agency determines that a FONSI is appropriate, and after preparation of the FONSI, the Agency will prepare or review a public notice announcing the availability of the FONSI and direct the applicant to publish the public notice in a newspaper(s) of general circulation, as described in § 1970.14(d)(2). In such case, the applicant must obtain an "affidavit of publication" or other such proof from all publications (or equivalent verification if other media were used) and must submit the affidavits and verifications to the Agency.

§ 1970.103 Supplementing EAs.

If the applicant makes substantial changes to a proposal or if new relevant environmental information is brought to the attention of the Agency after the issuance of an EA or FONSI, supplementing an EA may be necessary before the action has been implemented. Depending on the nature of the changes, the EA will be supplemented by revising the applicable section(s) or by appending the information to address potential impacts not previously considered. If an EA is supplemented, public notification will be required in accordance with § 1970.102(b)(7) and (8).

§ 1970.104 Finding of No Significant Impact.

The Agency may issue a FONSI or a revised FONSI only if the EA or supplemental EA supports the finding that the proposed action will not have a significant effect on the human environment. If the EA does not support a FONSI, the Agency will follow the requirements of subpart D of this part before taking action on the proposal.

(a) A FONSI must include:

(1) A summary of the supporting EA consisting of a brief description of the proposed action, the alternatives considered, and the proposal's impacts;

(2) A notation of any other EAs or EISs that are being or will be prepared and that are related to the EA;

(3) A brief discussion of why there would be no significant impacts;

(4) Any mitigation essential to finding that the impacts of the proposed action would not be significant;

(5) The date issued; and

(6) The signature of the appropriate Agency approval official.

(b) The Agency must ensure that the applicant has committed to any mitigation that is necessary to support a FONSI and possesses the authority and ability to fulfill those commitments. The Agency must ensure that mitigation, and, if appropriate, a mitigation plan that is necessary to support a FONSI, is made a condition of financial assistance.

(c) The Agency must make a FONSI available to the public as provided at 40 CFR 1501.4(e) and 1506.6.

(d) The Agency may revise a FONSI at any time provided that the revision is supported by an EA or a supplemental EA. A revised FONSI is subject to all provisions of this section.

§§ 1970.105–1970.150 [Reserved]

Subpart D—NEPA Environmental Impact Statements

§ 1970.151 General.

(a) The purpose of an EIS is to provide a full and fair discussion of significant environmental impacts and to inform the appropriate Agency decision maker and the public of reasonable alternatives to the applicant's proposal, the Agency's proposed action, and any measures that would avoid or minimize adverse impacts.

(b) Agency actions for which an EIS is required include, but are not limited to:

(1) Proposals for which an EA was initially prepared and that may result in significant impacts that cannot be mitigated;

(2) Siting, construction (or expansion), and decommissioning of major treatment, storage, and disposal facilities for hazardous wastes as designated in 40 CFR part 261;

(3) Proposals that change or convert the land use of an area greater than 640 contiguous acres;

(4) New electric generating facilities, other than gas-fired prime movers (gas-fired turbines and gas engines) or renewable systems (solar, wind, geothermal), with a rating greater than 50 average MW, and all new associated electric transmission facilities;

(5) New mining operations when the applicant has effective control (i.e., applicant's dedicated mine or purchase of a substantial portion of the mining equipment); and

(6) Agency proposals for legislation that may have a significant environmental impact.

(c) Failure to achieve compliance with this part will postpone further consideration of the applicant's proposal until the Agency determines that such compliance has been achieved or the applicant withdraws the application. If compliance is not achieved, the Agency will deny the request for financial assistance.

§ 1970.152 EIS funding and professional services.

(a) *Funding for EISs.* Unless otherwise approved by the Agency, an applicant must fund an EIS and any supplemental documentation prepared in support of an applicant's proposal.

(b) *Acquisition of professional services.* Applicants shall solicit and procure professional services in accordance with and through the third-party contractor methods specified in 40 CFR 1506.5(c), and in compliance with applicable state or local laws or regulations. Applicants and their officers, employees, or agents shall not engage in contract awards or contract administration if there is a conflict of interest or receipt of gratuities, favors or any form of monetary value from contractors, subcontractors, potential contractors or subcontractors, or other parties performing or to perform work on an EIS. To avoid any conflicts of interest, the Agency is responsible for selecting the EIS contractor and the applicant must not initiate any procurement of professional services to prepare an EIS without prior written approval from the Agency. The Agency reserves the right to consider alternate procurement methods.

(c) *EIS scope and content.* The Agency will prepare the scope of work for the preparation of the EIS and will be responsible for the scope, content and development of the EIS prepared by the contractor(s) hired or selected by the Agency.

(d) *Agreement Outlining Party Roles and Responsibilities.* For each EIS, an agreement will be executed by the Agency, the applicant, and each third-party contractor, which describes each party's roles and responsibilities during the EIS process.

(e) *Disclosure statement.* The Agency will ensure that a disclosure statement is executed by each EIS contractor. The disclosure statement will specify that the contractor has no financial or other interest in the outcome of the proposal.

§ 1970.153 Notice of Intent and scoping.

(a) *Notice of Intent.* The Agency will publish a Notice of Intent (NOI) in the FEDERAL REGISTER that an EIS will be prepared and, if public scoping meetings are required, the notice will be published at least 14 days prior to the public scoping meeting(s).

(1) The NOI will include a description of the following: the applicant's proposal and possible alternatives; the Agency's scoping process including plans for possible public scoping meetings with time and locations; background information if available; and contact information for Agency staff who can answer questions regarding the proposal and the EIS.

(2) The applicant must publish a notice similar to the NOI, as directed and approved by the Agency, in one or more newspapers of local circulation, or provide similar information through other distribution methods as approved by the Agency. If public scoping meetings are required, such notices must be published at least 14 days prior to each public scoping meeting.

(b) *Scoping.* In addition to the Agency and applicant responsibilities for public involvement identified in § 1970.14 and as part of early planning for the proposal, the Agency and the applicant must invite affected Federal, state, and local agencies and tribes to inform them of the proposal and identify the permits and approvals that must be obtained and the administrative procedures that must be followed.

(c) *Significant issues.* For each scoping meeting held, the Agency will determine, as soon as practicable after the meeting, the significant issues to be analyzed in depth and identify and eliminate from detailed study the

issues that are not significant, have been covered by prior environmental review, or are not determined to be reasonable alternatives.

§ 1970.154 Preparation of the EIS.

(a) The EIS must be prepared in accordance with the format outlined at 40 CFR 1502.10.

(b) The EIS must be prepared using an interdisciplinary approach that will ensure the integrated use of the natural and social sciences and the environmental design arts. The disciplines of the preparers must be appropriate to address the potential environmental impacts associated with the proposal. This can be accomplished both in the information collection stage and the analysis stage by communication and coordination with environmental experts such as those at universities; local, state, and Federal agencies; and Indian tribes.

(c) The Agency will file the draft and final EIS with the U. S. Environmental Protection Agency's (EPA) Office of Federal Activities.

(d) The Agency will publish in the FEDERAL REGISTER a Notice of Availability announcing that either the draft or final EIS is available for review and comment. The applicant must concurrently publish a similar announcement using one or more distribution methods as approved by the Agency in accordance with § 1970.14.

(e) Minimum public comment time periods are calculated from the date on which EPA's Notice of Availability is published in the FEDERAL REGISTER. The Agency has the discretion to extend any public review and comment period if warranted. Notification of any extensions will occur through the FEDERAL REGISTER and other media outlets.

(f) When comments are received on a draft EIS, the Agency will assess and consider comments both individually and collectively. With support from the third-party contractor and the applicant, the Agency will develop responses to the comments received. Possible responses to public comments include: Modifying the alternatives considered;

negotiating with the applicant to modify or mitigate specific project elements of the original proposal; developing and evaluating alternatives not previously given serious consideration; supplementing or modifying the analysis; making factual corrections; or explaining why the comments do not warrant further response.

(g) If the final EIS requires only minor changes from the draft EIS, the Agency may document and incorporate such minor changes through errata sheets, insertion pages, or revised sections to be incorporated into the draft EIS. In such cases, the Agency will circulate such changes together with comments on the draft EIS, responses to comments, and other appropriate information as the final EIS. The Agency will not circulate the draft EIS again; although, if requested, a copy of the draft EIS may be provided in a timely fashion to any interested party.

§ 1970.155 Supplementing EISs.

(a) A supplement to a draft or final EIS will be announced, prepared, and circulated in the same manner (exclusive of meetings held during the scoping process) as a draft and final EIS (see 7 CFR 1970.154). Supplements to a draft or final EIS will be prepared if:

(1) There are substantial changes in the proposed action that are relevant to environmental concerns; or

(2) Significant new circumstances or information pertaining to the proposal arise which are relevant to environmental concerns and the proposal or its impacts.

(b) The Agency will publish an NOI to prepare a supplement to a draft or final EIS.

(c) The Agency, at its discretion, may issue an information supplement to a final EIS where the Agency determines that the purposes of NEPA are furthered by doing so even though such supplement is not required by 40 CFR 1502.9(c)(1). The Agency and the applicant must concurrently have separate notices of availability published. The notice requirements must be the same as for a final EIS and the information supplement must be circulated in the same manner as a final EIS. The Agency will take no final action on any proposed modification discussed in the information supplement until 30 days after the Agency's notice of availability or the applicant's notice is published, whichever occurs later.

§ 1970.156 Record of Decision.

(a) The ROD is a concise public record of the Agency's decision. The required information and format of the ROD will be consistent with 40 CFR 1505.2.

(b) Once a ROD has been executed by the Agency, the Agency will issue a FEDERAL REGISTER notice indicating its availability to the public.

(c) The ROD may be signed no sooner than 30 days after the publication of EPA's Notice of Availability of the final EIS in the FEDERAL REGISTER.

§§ 1970.157–1970.200 [Reserved]

PART 1980—GENERAL

Subparts A–D [Reserved]

Subpart E—Business and Industrial Loan Program

AUTHORITY: 5 U.S.C. 301; 7 U.S.C. 1989.
Subpart E also issued under 7 U.S.C. 1932(a).

EDITORIAL NOTE: Nomenclature changes to part 1980 appear at 80 FR 9905, Feb. 24, 2015.

Subparts A–D [Reserved]

Subpart E—Business and Industrial Loan Program

SOURCE: 52 FR 6501, Mar. 4, 1987, unless otherwise noted.

§ 1980.401 Introduction.

(a) Direct Business and Industry (B&I) loans are disbursed by the Agency under this subpart. B&I loan guarantees are to be processed and serviced under the provisions of subparts A and B of part 4279 and subpart B of part 4287 of this title. Any processing or servicing activity conducted pursuant to this subpart involving authorized assistance to relatives, or business or close personal associates, is subject to the provisions of part 1900 subpart D of this chapter. Applicants for this assistance are required to identify any known relationship or association with any Agency employee.

(b) The purpose of the B&I program is to improve, develop or finance business, industry and employment and improve the economic and environmental climate in rural communities, including pollution abatement and control. This purpose is achieved through bolstering the existing private credit structure through guarantee of quality loans which will provide lasting community benefits. It is NOT intended that the guarantee authority be used for marginal or substandard loans or to "bail out" lenders having such loans.

(c) This subpart and its appendices (especially appendix I and appendix K) also contain regulations for Drought and Disaster (D&D) and Disaster Assistance for Rural Business Enterprises (DARBE) guaranteed loans authorized by section 331 of the Disaster Assistance Act of 1988 (Pub. L. 100–387) and section 401 of the Disaster Assistance Act of 1989 (Pub. L. 101–82). D&D loans must be to alleviate distress caused to rural business entities, directly or indirectly, by drought, hail, excessive moisture, or related conditions occurring in 1988, or to provide for the guarantee of loans to such rural business entities that refinance or restructure debt as a result of losses incurred, directly or indirectly, because of such natural disasters and are limited to a guarantee of principal only. DARBE loans must be to alleviate distress caused to rural business entities, directly or indirectly, by drought, freeze, storm, excessive moisture, earthquake, or related conditions occurring in 1988 or 1989, or to provide for the guarantee of loans to such rural business entities that refinance or restructure debt as a result of losses incurred, directly or indirectly, because of such natural disasters and within certain parameters guarantee both principal and interest.

(d) The B&I loan program is administered by the Administrator through a State Director serving each State. The State Director is the focal point for the program and the local contact person for processing and servicing activities, although this subpart refers in various places to the duties and responsibilities of other Rural Development employees.

(e) Throughout this subpart there appear Administrative provisions for the State Director, District Director, and County Supervisor. These provisions establish the internal duties, responsibilities and procedures to carry out the requirements of the program. These provisions are identified as "Administrative" and follow appropriate sections of this subpart.

(f) This subpart and its appendices also contains regulations for Business and Industry Disaster (BID) loans under the authority of the Dire Emergency Supplemental Appropriations Act, 1992, Public Law 102–368. This program provides B&I guarantees for loans needed as a result of natural disasters. Some of the requirements of this subpart are waived or altered for BID loans. The waivers and alterations are provided in § 1980.498 of this subpart.

[52 FR 6501, Mar. 4, 1987, as amended at 54 FR 4, Jan. 3, 1989; 54 FR 42483, Oct. 17, 1989; 55 FR 19245, May 8, 1990; 57 FR 45969, Oct. 5, 1992; 58 FR 229, Jan. 5, 1993; 61 FR 67633, Dec. 23, 1996]

§ 1980.402 Definitions.

(a) The following general definitions are applicable to the terms used in this subpart. *Adjusted tangible net worth.* Tangible balance sheet equity plus allowed tangible asset appreciation and subordinated owner debt.

Allowed tangible asset appreciation. The difference between the current net book value recorded on the financial statements (original cost less cumulative depreciation) of real property assets and the lesser of their current market value or original cost, where current market value is determined using an appraisal satisfactory to the Agency.

Area of high unemployment. An area in which a B&I loan guarantee can be issued, consisting of a county or group of contiguous counties or equivalent subdivisions of a State which, on the basis of the most recent 12-month average or the most recent annual average data, has a rate of unemployment 150 percent or more of the national rate. Data used must be those published by the Bureau of Labor Statistics, U.S. Department of Labor.

Biogas. Biomass converted to gaseous fuel.

Biomass. Any organic material that is available on a renewable or recurring basis including agricultural crops, trees grown for energy production, wood waste and wood residues, plants, including aquatic plants and grasses, fibers, animal waste and other waste materials, fats, oils, greases, including recycled fats, oils and greases. It does not include paper that is commonly recycled or unsegregated solid waste.

Borrower. A borrower may be a cooperative organization, corporation, partnership, trust or other legal entity organized and operated on a profit or nonprofit basis; an Indian Tribe on a Federal or State reservation or other Federally recognized tribal group; a municipality, county or other political subdivision of a State; or an individual. Such borrower must be engaged in or proposing to engage in improving, developing or financing business, industry and employment and improving the economic and environmental climate in rural areas, including pollution abatement and control.

Business and Industry Disaster Loans. Business and Industry loans guaranteed under the authority of the Dire Emergency Supplemental Appropriations Act, 1992, Public Law 102–368. These guaranteed loans cover costs arising from the direct consequences of natural disasters such as Hurricanes Andrew and Iniki and Typhoon Omar that occur after August 23, 1992, and receive a Presidential declaration. Also included are the costs to any producer of crops and livestock that are a direct consequence of at least a 40 percent loss to a crop, 25 percent loss to livestock, or damage to building structures from a microburst wind occurrence in calendar year 1992.

Commercially available. Energy projects utilizing technology that has a proven operating history, and for which there is an established industry for the design, installation, and service (including spare parts) of the equipment.

Community facilities. For the purposes of this subpart, community facilities are those facilities designed to aid in the development of private business and industry in rural areas. Such facilities include, but are not limited to, acquisition and site preparation of land for industrial sites (but not for improvements erected thereon), access streets and roads serving the site, parking areas extension or improvement of community transportation systems serving the site and utility extensions all incidental to site preparation. Projects eligible for assistance under Subpart A of Part 1942 of this chapter are not eligible for assistance under this subpart.

Development cost. These costs include, but are not limited to, those for acquisition, planning, construction, repair or enlargement of the proposed facility; purchase of buildings, machinery, equipment, land easements, rights of way; payment of startup operating costs, and interest during the period before the first principal payment becomes due, including interest on interim financing.

Disaster Assistance for Rural Business Enterprises. Guaranteed loans authorized by section 401 of the Disaster Assistance Act of 1989 (Pub. L. 101–82), providing for the guarantee of loans to assist in alleviating distress caused to rural business entities, directly or indirectly, by drought, freeze, storm, excessive moisture, earthquake, or related conditions occurring in 1988 or 1989, and providing for the guarantee of loans to such rural business entities that refinance or restructure debt as a result of losses incurred, directly or indirectly, because of such natural disasters. See this subpart and its appendices, especially Appendix K, containing additional regulations for these loans.

Drought and Disaster Guaranteed Loans. Guaranteed loans authorized by section 331 of the Disaster Assistance Act of 1988 (Pub. L. 100–387), providing for the guarantee of loans to assist in alleviating distress caused to rural business entities, directly or indirectly, by drought, hail, excessive moisture, or related conditions occurring in 1988, and providing for the guarantee of loans to such rural business entities that refinance or restructure debt as a result of losses incurred, directly or indirectly, because of such natural disasters.

Energy projects. Commercially available projects that produce or distribute energy or power and/or projects that produce biomass or biogas fuel.

Farmers Home Administration (FmHA). The former agency of USDA that previously administered the programs of this Agency. Many Instructions and forms of FmHA are still applicable to Agency programs.

Hurricane Andrew. A hurricane that caused damage in southern Florida on August 24, 1992, and in Louisiana on August 26, 1992.

Hurricane Iniki. A hurricane that caused damage in Hawaii on September 11, 1992.

Letter of conditions. Letter issued by Rural Development under Public Law 103–354 to a borrower setting forth the conditions under which Rural Development will make a direct (insured) loan from the Rural Development Insurance Fund.

Loan classification system. The process by which loans are examined and categorized by degree of potential for loss in the event of default.

Microburst wind. A violently descending column of air associated with a thunderstorm which causes straight-line wind damage.

Problem loan. A loan which is not performing according to its original terms and conditions or which is not expected in the future to perform according to those terms and conditions.

Public body. A municipality, political subdivision, public authority, district, or similar organization.

Qualified Intellectual Property. Trademarks, patents or copyrights included on current (within one year) audited balance sheets for which an audit opinion has been received that states the financial reports fairly represent the values therein and the reported value has been arrived at in accordance with GAAP standards for valuing intellectual property. The supporting work papers must be satisfactory to the Administrator.

Refinancing loan. A loan, all of the proceeds of which are applied to extinguish the entire balance of an outstanding debt.

Seasoned loan. A loan which:

(1) Has a remaining principal guaranteed loan balance of two-thirds or less of the original aggregate of all existing B&I guaranteed loans made to that business.

(2) Is in compliance with all loan conditions and B&I regulations.

(3) Has been current on the B&I guaranteed loan(s) payments for 24 consecutive months.

(4) Is secured by collateral which is determined to be adequate to ensure there will be no loss on the B&I guaranteed loan.

State. Any of the 50 States, the Commonwealth of Puerto Rico, the Virgin Islands of the United States, Guam, American Samoa, the Commonwealth of the Northern Mariana Islands, the Republic of Palau, the Federated States of Micronesia, and the Republic of the Marshall Islands.

Subordinated owner debt. Debt owed by the borrower to one or more of the owner(s) that is subordinated to debt owed by the borrower to the Agency or guaranteed by the Agency (aggregate B&I loan exposure) pursuant to a subordination agreement satisfactory to the Agency. The debt must have been issued in exchange for cash loaned to the borrower for the benefit of the borrower's business. The terms of the subordination agreement must provide that repayment will not commence until the earlier of the date all aggregate B&I loan exposure has been repaid or when a period of three consecutive years has passed during which the borrower has met all loan covenants and evidenced operating profit sufficient to commence partial repayment of this subordinated debt after giving effect to the annual debt service requirements of the aggregate B&I loan exposure. The partial repayment schedule in the case of the latter scenario is subject to annual Agency concurrence and may not be more accelerated than the rate of the debt repayment schedule in effect for the Agency's aggregate B&I loan exposure.

Tangible balance sheet equity. Total equity less the value of intangible assets recorded on the financial statements, as determined from balance sheets prepared in accordance with generally accepted accounting principles (GAAP), plus qualified intellectual property.

Typhoon Omar. A typhoon that caused damage in Guam on August 28, 1992.

Working capital. The excess of current assets over current liabilities. It identifies the relatively liquid portion of total enterprise capital which constitutes a margin or buffer for meeting obligations within the ordinary operating cycle of the business.

(b) Accounting terms not otherwise defined in this part shall have the definition ascribed to them under generally accepted accounting principles (GAAP).

[71 FR 33185, June 8, 2006]

§ 1980.403 Citizenship of borrowers.

Loans to individuals will be made or guaranteed only to those who are citizens of the United States or reside in the United States after being legally admitted for permanent residence. At least 51 percent of the outstanding interest in any corporation or organization-type applicant must be owned by those who are either citizens of the United States or reside in the United States after being legally admitted for permanent residence.

§ 1980.404 [Reserved]

§ 1980.405 Rural areas.

The business financed with a B&I loan must be located in a rural area. Loans to borrowers with facilities located in both rural and non-rural areas will be limited to the amount necessary to finance the facility located in the eligible rural area. Cooperatives that are headquartered in a non-rural area may be eligible for a B&I loan if the loan is used for a project or venture that is located in a rural area. Rural areas are any areas other than a city or town that has a population of greater than 50,000 inhabitants; and the urbanized area contiguous and adjacent to such a city or town, as defined by the U.S. Bureau of the Census. For the purpose of this section:

(a) The population figure is obtained from the most recent decennial Census of the United States (decennial Census). If the applicable population figure cannot be obtained from the most recent decennial Census, RD will determine the applicable population figure based on available population data; and

(b) An urbanized area means a densely populated territory as defined in the most recent decennial Census or other Agency-accepted data source if not defined in the most recent decennial Census.

[80 FR 9905, Feb. 24, 2015]

§§ 1980.406–1980.410 [Reserved]

§ 1980.411 Loan purposes.

Loans to borrowers with facilities located in both urban and rural areas will be limited to the amount necessary to finance the facility located in the eligible rural area.

(a) *Private entrepreneurs.* Loans may be for improving, developing or financing business, industry and employment and improving the economic and environmental climate, including pollution and abatement control, of rural areas, and may include but not be limited to:

(1) Business and industrial acquisitions, construction, conversion, enlargement, repair, modernization of development cost.

(2) Purchasing and development of land, easements, rights-of-way, buildings, facilities, leases or materials.

(3) Purchasing of equipment, leasehold improvements machinery or supplies.

(4) Pollution control and abatement including those in connection with farming and ranching operations.

(5) Transportation services incidental to industrial development.

(6) Startup costs and working capital.

(7) The financing of housing development sites located in open country or cities, towns or villages with populations not in excess of those eligible for Rural Development rural housing loans, provided the community demonstrates a need for additional housing to prevent a loss of jobs in the area, or to house families moving to the area as a result of new employment opportunities.

(8) Loans, other than for working capital or debt refinancing, for meat processing facilities and integrated meat and poultry operations. Loans may not be guaranteed for agricultural production as defined in § 1980.412(e); however, applicants who are in the business of processing, marketing or packaging of agricultural products, as well as agricultural production, may be eligible for loan assistance for that portion of the business other than agricultural production provided the agricultural production aspect is separate from the rest of the business; i.e., the production aspects are handled through

separate legal business entities or through maintenance of the accounting system in such a manner as to clearly identify the use of and future accounting of the loan proceeds and operation of the business.

(9) Interest (including interest on interim financing) during the period before the first principal payment becomes due or the facility becomes income producing, whichever occurs first.

(10) Feasibility studies.

(11) *Debt refinancing.* Lenders and Rural Development must provide as part of their loan analysis the reasons for refinancing and the file must be documented accordingly. Refinancing debts may be allowed in connection with viable projects when it is determined by the lender and Rural Development that it is necessary to create new or save existing jobs. Rural Development will consider any lender's exposure as it relates to this item and may adjust the guarantee percentage accordingly. Refinancing in accordance with this paragraph may be insured or guaranteed only when:

(i) It is necessary to spread substantial debt payment over a longer period of time thereby improving the business' net cash flow and working capital position consistent with the useful life of the asset(s) being refinanced, or

(ii) For payment of short-term debt when required in situations customarily financed over long periods of time (e.g., financing the purchase of real estate, machinery, or equipment with short-term debt or cash expenditures, when lenders would not extend reasonable longer terms to the business), or

(iii) It is necessary to place a permanent loan subsequent to an interim loan for financing the construction of the project.

(iv) It does not refinance subordinated owner debt; or

(v) (Except where the amount to be refinanced is owed directly to the Federal government or is Federally guaranteed) the amount to be refinanced by the Agency is a secondary part (less than 50 percent) of the overall loan requested.

(12) Reasonable fees and charges only as specifically listed below and disclosed on Form FD 449-1, "Application for Loan and Guarantee," or on an addendum to the application at the time the request is submitted to Rural Development for processing. Authorized fees include professional fees rendered by professionals generally licensed by individual State or accreditation Associations, such as Engineers, Architects, Lawyers, Accountants, and Appraisers. The amount of the fee will be what is reasonable and customary in the community or region where the project is located. For example, Architects and Engineers customarily charge fees based on a percentage of estimated project costs. Lawyers, Accountants, and Appraisers customarily charge for services on an hourly basis. Any fees for professional or expert services are to be fully documented and justified on the Form RD 449-1 and are subject to Rural Development review and approval before the application is presented to the Rural Development State Loan Review Board for action. The above approved fees and charges may be funded out of loan proceeds.

(13) Rural Development guarantee fee.

(14) Acquisition of membership and/or stocks, bonds, or debentures necessary to obtain a loan from Production Credit Associations, Banks for Cooperatives, Small Business Investment Companies, and other lenders, provided such acquisition is required of all their borrowers. However, a lender which requires membership fees in such organization or the purchase of securities issued by such organization will not use such proceeds to acquire, lease or improve property which does not benefit its members.

(15) Aquaculture including conservation, development and utilization of water for aquaculture. Aquaculture means the culture or husbandry of aquatic animals or plants by private industry for commercial purposes including the culture and growing of fish by private industry for the purpose of granting or augmenting publicly-owned and regulated stock of fish.

(16) *Energy projects.* Commercially available energy projects that produce biomass fuel or biogas as an output must have completed two operating cycles at design performance levels submitted to the Agency. Projects that

produce steam or electricity as an output must have met or exceeded acceptance test performance criteria submitted to the Agency and be successfully interconnected with the purchaser of the output. Performance or acceptance test requirements for all other energy projects will be determined by the Agency on a case by case basis. Financing for energy projects will only be allowed when the facility has been constructed according to plans and specifications and is producing at the quality and quantity projected in the application.

(b) *Public bodies.* See §§ 1980.481 and 1980.488.

[52 FR 6501, Mar. 4, 1987, as amended at 53 FR 45258, Nov. 9, 1988; 54 FR 28022, July 5, 1989; 71 FR 33187, June 8, 2006]

§ 1980.412   Ineligible loan purposes.

Loans may *not* be made or guaranteed if the funds are used:

(a) To pay off a creditor in excess of the value of the collateral.

(b) For distribution or payment to the owner, partners, shareholders or beneficiaries of the applicant or members of their families when such persons will retain any portion of their equity in the business.

(c) For projects in which such assistance exceeds $1 million and when direct employment increases more than 50 employees which is calculated to or is likely to result in the transfer from one area to another of any employment or business activity provided by the operations of the applicant. This limitation will not prohibit assistance for the expansion of an existing business entity through the establishment of a new branch, affiliate or subsidiary of such entity if the expansion will not result in an increase in the unemployment in the area of original location or in any other area where such entity conducts business operations unless there is reason to believe that such explanation is being established with the intention of closing down the operations of the existing business entity in the area of its original location or in any other area where it conducts such operations.

(d) For projects in which such assistance exceeds $1 million and when direct employment increased more than 50 employees which is calculated to or

likely to result in an increase in the production of goods, materials or commodities, or the availability of services or facilities in the area when there is not sufficient demand for such goods, materials, commodities, services or facilities to employ the efficient capacity of existing competitive commercial or industrial enterprises, unless such financial or other assistance will not have an adverse effect upon existing competitive enterprises in the area.

(e) For agricultural production which means the cultivation, production (growing), and harvesting, either directly or through integrated operations, of agricultural products (crops, animals, birds, and marine life, either for fiber or food for human consumption), and disposal or marketing thereof, the raising, housing, feeding (including commercial custom feedlots), breeding, hatching, control, and/or management of farm and domestic animals. Exceptions to this definition are:

(1) Aquaculture as identified under eligible purposes.

(2) Commercial nurseries primarily engaged in the production of ornamental plants and trees and other nursery products such as bulbs, florists' greens, flowers, shrubbery, flower and vegetable seeds, sod, and the growing of vegetables from seed to the transplant stage.

(3) Forestry which includes establishments primarily engaged in the operation of timber tracts, tree farms, forest nurseries, and related activities such as reforestation.

(4) Loans for livestock and poultry processing as identified under eligible purposes.

(5) The growing of mushrooms or hydroponics.

(f) For the transfer of ownership of a business unless the loan will keep the business from closing, or prevent the loss of employment opportunities in the area, or provide expanded job opportunities.

(g) For financing community antenna television services or facilities.

(h) Charitable and educational institutions, churches, organizations affiliated with or sponsored by churches, and fraternal organizations.

(i) For lending and investment institutions and insurance companies.

(j) For assistance to government employees and military personnel who are directors, officers or have a major ownership of 20 percent or more in the business.

(k) For any legitimate business activity when more than 10 percent of the annual gross revenue is derived from legalized gambling activity.

(l) For any illegal business activity.

(m) For hotels, motels, tourist homes, or convention centers.

(n) For any tourist, recreation or amusement facility.

(o) For any line of credit.

Administrative

Par (c) and (d). The State Director will review the criteria in §1980.412(c) and (d) and make a written determination with supporting data and reasons as to the determinations. Such review must be independent of the Department of Labor certification. The State Director will make sure the loan file contains these determinations as part of the loan analysis prior to the issuance of the Conditional Commitment for Guarantee.

[52 FR 6501, Mar. 4, 1987, as amended at 53 FR 45258, Nov. 9, 1988]

§1980.413 Transactions which will not be guaranteed.

(a) The following transactions will not be guaranteed by the Agency:

(1) The guarantee of lease payments.

(2) The guarantee of loans made by other Federal agencies. This does not preclude the guaranteeing of loans made by the Bank for Cooperatives, Federal Land Bank, or Production Credit Association.

(3) The guarantee or making of any B&I loans(s), to any one borrower, when the total amount of the B&I loans(s) requested plus the outstanding balance of any existing B&I loan(s) is in excess of $10 million.

(b) Guaranteeing of loans involved in tax-exempt obligations under §1980.23 of subpart A of this part.

Administrative

The State Director will consider the overall State allocations of funding authority in recommending loans for processing. Loan requests which fall within Small Business Administration (SBA) authority should continue to be referred to SBA. If the State Director decides to process SBA size loans, the loan file must be fully documented as to the reasons for such actions.

[52 FR 6501, Mar. 4, 1987, as amended at 53 FR 40401, Oct. 17, 1988]

§1980.414 Fees and charges by lender and others.

[See Subpart A, §1980.22]

(a) All fees and charges must be specifically documented and justified on the Form RD 449-1 or on an addendum to the application at the time the loan request is submitted to Rural Development for processing. Allowable fees will be those reasonably and customarily charged borrowers in similar circumstances in the ordinary course of business and are subject to Rural Development review and approval.

(b) Packaging fees include services rendered by the lender or others in connection with preparation of the application and seeing the project through to final decision. These services may or may not be performed by an investment banker. If an investment banker provides needed assistance in addition to the packaging of the loan, additional charges may be added to the packaging fee. The maximum allowable packaging fees are 2 percent of the total principal amount of the loan up to $1 million and on all amounts over $1 million, an additional one-fourth percent up to total maximum fee of $50,000. Packaging fees, investment banker fees and other fees and charges not specifically provided for in this section are permitted subject to Rural Development review and approval. Loan proceeds may be used to pay fees as specifically authorized under §§1980.411(a)(12) and (13). Packaging fees, investment banker fees, and any other fees or charges shall not be paid from loan proceeds.

[52 FR 6501, Mar. 4, 1987, as amended at 53 FR 45258, Nov. 9, 1988]

§§1980.415–1980.418 [Reserved]

§1980.419 Eligible lenders.

[See Subpart A, §1980.13.]

Administrative

A. *Par (a) of subpart A, §1980.13* requires National Office approval for any variations.

B. *Par (b)(4) of subpart A, §1980.13,* State Director submits information to National Office with recommendations.

C. With prior written approval of the Rural Development National Office, a new eligible lender may be substituted for the original lender provided the new lender agrees to assume all original loan requirements including liabilities, servicing responsibilities and acquiring legal title to the unguaranteed portion of the loan. Such approval will be granted by the National Office only when a lender discontinues lending operations or other extreme situations require a substitution of lender. If approved by the National Office, the State Director will submit to the Finance Office Form RD 1980–42. "Notice of Substitution of Lender."

§ 1980.420 Loan guarantee limits.

The percentage of guarantee, up to the maximum allowed by this section, is a matter of negotiation between the lender and Rural Development.

(a) For loans of $2 million or less, the maximum percentage of guarantee is 90 percent.

(b) For loans over $2 million but not over $5 million, the maximum percentage of guarantee is 80 percent.

(c) For loans in excess of $5 million, the maximum percentage of guarantee is 70 percent.

(d) Lenders and borrowers will propose the percentage of guarantee. Rural Development informs lenders and borrowers in writing on Form RD 449–14 of any percentage of guarantee less than proposed by the lender and borrower, and the reasons therefore. Rural Development determines the percentage of guarantee after considering all credit factors involved, including but not limited to:

(1) *Borrower's management.* The borrower's management, and when appropriate, equity capital, history of operation, marketing plan, raw material requirements, and availability of necessary supporting utilities and services;

(2) *Collateral.* Collateral for the loan;

(3) *Financial condition.* Financial condition of borrower or borrower's principals, if appropriate;

(4) *Lender's exposure.* The lender's exposure before and after the loan, and any applicable limits on the lender's lending authority; and

(5) *Trends and conditions.* Current trends and economic conditions.

§§ 1980.421–1980.422 [Reserved]

§ 1980.423 Interest rates.

(a) *Guaranteed loans.* Rates will be negotiated between the lender and the borrower. They may be either fixed or variable as long as they are legal. Interest rates will be those rates customarily charged borrowers in similar circumstances in the ordinary course of business and are subject to Rural Development review and approval. Should any part of the loan(s) be sold by the lender, Rural Development, in its analysis, will take into consideration in approving the lender's interest rate, the rate at which guaranteed loans are being sold or traded in the secondary market.

(1) A variable interest rate must be a rate that is tied to a base rate published periodically in a recognized national or regional financial publication specifically agreed to by the lender and borrower. The variable interest rate may be adjusted at different intervals during the term of the loan but the adjustments may not be more often than quarterly. The intervals between interest rate adjustments will be specified in the Loan Agreement. The lender must incorporate within the variable rate promissory note at loan closing, the provision for adjustment of payment installments coincident with an interest rate adjustment. This will assure that the outstanding principal balance is properly amortized within the prescribed loan maturity to eliminate the possibility of a balloon payment at the end of the loan.

(2) Under a Memorandum of Understanding between Rural Development and the Farm Credit Administration dated September 25, 1974, the interest rate on loans made by the Bank for Cooperatives, Federal Land Banks and Production Credit Associations may be a variable rate based on their administrative and borrowing costs.

(3) Any change in the interest rate between the date of issuance of the Form RD conditional Commitment For Guarantee," and before the issuance of the Loan Note Guarantee must be approved by the State Director. Approval of such change will be shown on an amendment to Form RD 449–14.

(4) It is permissible to have one interest rate on the guaranteed portion of the loan and another interest rate on the unguaranteed portion of the loan, provided the lender and borrower agree and:

(i) The rate on the unguaranteed portion does not exceed that currently being charged on loans of similar size and purpose for borrowers under similar circumstances.

(ii) The rate on the guaranteed portion of the loan will not exceed the rate on the unguaranteed portion.

(5) When multi-rates are used, the lender will provide Rural Development with the overall effective interest rate for the entire loan.

(6) The borrower, lender and holder (if any) may collectively effect a permanent reduction in the interest rate of their B&I guaranteed loan at any time during the life of the loan upon written agreement by these parties. Rural Development must be notified by the lender, in writing, within 10 calendar days of the change. If the guaranteed portion has been repurchased by Rural Development, then Rural Development is a holder and must affirm or reject interest rate change proposals. When Rural Development is a holder, it will concur in such interest rate change only when it is demonstrated to Rural Development that the change is a more viable alternative than initiating or proceeding with liquidation of the loan or continuing with the loan in its present state and that the Government's financial interests are not adversely affected. Factors which will be considered in making such determination will include whether the proposed interest rate will be below the Government's cost of borrowing money, whether continuing with the loan would realistically promote or enhance rural development and employment in rural areas, whether the monetary recovery would be increased by proceeding immediately to liquidation, if applicable, or allowing the borrower to continue at a reduced interest rate, and whether an in-depth financial analysis by the lender reasonably indicates that the business would be successful at a lower interest rate and reasonably indicates that the borrower could make the reduced payment and pay off

amounts in arrears, if any. The Rural Development will reflect the documentation of the interest rate change decision.

(i) Fixed rates cannot be changed to variable rates to reduce the interest rate to the borrower unless the variable rate has a ceiling which is less than the original fixed rate.

(ii) Variable rates can be changed to reduced fixed rates. In a final loss settlement, when qualifying rate changes were made with the required written agreements and notification, the interest will be calculated for the periods the given rates were in effect, except that interest claimed on a loan which originated at a variable rate can never exceed the amount which would have been eligible for claim had the variable interest remained in force. The lesser cost to the Government will always prevail. The lender must maintain records which adequately document the accrued interest claimed.

(iii) The lender is responsible for the legal documentation of interest changes by an allonge attached to the promissory note(s) or any other legally effective amendment of the rate(s); however, no new note(s) may be issued.

(7) No increases in interest rates will be permitted under the B&I loan guarantee except the normal fluctuations in approved variable interest rate loans.

(b) *Insured loans.* (1) Loans for other than those in paragraph (b)(2) of this section will bear interest at a rate prescribed by Rural Development, and will be announced periodically. The interest rate for insured loans will be the rate in effect at the time the loan is approved or at the time the loan is closed, whichever rate is lower.

(2) Loans to public bodies, nonprofit associations and Indian Tribes used to finance community facilities will bear interest at the rate prescribed in RD Instruction 440.1, Exhibit B (available in any Rural Development Office).

Administrative

Par (a)(6) and (a)(7). (Added 4–26–85, SPE-CIAL PN.) The Director will notify the Finance Office of any interest rate reduction by using Form RD 1980–47, "Guaranteed Loan Borrower Adjustments." The State Director will make corrections to the Rural Community Facility Tracking System (FCFTS) reflecting the interest rate change. The Rural Development loan file, as well as the attachments to the copy of the promissory note in the file, will be documented by the State Director to reflect any change in the interest rate.

[52 FR 6501, Mar. 4, 1987, as amended at 54 FR 28022, July 5, 1989]

§ 1980.424   Term of loan repayment.

(a) Principal and interest on the loan will be due and payable as provided in the promissory note except, any interest accrued as the result of the borrower's default on the guaranteed loan(s) over and above that which would have accrued at the normal note rate on the guaranteed loan(s) will not be guaranteed by Rural Development. The lender will structure repayments as established in the loan agreement between the lender and borrower. Ordinarily, such installments will be scheduled for payment as agreed upon by the lender and applicant but on terms that reasonably assure repayment of the loan. However, the first installment to include a repayment of principal may be scheduled for payment after the project is operable and has begun to generate income, but such installment will be due and payable within three years from the date of the promissory note and at least annually thereafter. Interest will be due at least annually from the date of the note. Ordinarily, monthly payments will be expected, except for seasonal-type businesses.

(b) The maximum time allowable for final maturity for an Rural Development guaranteed B&I loan will be limited to thirty (30) years for land, buildings and permanent fixtures; the usable life of the machinery and equipment purchased with loan funds, but not to exceed fifteen (15) years; and seven (7) years for the working capital portion of the loan. The term for a loan that is being refinanced may be based on the collateral the lender will take to secure the loan.

(c) The maximum time allowable for final maturity of an Rural Development insured loan for community facilities will not exceed forty (40) years.

(d) Rural Development will not guarantee any loan in which the promissory note or any other document provides for the payment of interest upon interest.

Administrative

It is permissible for lenders to structure the borrower's financial proposal under the multi-note option as provided for in paragraph III A.2. of Form RD 449–35, "Lender's Agreement," in the following ways:

A. To treat the entire financial package of the borrower as one loan (i.e., loan purposes may include one or any combination of working capital, machinery and equipment or real estate) provided:

1. The loan is amortized to provide repayment of the working capital portion within the 7 years, the machinery and equipment portion within useful life or 15 years, whichever is less, and real estate portion within 30 years.

2. One note represents the unguaranteed portion of the loan. It is permissible to issue as many as 10 notes or the guaranteed portion of the loan.

3. A Form RD 449–34, "Loan Note Guarantee," is attached to all notes, including the unguaranteed note.

4. One interest rate (either variable or fixed) is used for the entire loan or one interest rate is used on the guaranteed portion and a different interest rate is used on the unguaranteed portion, subject to the requirements and conditions found in § 1980.423 of this subpart.

5. One of each of the following Forms: RD 449–14, RD 1940–3, "Request for Obligation of Funds—Guaranteed Loans," RD 449–35, and RD 1980–19, "Guaranteed Loan Closing Report," is used.

B. To treat the financial package of the borrower as separate loans that are processed as a single application provided:

1. A separate loan is made for each purpose (i.e., working capital, machinery and equipment or real estate). As an example, a working capital loan could be structured as follows:

One note for $XXXX at X% interest due in 7 years representing the unguaranteed portion of the loan, and

Up to 10 notes for $XXXX at X% interest due in 7 years representing the guaranteed portions of the loan.

2. A Form RD 449–34 is attached to all notes, including the unguaranteed note.

3. A different interest rate may be used on the guaranteed and unguaranteed portions of the loan, subject to the requirements and conditions found in § 1980.423 of this subpart.

4. Separate Forms RD 449–14, 1940–3, 449–35, and 1980–19 are required for each loan. If you have two loans, one for working capital and another for real estate, then a set of these forms will be required for each loan.

C. Form RD 449–36, "Assignment Guarantee Agreement," will never be used when the multi-note option is utilized.

D. Par. (b). The State Director will assure that the loan officer reviewing the application fully evaluates the useful life of the collateral offered for the loan when determining maturities for the loan. Loan requests for the maximum maturities could result in collateral obsolescence prior to full repayment of the indebtedness. The loan file must be documented to support the maturity granted for the loan.

[52 FR 6501, Mar. 4, 1987, as amended at 56 FR 8271, Feb. 28, 1991]

§ 1980.425 Availability of credit from other sources.

(a) Inability to obtain credit elsewhere is not a requirement for guaranteed assistance under this subpart.

(b) To be eligible for an insured loan under this subpart, the borrower must be unable to obtain the required credit from private or cooperative sources at reasonable rates and terms, taking into consideration prevailing private and cooperative rates and terms in the community in or near the borrower's location(s) for loans for similar purposes and period of time. The borrower's inability to obtain such credit elsewhere will be determined in accordance with subpart A of part 1942 of this chapter.

§§ 1980.426–1980.431 [Reserved]

§ 1980.432 Environmental review requirements.

[See subpart A, § 1980.40 and 7 CFR part 1970.] *Administrative*

Loans made under this part must be in compliance with the environmental review requirements in accordance with 7 CFR part 1970.

[81 FR 11047, Mar. 3, 2016]

§ 1980.433 Flood or mudslide hazard area precautions.

(See subpart A, § 1980.42.)

Administrative

The State Director is responsible for determining if a project is located in a special flood or mudslide hazard area. Refer to subpart B of part 1806 of this chapter [RD Instruction 426.2].

§ 1980.434 Equal opportunity and nondiscrimination requirements.

(See subpart A § 1980.41.)

Administrative

The State Director will assure that equal opportunity and nondiscrimination requirements are met. If there is indication of noncompliance with these requirements, such facts will be reported by the Compliance Reviewing Officer or Rural Development Official in writing to the Administrator, ATTN: Equal Opportunity Officer.

§§ 1980.435–1980.440 [Reserved]

§ 1980.441 Borrower equity requirements.

(a) A minimum of 10 percent tangible balance sheet equity will be required for existing businesses at loan closing. A minimum of 20 percent tangible balance sheet equity will be required for new businesses at loan closing. For energy projects, the minimum tangible balance sheet equity requirement range will be between 25 percent and 40 percent. Criteria for considering the minimum equity required for an individual application will be based on: existing businesses with successful financial and management history vs. start-up businesses; personal/corporate guarantees offered; contractual relationships with suppliers and buyers; credit rating; and strength of the business plan/feasibility study. Where the application is a request to refinance outstanding Federal direct or guaranteed loans, without any new financing, the equity requirement may be determined using adjusted tangible net worth. An application that combines a refinancing loan or guarantee request with a new loan or guarantee request is subject to the standard, unadjusted, equity requirement except as provided in paragraphs (a)(1) or (a)(2) of this section. Increases or decreases in the equity requirements may be imposed or granted as follows:

(1) A reduction in the equity requirement for existing businesses may be permitted by the Administrator. In order for a reduction to be considered, the borrower must furnish the following:

(i) Collateralized personal and corporate guarantees, including any parent, subsidiary, or affiliated company, when feasible and legally permissible, and

(ii) Pro forma and historical financial statements that indicate the business to be financed meets or exceeds the median quartile (as identified in the Risk Management Association's Annual Statement Studies or similar publication) for the current ratio, quick ratio, debt-to-worth ratio, debt coverage ratio, and working capital.

(2) The approval official may require more than the minimum equity requirements provided in this paragraph if the official makes a written determination that special circumstances necessitate this course of action.

(b) The equity requirement must be met in the form of either cash or tangible earning assets contributed to the business and reflected on the balance sheet.

(c) The equity requirement must be determined using balance sheets prepared in accordance with GAAP and met upon giving effect to the entirety of the loan in the calculation, whether or not the loan itself is fully advanced, as of the date the loan is closed; a certification to this effect is required of all guaranteed lenders.

(d) The modified formula for determining whether the equity requirement is met, "adjusted tangible net worth," may be used only in cases where the guarantee requested is for a loan, the proceeds of which are to be used entirely to refinance a debt owed to the Federal government or Federally guaranteed debt. In all other situations, the equity requirement must be determined using tangible net worth.

[71 FR 33187, June 8, 2006]

§ 1980.442　Feasibility studies.

A feasibility study by a recognized independent consultant will be required for all loans, except as provided in this paragraph. The cost of the study will be borne by the borrower and may be paid from funds included in the loan. The loan approval official may make an exception to the requirement of a feasibility study for loans to existing businesses when the financial history of the business, the current financial condition of the business, and guarantees or other collateral offered for the loan are sufficient to protect the inter-

est of the lenders and Rural Development. Rural Development will thoroughly document the justification for the exception to the feasibility study for such businesses. An acceptable feasibility study should include but not be limited to:

(a) *Economic feasibility.* Information related to the project site, availability of trained or trainable labor; utilities; rail, air and road service to the site; and the overall economic impact of the project.

(b) *Market feasibility.* Information on the sales organization and management, nature and extent of market area, marketing plans for sale of projected output, extent of competition and commitments from customers or brokers.

(c) *Technical feasibility.* Technical feasibility reports shall be prepared by individuals who have previous experience in the design and analysis of similar facilities and/or processes as are proposed in the application. The technical feasibility reports shall address the suitability of the selected site for the intended use, including an environmental impact analysis. The report shall be based upon verifiable data and contain sufficient information and analysis so that a determination may be made on the technical feasibility of achieving the levels of income and/or production that are projected in the financial statements. The report shall also identify any constraints or limitations in these financial projections and any other facility or design related factors which might affect the success of the enterprise. The report shall also identify and estimate project operating and development costs and specify the level of accuracy of these estimates and the assumptions on which these estimates have been based. For the purpose of the technical feasibility reports, the project engineer or architect may be considered an independent party provided the principals of the firm or any individual of the firm who participates in the technical feasibility report does not have a financial interest in the project, and provided further that no other individual or firm with the expertise necessary to make such a determination is reasonably available to perform the function.

(d) *Financial feasibility.* An opinion on the reliability of the financial projections and the ability of the business to achieve the projected income and cash flow. An assessment of the cost accounting system, the availability of short-term credit for seasonal business and the adequacy of raw material and supplies.

(e) *Management feasibility.* Evidence that continuity and adequacy of management has been evaluated and documented as being satisfactory.

Administrative

Rural Development loan approval officials will be selective in approving borrowers for new business ventures involved in unproven products, services, or markets. Should such businesses be considered, additional equity will usually be required.

[52 FR 6501, Mar. 4, 1987, as amended at 58 FR 40039, July 27, 1993]

§1980.443 **Collateral, personal and corporate guarantees and other requirements.**

(a) *Collateral.* (1) The lender is responsible for seeing that proper and adequate collateral is obtained and maintained in existence and of record to protect the interest of the lender, the holder, and Rural Development.

(2) Collateral must be of such a nature that repayment of the loan is reasonably assured when considered with the integrity and ability of project management, soundness of the project, and applicant's prospective earnings. Collateral may include, but is not limited to the following: Land, buildings, machinery, equipment, furniture, fixtures, inventory, accounts receivable, cash or special cash collateral accounts, marketable securities and cash surrender value of life insurance. Collateral may also include assignments of leases or leasehold interest, revenues, patents, and copyrights.

(3) All collateral must secure the entire loan. The lender will not take separate collateral to secure only that portion of the loan or loss not covered by the guarantee. The lender will not require compensating balances or certificates of deposit as a means of eliminating the lender's exposure on the unguaranteed portion of the loan. However, compensating balances as used in the ordinary course of business may be used.

(4) Release of collateral of a going concern is based on a complete analysis of the proposal.

(i) Release of collateral prior to payment-in-full of the Rural Development guaranteed debt must be requested by the lender and concurred with by the State Director as prescribed in §1980.469 Administrative D.2 of this subpart subject to the following conditions:

(A) Collateral taken initially or subsequently may not be released prior to the payoff, in full, of the loan balance without adequate consideration for the value of that collateral. Adequate consideration may include, but is not limited to:

(*1*) Application of the net proceeds from the sale of the collateral to the note in inverse order of maturity. All or part of the total proceeds, if approved by the Administrator, may be applied to the payment of current or delinquent principal and interest on the note; or

(*2*) Use of the net proceeds from the sale of collateral to purchase collateral of equal or greater value for which the lender will obtain a first lien position; or

(*3*) Application of net proceeds from the sale of collateral to the borrower's business operations in such a manner that enhancement of the borrower's debt service ability can be clearly demonstrated; for example, the payoff or reamortization of the loan as the result of a large extra payment which reduces subsequent installments on the loan; or

(*4*) Assurance to Rural Development that the release of collateral will contribute to the project's success thereby furthering the goals of the B&I program to show why the release of collateral will contribute to the success of the borrower and repayment of the loan; and

(B) Rural Development must not be adversely affected by the release of collateral; and

(C) If the release of collateral does not involve a reduction of the Rural Development guaranteed debt equal to the net proceeds of the disposition of the collateral, then it must be determined that the remaining collateral is sufficient to provide for the recovery of the Rural Development guaranteed loan(s).

(ii) Sale of collateral of a going concern to the borrower, borrower's stockholder(s) or officer(s), the lender or lender's stockholder(s) or officer(s) must be based on an arm's-length transaction with the concurrence of Rural Development.

(b) *Personal and corporate guarantees.* (1) Unconditional personal/corporate guarantees (i.e., absolute guarantees of full and punctual payment and performance by the borrower) from owners or major stockholders as determined by Rural Development and all partners of partnerships (except for limited partnerships) unless restricted by law *will* be required unless exempted as provided for in paragraph (b)(2) of this section. Guarantees of parent, subsidiaries, or affiliated companies and/or secured guarantees may also be required. Rural Development is not a co-guarantor with the personal or corporate guarantors. The personal and corporate guarantees are part of the collateral for the loan.

(2) An exception to the requirement for personal or corporate guarantees may be made by Rural Development when requested by the lender and if:

(i) The borrower has a satisfactory and current (not over 90 days old) credit report, proven management, evidence of the market necessary to support projections, profitable historical performance of no less than 3 years, abundant collateral to protect the lender and Rural Development, sufficient cash flow to service its debts and meets key industry standards such as those of Robert Morris Associates, Dunn and Bradstreet or the like; or

(ii) The borrower's stock is widely enough held so that no one individual can exercise control. Examples of control would include but are not limited to: Holding sufficient proxies and maintaining sufficient family or special interest voting blocks; or

(iii) A borrower which has a parent, subsidiary, or affiliate which is legally restricted from guaranteeing, or if the guarantee would conflict with existing contractual obligations. Examples of existing contractual obligations include but are not limited to restric-

tions in loan agreements or in credit lines which may preclude guaranteeing.

(3) No guarantees are required from any partners in a limited partnership.

(4) As a general rule, stockholders of publicly traded corporations will not be required to guarantee. However, such guarantees can be required from some of the stockholders where such guarantees are determined necessary to adequately protect the interest of the Government.

(5) If the guarantee would conflict with existing contractual restrictions, the Administrator will have the authority to grant exceptions to the above restrictions upon a finding by the Administrator that such a guarantee is not necessary to adequately protect the Government's interest. Relief would only be granted as to contractual restrictions existing at the time the lender filed an application with Rural Development.

(6) Unsecured personal guarantees, while collateral, will not be considered for purposes of adequacy of security. Personal guarantees will be secured by collateral when business collateral offered is determined by Rural Development to be insufficient or when the borrower's credit does not meet the program's normal requirements or anytime the lender deems such security should be taken.

(7) Guarantors of borrowers will:

(i) In the case of personal guarantees, provide current financial statements (not over 60 days old at time of filing), signed by the guarantors, which make a clear disclosure of community or homestead property.

(ii) in the case of corporate guarantees, provide current financial statements (not over 90 days old at time of filing), certified by an officer of the corporation.

(iii) When applicable, provide written evidence to Rural Development of their inability to provide a guarantee because of existing contractual arrangements or legal restrictions.

(c) *Other requirements.* (1) The lender will ascertain that no claim or liens of laborers, material men, contractors, subcontractors, suppliers of machinery and equipment or other parties are against the collateral of the borrower, and that no suits are pending or threatened that would adversely affect the collateral of the borrower when the security instruments are filed.

(2) Hazard insurance with a standard mortgage clause naming the lender as beneficiary will be required on every loan in an amount that is at least the lesser of the depreciated replacement value of the property being insured or the amount of the loan. Hazard insurance includes fire, windstorm, lightning, hail, business interruption, explosion, riot, civil commotion, aircraft, vehicle, marine, smoke, builder's risk, public liability, property damage, flood or mudslide or any other hazard insurance that may be required to protect the collateral.

(3) Ordinarily, life insurance, which may be decreasing term insurance, is required for the principals and key employees of the borrower and will be assigned or pledged to the lender. A schedule of life insurance available for the benefit of the loan will be included as part of the application.

(4) Workman's compensation insurance is required in accordance with State law.

Administrative

A. *Par (a)(2).* Rural Development's credit analysis of collateral will consist of the following:

1. Little or no value will be assigned to unsecured personal or corporate guarantees.

2. A maximum of 80 percent of current market value will be given to real estate. Special purpose real estate should be assigned less value.

3. Rural Development at its option may permit a maximum of 60 percent of book value to be assigned to acceptable accounts receivable; however, all accounts over 90 days past due, contra accounts, affiliated accounts and other accounts deemed, by the Rural Development official, not to be collateral will be omitted. Calculations to determine the percentage to be applied in the analysis are to be based on the realizable value of the accounts receivable taken from a current aging of accounts receivable from the borrower's most recent financial statement.

4. A maximum of 60 percent of book value will be assigned to inventory.

5. Collateral value assigned to machinery and equipment, furniture and fixtures will be based upon its marketability, mobility, useful life and alternative uses, if any.

B. *Par (b).* The State Director will assure that the collateral values and personal and corporate guarantees are fully reviewed, analyzed and the loan file is documented as to the facts and reasons for decisions reached.

§1980.444 **Appraisal of property serving as collateral.**

(a) Appraisal reports prepared by independent qualified fee appraisers will be required on all property that will serve as collateral. In the case of loans two million dollars or less, the State Director may modify this requirement by permitting the appraisal to be made by a qualified appraiser on the lender's staff with experience appraising the type of collateral involved. The appraisers will give their opinion regarding the current market value of the collateral and the purpose for which the appraisal will be used. The lender will be responsible for assuring that appropriate appraisals are made.

(b) The lender will be responsible for determining that appraisers have the necessary qualifications and experience to make the appraisals. The lender will consult with Rural Development for its recommendations before having the appraisal made.

(c) The lender will determine that the fees or charges of appraisers are reasonable.

(d) Independent appraisals will be made in accordance with the accepted format of the industry and those prepared by the lender in accordance with its policy and procedures. All appraisals will become part of the application. (See §1980.541(i)(6) of this subpart.)

(e) If a subsequent loan request is made within 3 years from the date of the most recent borrower's appraisal report, and there is no significant change in collateral, then the Rural Development State Director in his/her discretion, and if the lender agrees, may use the existing appraisal report in lieu of having a new appraisal prepared.

[52 FR 6501, Mar. 4, 1987, as amended at 53 FR 40401, Oct. 17, 1988]

§1980.445 **Periodic financial statements and audits.**

All borrowers will be required to submit periodic financial statements to the lender. Lenders must forward copies of the financial statements and the lender's analysis of the statements to the Agency.

(a) *Audited financial statements.* Except as provided in paragraphs (d) and (e) of this section, all borrowers with a total principal and interest loan balance for loans under this subpart, at the end of the borrower's fiscal year, of more than $1 million must submit annual audited financial statements. The audit must be performed in accordance with generally accepted accounting principles (GAAP) and any other requirements specified in this subpart.

(b) *Unaudited financial statements.* For borrowers with a loan balance (principal plus interest at year-end) of $1 million or less, the Agency will require annual financial statements which may be statements compiled or reviewed by an accountant qualified in accordance with the publication "Standards for Audit of Governmental Organizations, Programs, Activities and Functions" instead of audited financial statements.

(c) *Internal financial statements.* The Agency may require submission of financial statements prepared by the borrower at whatever frequency is determined necessary to adequately monitor the loan. Quarterly financial statements will be required on new business enterprises or those needing close monitoring.

(d) *Minimum requirements.* This section sets out minimum requirements for audited and unaudited financial statements to be submitted to the Agency. If specific circumstances warrant, the Agency may require audited financial statements or independent unaudited financial statements in excess of the minimum requirements. For example, loans that depend heavily on inventory and accounts receivable for collateral will normally be audited, regardless of the size of the loan. Nothing in this section shall be considered an impediment to the lender requiring financial statements more frequently than required by the Agency or requiring audited financial statements when the Agency would accept unaudited financial statements.

(e) *Public bodies and nonprofit corporations.* Notwithstanding other provisions of this section, any public body or nonprofit corporation that receives a guarantee of a loan that meets the thresholds established by 2 CFR part 200, subpart F, as codified by 2 CFR 400.1, must provide an audit for the fiscal year of the borrower in which the Loan Note Guarantee is issued. If the loan is for development or purchases made in a previous fiscal year through interim financing, an audit will also be provided for the fiscal year in which the development or purchases occurred. Any audit provided by a public body or nonprofit corporation required by this paragraph will be considered adequate to meet the requirements of this section for that year.

§§ 1980.446–1980.450 [Reserved]

§ 1980.451 Filing and processing applications.

(a) *Borrowers' and lenders' contact.* Borrowers and lenders desiring Rural Developmentassistance as provided in this subpart may file preapplications or applications with the County Supervisor or District Director servicing the area in which the project is to be located. In either case, the requirements of § 1980.46 of subpart A of this part must be met. The County Supervisor or District Director receiving the request for assistance will promptly notify the State Director of the nature and facts of the request. The Rural Development State Director will promptly arrange an early meeting with the borrower and lender representatives to discuss assembly, preparation and processing of preapplications and applications. The State Director may call upon the County Supervisor and District Director to assist the State Office in any way necessary.

(b) *Applications from cooperatives.* Borrowers eligible for loans from the Bank for Cooperatives will be encouraged to obtain guaranteed loans from that source since the Bank for Cooperatives is experienced in making and servicing such loans and can provide substantial counsel to the applicant. Applications must be submitted to the Bank for Cooperatives as a test for credit elsewhere when an insured loan is being considered. (See RD Instruction 2000–Q available in any Rural Development office for Memorandum of Understanding between Rural Development and Farm Credit Administration.)

(c) *Borrowers eligible for Small Business Administration (SBA) assistance.* All borrowers for loan guarantees eligible for SBA assistance will be advised by Rural Development at the time of receipt of the preapplication of the availability of such assistance and will be encouraged to apply to that agency. (See RD Instruction 2000–P available in any Rural Development office for Memorandum of Understanding between SBA and Rural Development).

(d) *Loan Priorities.* Applications and preapplications received by Rural Development will be considered in the order received; however, for the purpose of assigning priorities as described in paragraph (d)(3) of this section, Rural Development will compare an application to other pending applications.

(1) Rural Development will cooperate fully with appropriate State agencies in guaranteeing and insuring loans in a manner which will assure maximum support of the State's strategies for development of its rural areas.

(2) When applications on hand otherwise have equal priority, the applications from a veteran will have preference. A veteran is a person who has been discharged or released from the active forces of the United States Army, Navy, Air Force, Marine Corps, or Coast Guard under conditions other than dishonorable and who served on active duty in such forces:

(i) During the period April 6, 1917, though March 31, 1921;

(ii) During the period of December 7, 1941, through December 31, 1946;

(iii) During the period of June 27, 1950, through January 31, 1955; or

(iv) For a period of more than 180 days, any part of which occurred after January 31, 1955; but on or before May 17, 1975. Discharges under conditions other than dishonorable include "clemency discharges."

(3) Priorities will be assigned by Rural Development to eligible applications on the basis of a point system that takes into account project location, the creation and saving of jobs, the cost at which those jobs would be created or saved, seasonal and part-time job impact, and leveraging of

Rural Development assistance. The application and supporting information submitted with it will be used to determine an eligible proposed project's priority for available funds or guarantee authority. The priorities described in this paragraph will be used by Rural Development to score projects. A copy of the calculation of the score should be placed in the case file for future reference.

(i) *Location priorities.* The priority score for location will be the score for the highest-ranked category in which the project fits. If the location does not fit one of these categories, its receives no points for location. The categories, and their point scores, are:

(A) Located in a city or area under 25,000 population (10 points).

(B) Located in a city or area under 25,000 population that is in an area of high unemployment as of the date of application (20 points).

(C) Located in an area of high unemployment as of the date of application, provided the borrower certifies in writing to the State Director in simple narrative or letter form that the project will employ on a permanent, full-time basis (providing at its own cost such training or retraining as may be needed) persons (numbering no fewer than 25 percent of the project's employment) who are members of displaced farm families which recently derived from farming or ranching the majority of their combined incomes but are no longer actively engaged in farming or ranching as operators or employees (35 points).

(ii) *Jobs priorities.* The priority score for jobs created and/or saved is the score for the highest-ranked category in which the project fits. If the project does not fit one of these categories, it receives no points for jobs. The categories, and their point scores, are:

(A) Project will contribute to the overall economic stability of the project area and generate permanent jobs beyond the entrepreneur and the entrepreneur's household (10 points).

(B) Project will contribute to the overall economic stability of the project area and will employ on a permanent, full-time basis a number of persons that is significant in the context of the area's economy (20 points).

(C) Project will contribute to the overall economic stability of the project area, will employ on a permanent, full-time basis a number of persons that is significant in the context of the area's economy, and will retain in that area a significant number of jobs that would otherwise be lost (35 points)

(iii) *Job cost priorities.* The priority score for the project's cost per job is the score for the highest-ranked category in which the project fits. First, divide the amount of the Rural Development guaranteed loan by the number of jobs created or saved. This will result in the cost per job. Count only full-time jobs. Part-time jobs may be reduced to a fraction of a full-time job and counted. For example, a 20-hour-per-week job, or a job that is full-time for six months per year, is one-half of a job. Second, determine the State's nonmetropolitan household income as described in § 1980.451(d)(3)(vi). Third, divide the cost per job by the State's nonmetropolitan household income. For example, if the cost per job is $10,000 and the State's nonmetropolitan household income is $20,000, the result will be 0.5. The categories, and their point scores are:

(A) Loans on which the result is greater than 1.5 but less than 2.0 (5 points).

(B) Loans on which the result is from 1.0 to 1.5 (15 points).

(C) Loans on which the result is less than 1.0 (25 points).

If the result exceeds 2.0, a high cost per job in that State, no points are received for job cost.

(iv) *Additional Points.* There shall be added to the score the points indicated for any and all of the following criteria met by the project.

(A) FmHA or its successor agency under Public Law 103–354 guaranteed loan is less than 50 percent of project cost (5 points).

(B) Percentage of guarantee is 10 or more percentage points less than the maximum allowable for a loan of its size (5 points).

(C) Project will, in addition to any permanent full-time jobs, create a significant number of part-time or seasonal jobs that will provide additional income to underemployed residents of the project area without their having to give up any present part-time or seasonal jobs (10 points).

(v) *Administrative Points.* The State Director may assign up to 20 points to an application in addition to those points scored under § 1980.451(d)(3) (i) through (iv). These administrative points are intended to be assigned by a State Director only in cases of unforeseen exigencies, emergencies, benefits to other FmHA or its successor agency under Public Law 103–354-assisted projects (including the limiting of financial risks affecting FmHA or its successor agency under Public Law 103–354 loans and loan guarantees) or the loss of financing if Rural Development funds are not committed in a timely fashion. They may also be assigned in cases in which the project's goods or services are essential to other Federally assisted projects and activities in the area or to the successful implementation of an economic development strategy for the area that is sponsored and/or operated by an agency of the Federal or State government. An explanation for the assigning of these points by the State Director will be appended to the calculation of the project score maintained in the case file. If an application is considered in the National Office, the Administrator may also assign up to 20 points. An assignment of points by the Administrator will be by memorandum, stating the Administrator's reasons, and that memorandum will be appended to the calculation of the project score maintained in the case file. In assigning priorities to applications and in selecting projects for funding, Rural Development will consider State development strategies. Funds (guarantee authority) allocated for use as prescribed in this regulation are to be considered for use by Indian tribes within the State regardless of whether State development plans include Indian reservations within the State's boundaries. It is essential that Indians residing on such reservations have equal opportunity to participate in any benefits of these programs.

(vi) *Indexation.* When current, annual data are not available to determine a State's nonmetropolitan household income for purposes of the calculations described in paragraph (d)(3)(iii) of this section, indexation of census data is necessary. The State Director will use the figure from the most recent decennial census of the United States, increased by a factor representing the increase since the year of that census in the Consumer Price Index ("CIP-U"). That factor shall be furnished annually by the National Office, Rural Development.

(e) *Filing preapplications and applications.* Borrowers or lenders may file preapplications described in paragraph (f) of this section if they desire an expression of Rural Development interest prior to assembling the complete application and request for Loan Note Guarantee or they may present the complete application, in one package, including the material required in paragraphs (f), (i), (j), and (k) of this section.

(f) *Preapplications.* Applicants may file preapplications with the County, District, or State Office including:

(1) A letter prepared by the borrower and the lender which shall include:

(i) Borrower's name, address, contact person and telephone number.

(ii) Amount of loan request.

(iii) Name of the proposed lender, address, contact person, and telephone number.

(iv) Brief description of the projects, products and services provided.

(v) Type and number of employment opportunities and unemployment rate where the project will be located.

(vi) Amount of borrower's equity and guarantees offered.

(vii) Anticipated loan maturity and interest rates.

(viii) Availabiity of raw materials and supplies.

(ix) If a corporation, names and addresses of borrower's parent, affiliates and/or subsidiary firms and a brief description of relationship, products and ownership among borrower, parent, affiliates and subsidiary firms.

(2) Form RD 449–22, "Certification of Non-Relocation and Market and Capacity Information Report."

(3) Form RD 449–4, "Statement of Personal History," for a proprietor (owner), each partner, officer, director, key employee and stockholders holding 20 percent or more interest in the borrower except for those corporations listed on a major stock exchange and for those so listed if required by Rural Development. Forms RD 449–4 are not required to be submitted for elected officials and appointed officials in connection with loan applications from public bodies. Failure to report full, complete and accurate information on the Statement of Personal History may result in Rural Development's not making or guaranteeing the loan. Whenever possible, a local, regional, or national credit report, furnished by the lender, will be used to verify data on Form RD.

(4) A record of any pending or final regulatory or legal (civil or criminal) action against the borrower, parent, affiliate, proposed guarantors, subsidiaries, principal stockholders, officers and directors.

(5) For existing businesses, a current balance sheet, and latest profit and loss statement (not more than 60 days old) and financial statements including parent, affiliate and subsidiary firms, for at least the last 3 years or more if necessary for a thorough evaluation.

(6) A detailed projection of gross revenue, net earnings and cash flow statements for 3 years including assumptions upon which such forecasts are based.

(7) Sales projections indicating the percent of the national and local market the business expects to obtain.

(8) Intergovernmental consultation should be carried out in accordance with 2 CFR part 415, subpart C, "Intergovernmental Review of Department of Agriculture Programs and Activities."

(g) *Preliminary determination by Rural Development.* If preparation information indicates the project will not meet Rural Development's minimum credit standards for a sound loan, is ineligible, does not have sufficient priority or that funds or guarantee authority are not available for the project, Rural Development will so inform the lender. The lender will be notified in writing with all reasons for the decision indicated. If it appears that the project is eligible, has sufficient priority, is economically feasible and loan guarantee authority is available, Rural Development will inform the lender and borrower in writing and request that they complete the application.

(h) *Department of Labor certifications.* Rural Development will submit Form RD 449–22 to the Department of Labor for the necessary certification that the proposal will not be in conflict with § 1980.412(c) and (d).

(i) *Content of Applications:*

(1) Form RD 449–1.

(2) Form RD 449–2.

(3) Environmental review documentation as required in accordance with 7 CFR part 1970.

(4) Architectural or engineering plans, if applicable.

(5) Cost estimates and forecasts of contingency funds to cover inflation or project changes.

(6) Appraisal reports.

(7) For existing businesses a pro forma balance sheet at startup and for at least three additional projected years, indicating the necessary startup capital, operating capital and short-term credit based on financial statements for the last three years, or more (if available); and projected cash flow and earnings statements for at least three years supported by a list of assumptions showing the basis for the projections. The business should submit a current balance sheet with a debt schedule of any debts to be refinanced and an income statement to Rural Development, through the lender, every 90 days from the time the application is filed with the lender to the time of issuance of the Loan Note Guarantee. If debt refinancing is requested, a debt schedule is prepared (correlated to the latest balance sheets) reflecting the debts to be refinanced including the name of the creditor, the original loan amount and loan balance, date of loan, interest rate, maturity date, monthly or annual payments, payment status and collateral that secured such loans.

(8) For new businesses, a pro forma balance sheet at startup and for the next three years, project cash flow (monthly first year, quarterly for two additional years) and projected earnings statements for three years supported by a list of assumptions showing the basis for the projections.

(9) Any credit reports obtained by the lender or Rural Development on the borrower, its principals and parent, affiliate and subsidiary firms.

(10) Form RD 400–1, "Equal Opportunity Agreement," if construction costing more than $10,000 is involved.

(11) Copies of building permits, if applicable, and any necessary certifications and recommendations of appropriate regulatory or other agency having jurisdiction over the project including any pollution control agency.

(12) Personal and corporate financial statements of those guarantors named in § 1980.443.

(13) Proposed loan agreement. (See paragraph VII of Form RD 449–35). Loan agreements between the borrower and lender will be required. The final executed loan agreement must include the Agency requirements as set forth in the Form RD 449–14 including the requirements for periodic financial statements in accordance with § 1980.445. The loan agreement must also include, but is not limited to, the following:

(i) Prohibition against assuming liabilities or obligations of others.

(ii) Restrictions on dividend payments.

(iii) Limitation on purchase or sale of equipment and fixed assets.

(iv) Limitations on compensation of officers and owners.

(v) Minimum working capital requirements.

(vi) Maximum debt to net worth ratio.

(vii) Restrictions concerning consolidations, mergers or other circumstances.

(viii) Limitations on selling the business without concurrence of the lender and Rural Development.

(ix) Repayment and amortization of the loan.

(x) List of collateral for the loan including a list of persons and/or corporations guaranteeing the loan with a schedule for providing the lender and Rural Development with personal and/or corporate financial statements. (See § 1980.443)

(14) A complete feasibility study when required. (See § 1980.442)

(15) Any additional information required by Rural Development.

(16) For companies listed on major stock exchanges and/or subject to the Securities and Exchange Commission regulations, a copy of Form 10–K, "Annual Report Pursuant to section 13 or 15 D of the Act of 1934."

(17) Documented evidence that the project is located within or without special flood or mudslide hazard areas.

(18) Notices of compliance with the Privacy Act of 1974.

(i) If the borrower is acting in a personal capacity and not as an entrepreneur for such entities as proprietorships, partnerships, or corporations, and Rural Development solicits personal information for him/her, the individual will be provided Form RD 410–9, "Statement Required by the Privacy Act."

(ii) If Rural Development desires to obtain information concerning an individual from any source, Rural Development will provide such source with Form RD 410–10, "Privacy Act Statement to References."

(19) On any request for refinancing of existing loan(s) as authorized under § 1980.411(a)(11), the lender is required, as a minimum, to obtain the previously held collateral as security for the guaranteed loan(s). Additional collateral will be required by Rural Development when refinancing of unsecured or undersecured loans is unavoidable in order to accomplish the necessary strengthening of the firm's current position.

(j) *Use of forms.* Rural Development numbered forms will be used where shown in both preapplications and applications. Otherwise, lenders should use their forms, real estate mortgages, security instruments and other agreements, provided such forms do not contain any provisions that are in conflict or are inconsistent with provisions of the subpart.

(k) *Certificate of need.* If the loan request is for health care facilities (e.g., hospitals or nursing homes), a "Certificate of Need" will be obtained by the borrower from the appropriate regulatory or other agency having jurisdiction over the project and submitted to Rural Development by the lender. If a significant part of the project's income will be from third party payors, (e.g., medicare or medicaid), the project will be designed and operated in a manner necessary to meet the requirements of the third-party payors.

Administrative

A. *The State Director:*

1. Determines if material and information submitted is completed and signed by the appropriate party in the appropriated capacity.

2. May request the comments and recommendations of the County Supervisor and District Director. Such comments will include but are not limited to the following: Community attitude toward project; a summary of comments regarding the proposal by the lender, county leaders and other interested parties; whether the project is likely to result in the need for additional community facilities such as schools, water, sewer and health care services, and if so, the community's plan for providing such facilities; availability of any required additional labor force and training plans for such force, if needed; an economic forecast of the effect on the community should the project fail, if financed.

3. Will furnish all individuals acting in a personal capacity at the time of filing a preapplication or application and two copies of Form RD 410–9. The individual will sign both copies, retaining one and providing Rural Development with the other copy which becomes a part of the loan file.

4. Will provide any source whom Rural Development obtains information concerning an individual with two copies of Form Rural Development 410–10. The source will sign both copies, retain one and provide Rural Development with the other copy which becomes a part of the loan file.

5. Will input the necessary data via terminal screens into the Rural Community Facility Tracking System (RCFTS). The RCFTS data structure consists of 3 sets: Applicant/Borrower (BOR), Facility (FAC), and Loan/Grant Request (LGR) sets. There are multiple screens for the BOR and LGR sets. The State Director may, if he/she so desires, prepare a Form RD 2033–34, "Management System Card—Business and Industry," in accordance with RD Instruction 2033–F.

6. Will forward immediately to the National Office on all projects.

(a) Form RD 449–22 (7 copies) for loans over $1 million and when direct employment increases more than 50 employees.

(b) For insured loans where the borrower leases facilities to another, submit Form RD 449–22 for such borrower. The lessor(s) will also be required to provide Form RD 449–22. Subsequent loan requests require resubmission of Form RD 449–22.

(c) A local, national or regional credit report and Form RD 449–4 for all loans over one million dollars or for loans, regardless of size, when the State Director believes a character evaluation check is advisable.

NOTE: Forms RD 449-22 and RD 449-4 should *only* be processed if a *complete* preapplication or application has been received.

B. Miscellaneous Administrative provisions:

1. *Par (f)*. Preapplications are not to be accepted or processed unless a lender has agreed in writing to finance the proposal. The preapplication letter is a joint letter prepared by the borrower and lender.

2. *Par (g)*. Upon receipt of all preapplications in excess of $5 million, the State Director will transmit to the National Office the material required under paragraph (f)(1), (f)(4) and (f)(5) of this section together with recommendations and observations an analysis of the quality and permanency of the employment opportunities involved in the project. The National Office will review the proposed project in relation to objectives, priorities and intent of the program and will advise the State Director. After receiving the National Office advice or for loans less than $5 million, the State Director will inform the borrower of the decision.

3. *Par (i)*. State Director submits a transmittal letter with recommendations on loan applications requiring National Office review. Included are:

(a) Loan file.

(b) Form RD 449-29, "Project Summary—Business Industrial Loan Division," including State Director's a spread sheets, financial history and projections (use attachments to Project Summary if necessary).

(c) Proposed Form RD 449-14.

(d) Copy of Rural Development State Loan Review Board Minutes.

(e) Notification of required financial and other reports, their frequency, due dates and fiscal yearend.

4. *Par (i)(9), Credit reports.*

(a) The National Office has a contract to provide credit reports for preapplications, applications, and in instances after the loan(s) is made, where a credit report is needed.

(b) States should first try to have the lender provide such a report because credit reports are the responsibility of the lender.

(c) Any state needing a credit report should telephone the National Office, Director, B&I, and give the name of the business and the city and State location. The report will be mailed to the State the same day, if possible.

5. *File documentation*. Applications will be organized in a loan file in accordance with RD Instruction 2033-A (available in any Rural Development office.) An 8-position folder with tabs will be utilized.

The State Director may supplement the Position Guides to include specific legal requirements within their State. If the lender prepares a complete application package, it may accompany the docket provided the docket is organized in a binder, indexed and tabbed. Feasibility studies should be kept separate. It is the responsibility of Rural Development employees who work on applications or servicing actions to add to the correspondence section of the loan file (also known as the running record) a written report of any field visits, meetings, telephone conversations and memorandums covering decisions or reasons for Rural Development's actions on the cases. Particular attention must be given to this requirement on cases that become delinquent or problems in order that Rural Development position will be defensible in the event of an adverse action.

6. *Par (i), (13), Audit agreements and requirements*. Rural Development urges the use of a written agreement between the lender and borrower to assure that there is no misunderstanding concerning Rural Development audit requirements.

7. *Par (i), Forms and documents found in loan docket*. The following table is a guide to forms and documents used in completing an application and loan docket. The filing position within the 8 position folder is shown on the right. Some of these items may not be applicable for a particular loan. However, a complete loan docket may need to include items in addition to the following:

DESCRIPTION OF RECORD OR FORM NUMBER AND TITLE

		Filing position
AD-425	Contractor's Affirmative Action Plan For Equal Employment Opportunity	1
RD 400-1	Equal Opportunity Agreement	6
RD 400-3	Notice to Contractors and Applicants	6
RD 400-4	Assurance Agreement	3
RD 400-6	Compliance Statement	6
RD 410-8	Applicant Reference Letter	3
RD 410-9	Statement Required by the Privacy Act	3
RD 410-10	Privacy Act Statement to References	3
RD 424-12	Inspection Report	6
RD 1940-3	Request for Obligation of Funds—Guaranteed Loans; Filing Position 2	2
RD 1970-1	Environmental Checklist for Categorical Exclusions	3
	Environmental Reports	3
	Environmental Assessments	3
	Environmental Impact Statements	3
RD 440-57	Acknowledgement of Obligated Funds/Check Request	2

DESCRIPTION OF RECORD OR FORM NUMBER AND TITLE—Continued

		Filing position
RD 449–1	Application for Loan and Guarantee	3
RD 449–2	Statement of Collateral	5
RD 449–4	Statement of Personal History	3
	Loan Closing Opinion of Lender's Legal Counsel	....

[52 FR 6501, Mar. 4, 1987, as amended at 53 FR 40401, Oct. 17, 1988; 53 FR 45258, Nov. 9, 1988; 55 FR 26199, June 27, 1990; 56 FR 8271, Feb. 28, 1991; 61 FR 18495, Apr. 26, 1996; 76 FR 80731, Dec. 27, 2011; 79 FR 76012, Dec. 19, 2014; 80 FR 9906, Feb. 24, 2015; 81 FR 11047, Mar. 2, 2016; 81 FR 26667, May 4, 2016]

§ 1980.452 Rural Development evaluation of application.

Rural Development will evaluate the application and make a determination whether the borrower is eligible, the proposed loan is for an eligible purpose and that there is reasonable assurance of repayment ability, sufficient collateral and sufficient equity and the proposed loan complies with all applicable statutes and regulations. If Rural Development determines it is unable to guarantee the loan, the lender will be informed in writing. Such notification will include the reasons for denial of the guarantee. If Rural Development is able to guarantee the loan, it will provide the lender and the borrower with Form RD 449–14, listing all requirements for such guarantees. Rural Development will include in the requirements of the Conditional Commitment for Guarantee a full description of the approved use of guaranteed loan funds as reflected in the Form RD 449–1. The Conditional Commitment for Guarantee may not be issued on any loan until the State Director has been notified by the National Officer that the Statements of Personal History(s) have been processed and cleared. Rural Development State Directors are the only persons authorized to execute Form RD 449–14.

Administrative

State Director evaluates the application and considers:

A. Rural area determinations. (See § 1980.405 of this subpart.)

B. Community impact of the proposal which includes:

1. Number of businesses and industries in the town or city.

2. Employment impact upon the community.

3. Availability of skilled and unskilled labor and permanency of employment opportunities.

4. Vocational and educational facilities to provide skilled labor, if applicable.

5. Policies of applicant regarding unemployment, lay-offs, wage scales, etc.

C. If debt refinancing is requested, consider in accordance with § 1980.411(a)(11) of this subpart and:

1. A complete review will be made to determine whether it is essential to restructure the company's debts on a schedule that will allow the business to operate successfully rather than merely guaranteeing an unsound loan.

(a) Obtain a borrower's complete debt schedule. Schedule should agree with borrower's latest balance sheet.

(b) Determine from lender if the borrower's present loan(s) is on the lender's regulatory examiner's report and if so determine the loan classification.

(c) Analyze lender's liability ledger on the borrower, individual customer credit file, installment Loan Ledger Card or Computer printouts and other credit reports.

(d) The percentage of guarantee should be adjusted to assure that the lender does not bring its previously existing unguaranteed exposure under the guarantee.

(e) Any special servicing requirements should be identified and included in the Conditional Commitment for Guarantee.

D. Applications will be analyzed by an Rural Development State Loan Review Board before execution of Form RD 449–14. When analyzing the B&I loan request, the State Loan Review Board will specifically address the issue of the guarantee percentage to be approved. Consideration of reducing the maximum guarantee to less than 90 percent is appropriate when the loan has sufficient strength to warrant further participation by the private sector or refinancing of existing lender debts to the borrower is involved. Ordinarily, B&I loan guarantees should be structured so that the lender bears a significant portion of the risk of loss from a default. "Significant" means equal to or greater than 20 percent of the loss stemming from default. All review board meetings will be fully documented, including the review and decision concerning the guarantee percentage, and will be signed by those Rural Development employees serving on the board. A copy of such documentation will be retained in the loan file.

1. Generally, the review board consists of the State Director as Chairperson, Community and Business Program Chief or the Business and Industry Chief (Loan Specialist) and either the Community Programs Chief, Rural Housing Chief, or Farmer Programs Chief, as appropriate.

2. The State Director may wish to contact non-Rural Development sources for expertise, such as banker or other lenders, industrial development specialists from state commissions, academicians, certified public accountants, tax attorneys, successful business and professional lenders, management consultants and officials from other Federal agencies. Outside resource consultants may be reimbursed only for their travel costs (transportation and subsistence). (See RD Instruction 2036-A which is available in any Rural Development Office).

3. The Rural Housing Loan Chief will be a member of the Rural Development State Loan Review Board if a site development loan (see §1980.411(a)(7) of this subpart) is being considered. The Community and Business Programs Chief (Loan Specialist) will be a member if a loan for facilities of the type financed under the provisions of Subpart A of Part 1942 of this chapter is being considered. The Farmer Programs Chief will be a member of the board if a project, the success of which is dependent on the production of agricultural products, is being considered. If the proposed project covers more than one program area, all the chiefs for those programs involved will be members of the board. If the approval of an application for a B&I loan may result in benefiting or hindering other Rural Development programs, the review board will determine whether the making of such loan or guarantee is likely to result in embarrassment for Rural Development as a result of a possible conflict of interest whereby other parties may accuse the agency of giving loan preference to housing borrowers (in the case of site development) or producers (in the case of agricultural processing plants) or other Rural Development programs.

4. The State Director may request the County Supervisor and/or District Director to attend the review board meeting whenever it is determined they may have special knowledge of the proposed loan which may affect the board's decision.

5. Prior to submission of a B&I guaranteed loan(s) request to the National Office for loan processing review and prior to loan approval, the appropriate loan processing official must visit the project site and discuss the loan proposal with the lender and bor-

rower. In the event there are multiple project sites the official should visit a representative sample of project sites to develop deeper understanding of the project operation. For businesses without a developed project site a visit is not necessary; however, a visit with the lender and borrower is still required. The findings of the visit should be documented in the loan docket submitted to the National Office.

6. The State Director will prepare an original and two copies of Form RD 1940-3 for each loan to be obligated. Also, for each initial loan, Form RD 1980-50, "Add, Delete, or Change Guaranteed Loan Borrower Information," will be prepared. The State Director will sign the original and one copy and conform the second copy. Form RD 1940-3 will not be mailed to the Finance Office. Notice of approval to lender will be accomplished by providing or sending the lender the signed copy of Form RD 1940-3 and Form RD 449-14 six working days from the date funds are reserved, unless an exception is granted by the National Office. The State Director or designee will record the actual date of lender notification on the original of the RD 1940-3 and retain the original of the form as a permanent part of the Rural Development case file. The State Director may retain the remaining conformed copy of Form RD 1940-3. The State Director or designee will use the State Office terminal to request reservation/obligation of funds. Use of the telephone for the reservation/obligation of funds is restricted to those instances when the State Office terminal is inoperative. Form RD 1980-50 will be prepared and distributed for initial loans only.

a. Immediately after contacting the Finance Office, the requesting official will furnish the requesting office's security identification code. Failure to furnish the security code will result in rejection of the request for reservation of authority. After the security code is furnished, all pertinent information contained on Form RD 1940-3 will be furnished to the Finance Office. Upon receipt of the telephone request for reservation of authority, the Finance Office will record all information necessary to process the request for reservation in addition to the date and time of the request.

b. The individual making the telephone request will record the date and time of the telephone request and place his/her signature in section 35 of Form RD 1940-3.

c. The Finance Office will terminally process telephone reservation requests. Those requests for reservation received before 2:30 p.m. Central Time, to the extent possible, will be processed on the date received; however, there may be instances in which the reservation will be processed on the next working day.

d. Each working day the Finance Office will notify the State Office by telephone of all projects for which authority was reserved during the previous night's processing cycle and the date of obligation. If authority cannot be reserved for a project, the Finance Office will notify the State Office that authority is not available within the State allocation. The obligation date will be the date of the request for reservation of authority which is being processed in the Finance Office. The Finance Office will mail to the State Director Form RD 440–57, "Acknowledgment of Obligated Funds/Check Request," prepared in duplicate, confirming the reservation of authority with the obligation date inserted as required by item No. 9 on the FMI for Form RD 440–57. Immediately after notification by telephone of the reservation of authority, the State Director will call the Legislative Affairs and Public Information staff in the National Office as required by RD Instruction 2015–C (available in any Rural Development office).

e. See RD Instruction 2015–C (available in any Rural Development office) for notification procedures.

7. State Director notifies the lender and borrower if he/she will not issue the Form RD 449–14.

[52 FR 6501, Mar. 4, 1987, as amended at 53 FR 45258, Nov. 9, 1988; 56 FR 8271, Feb. 28, 1991; 79 FR 55967, Sep. 18, 2014]

§ 1980.453 Review of requirements.

(a) Immediately after reviewing the conditions and requirements in Form RD 449–14 the lender and applicant should complete and sign the "Acceptance of Conditions," and return a copy to the Rural Development State Director. If certain conditions cannot be met, the lender and borrower may propose alternate conditions to Rural Development.

(b) If the lender indicates in the "Acceptance of Conditions" that it desires to obtain a Loan Note Guarantee and subsequently decides at any time after receiving a conditional commitment that it no longer wants a Loan Note Guarantee, the lender will immediately advise the Rural Development State Director.

Administrative

A. The State Director will negotiate with the lender and proposed borrower any changes made to the initially issued or proposed Form RD 449–14. For loans requiring National Office concurrence, a copy of Form RD 449–14 and any amendments thereto will be included when the loan file is submitted to the National Office for review. When the National Office recommends modifications

or additions to Form RD 449–14, the State Director will further negotiate these recommendations with the lender and proposed borrower. If, as a result of these further negotiations, the lender, proposed borrower or State Director presents alternate conditions which would result in a change in the scope of the proposed project and if the loan exceeds the State Director's loan approval authority, the State Director will submit these conditions by memorandum to the National Office for consideration with a copy of the revised Form RD 449–14 and any amendments thereto. If the loan is within the State Director's loan approval authority, the State Director may approve such changes.

B. On loan applications within the State Director's loan approval authority, the State Director will submit to the National Office, Business and Industry Division, within 30 days after the Form Rural Development 449–14 has been accepted:

1. A copy of Form RD 449–29.

2. A copy of Form RD 449–14 is accepted by the lender and borrower.

2. A copy of Rural Development State Loan Review Board Minutes.

4. Notification of required financial and other reports, their frequency, due dates and fiscal year-end.

5. A copy of the proposed loan agreement between the lender and the borrower.

6. When debt refinancing is involved, a copy of the justification for the refinancing.

7. The cover memorandum should indicate whether the Form RD 449–34 has been issued. If the Loan Note Guarantee has been issued, enclose a copy of the Lender Certification required by § 1980.60(a) of subpart A of this part, and, if not, a proposed date for issuance of the Form RD 449–34.

[52 FR 6501, Mar. 4, 1987, as amended at 54 FR 28022, July 5, 1989; 57 FR 4359, Feb. 5, 1992]

§ 1980.454 Conditions precedent to issuance of the Loan Note Guarantee.

In addition to compliance with the requirements of § 1980.60 of subpart A of this subpart, compliance with the following provisions are required prior to issuance of the Loan Note Guarantee.

(a) *Transfer of lenders.* The Rural Development State Director may approve a substitution of a new eligible lender in place of a former lender who holds an outstanding Conditional Commitment for Guarantee (where the Loan Note Guarantee has not yet been issued and the loan is within the State Director's loan approval authority) provided there are no changes in the borrower's ownership or control, loan purposes, scope of project and loan conditions in the Form RD 449–14 and the loan agreement remains the same. To effect such a substitution, the former lender will provide Rural Development with a letter stating the reasons it no longer desires to be a lender for the project. For loans in excess of the State Director's loan approval authority, National Office concurrence is required. The State Director will submit a recommendation concerning the transfer of lenders along with the lender's letter stating the reasons it no longer desires to be a lender for the project. The substituted lender will execute a new Part "B" of Form RD 449–1. If approved by Rural Development, the State Director will issue a letter or amendment to the original Form RD 449–14 reflecting the new lender and the new lender will acknowledge acceptance of the letter or amendment in writing.

(b) *Substitution of borrowers.* Rural Development will not issue a Loan Note Guarantee to the lender who is in receipt of a Form RD 449–14 with an obligation in a previous fiscal year if the originally approved borrower (including changes in legal entity) or owners are changed. The only exception to this provision prohibiting a change in the legal entity's form of ownership is when the originally approved borrower or owner is replaced with substantially the same individuals with substantially the same interests, as originally approved and identified in the Form RD 449–1, item 15. All requests for exceptions must be approved by the Rural Development National Office.

(c) *Changes in terms and conditions in Form RD 449–14.* It is the intent of Rural Development that once the Form RD 449–14 is issued and accepted by the lender, the commitment is not to be modified as to the scope of the project, overall facility concept, project purpose, use of proceeds or terms and conditions. Should changes be requested by the lender, the State Director will negotiate with the lender and proposed borrower any proposed changes to the originally accepted Form RD 449–14. If, as a result of these negotiations, the lender, proposed borrower or State Director presents alternate conditions which would result in a change in the scope of the project, and if the loan exceeds the State Director's loan approval authority, the State Director will submit these changes in the conditions by memorandum to the National Office for consideration with a copy of the revised Form FmHA or its successor agency under Public Law 103–354 449–14 and any amendments thereto. Changes to the conditional commitment may be approved by the State Director for loans within their loan approval authority.

(d) *Additional requirements for B&I guaranteed loans.* All B&I borrowers and lenders, as applicable, must comply with Appendix D, paragraphs (I) (A) and (B); (II)(A) through (II)(A)(2)(g)(1); (II) (B) and (C); (III) (A), (B), (C), (D), and (E).

(e) *Preguarantee review.* Coincident with, or immediately after loan closing, the lender will contact Rural Development and provide those documents and certifications required in §§ 1980.60 and 1980.61 of subpart A of this part. Only when the Rural Development B&I or C&BP Chief or Loan Specialist, as required in paragraph B. (Administrative) of this section, is satisfied that all conditions for the guarantee have been met will the Loan Note Guarantee be executed.

(f) *Loan closing.* When loan closing plans are established, the lender will notify Rural Development.

(g) *Closing of working capital loans.* The State Director will not issue a Loan Guarantee for a working capital loan prior to the completion of all proposed construction for the project. Working capital loan funds will not be used to pay short-term notes.

Administrative

A. *The State Director reviews:* 1. [Reserved]

2. Plans for inspections made on construction projects. These should be coordinated with the lender and borrower. Form RD 424–12, "Inspection Reports," may be used by the State Engineer or Architect who will make an inspection of the projects which involve substantial construction. The inspection shall be completed prior to the issuance of the Loan Note Guarantee to assure all construction is complete. The State Loan Specialist or Chief may also participate in the inspections.

3. Cost overruns, if any, and how they will be met. State Directors may approve cost overruns for projects in any amount or percentage within their loan approval authority

not to exceed 10 percent in loan amounts between $1 million and $10 million.

4. Basic credit requirements of all loans.

B. In all cases, the Program Chief or the B&I Loan Specialist will conduct a preguarantee review before issuance of the Loan Note Guarantee to assure that all requirements of the application, Conditional Commitment for Guarantee and Loan Agreement have been met including the required certifications using language specified by the regulations, and will provide such verification in the loan file, including arrangements for annual audit reports. In the conduct of this review, all requirements of §1980.60(a) of Subpart A of this part will be reviewed and special attention should be paid to reviewing current financial statements of the borrower to assure that no adverse change has taken place. The District Director may participate in the review.

C. The State Director or any other Rural Development personnel shall not sign any documents other than those specifically provided for in Subparts A or E of this part. No certificates shall be signed except the "Certificate of Incumbency and Signature" as set forth as Appendix B of this subpart.

D. *Par (a) Transfer of Lender.* The State Director will analyze all requests for substituted lenders including the servicing capability, eligibility and experience of the new lender before the request is approved. If approved, notify the Finance Office of the change using Form RD 1980–42, Do not deobligate and reobligate the loan if the Form RD 449–14 was issued in a previous fiscal year.

E. *Par (b) Substitution of borrowers.* The State Director will review any request for exceptions to substitution of borrowers and forward such requests with a memorandum of facts and recommendations to the National Office for a decision. The National Office will not approve any request where the legal entity is changed, such as from a corporation to a partnership, etc., or if the ownership changes more than 20 percent.

F. *Par (c) Changes in terms and conditions in Form RD 449–14.* The State Director will review any request for changes to Form RD 449–14. Only those changes which do not materially affect the project, its capacity, employment, original projections or credit factors may be approved. Changes in legal entities or where tax considerations are the reason for change will not be approved when modifying any loan guarantee or conditions of guarantee. State Directors may approve these changes in terms and conditions if the loan is within the State Director's loan approval authority and the change will not result in a major change in the scope of the project. Changes in terms and conditions for loans in excess of the State Director's loan approval authority, must be submitted to the National Office with a memorandum of facts and recommendations for review and concurrence.

In order to identify the number and types of action taken, the following procedures are to be followed when requests of this type are approved by Rural Development.

1. Start with the number 1 when the first modification is approved and enter this number in the upper right hand corner of the Letter of Concurrence and on the related "Modification or Administration Action" sheet.

2. Next to the modified wording on the work copy of the Conditional Commitment for Guarantee and the Term Loan Agreement or any form which has been modified, pencil in a short cross reference to the modification and identify the number given it.

3. File the copies of the "Modification or Administrative Action" sheet and related Letters of Concurrence numerically in the docket directly on top of the affected original documents of conditions.

4. This order of recordkeeping should include any requests which were declined by the National Office.

[52 FR 6501, Mar. 4, 1987, as amended at 53 FR 26413, July 12, 1988; 57 FR 4359, Feb. 5, 1992; 61 FR 18495, Apr. 26, 1996]

§§ 1980.455–1980.468 [Reserved]

§ 1980.469 Loan servicing.

The lender is responsible for loan servicing and for notifying the Rural Development (RD) of any violations in the Lender's Loan Agreement. (See Paragraph X of Form Rural DevelopmentRD 449–35).

(a) All B&I guaranteed loans in the lender's portfolio will be classified by the lender as soon as it is notified by the State Office to do so and again whenever there is a change in the loan which would impact on the original classification. The State Director will notify the lender of this requirement for all existing loan guarantees, when new Loan Note Guarantees are issued to a lender and/or when the State Office becomes aware of a condition that would affect the classification and justification of the classification will be sent to the State Office. The loans will be classified according to the following criteria:

(1) *Substandard Classifications.* Those loans which are inadequately protected by the current sound worth and paying

capacity of the obligor or of the collateral pledged, if any. Loans in this category must have a well defined weakness or weaknesses that jeopardize the payment in full of the debt. If the deficiencies are not corrected, there is a distinct possibility that the lender and Rural Development will sustain some loss.

(2) *Doubtful Classification.* Those loans which have all the weaknesses inherent in those classified Substandard with the added characteristics that the weaknesses make collection or liquidation in full, based on currently known facts, conditions and values, highly questionable and improbable.

(3) *Loss Classifications.* Those loans which are considered uncollectible and of such little value that their continuance as bankable loans is not warranted. Even though partial recovery may be effected in the future, it is not practical or desirable to defer writing off these basically worthless loans.

(b) There is a close relationship between classifications; and no classifications category should be viewed as more important than the other. The uncollectibility aspect of Doubtful and Loss classifications are of obvious importance; however, the function of the Substandard classification is to indicate those loans that are unduly risky which may result in future claims against the B&I guarantee.

(c) Substandard, Doubtful and Loss are adverse classifications. There are other classifications for loans which are not adversely classified but which require the attention and followup of the lenders and Rural Development. These classifications are:

(1) *Special Mention Classification.* Those loans which do not presently expose the lender and Rural Development to a sufficient degree of risk to warrant a Substandard classification but do possess credit deficiencies deserving the lender's close attention. Failure to correct these deficiencies could result in greater credit risk in the future. This classification would include loans that the lender is unable to supervise properly because of a lack of expertise, an inadequate loan agreement, the condition of or lack of control over the collateral, failure to obtain proper documentation or any other deviations

from prudent lending practices. Adverse trends in the borrower's operation or an imbalanced position in the balance sheet which has not reached a point that jeopardizes the repayment of the loan should be assigned to this designation. Loans in which actual, not potential, weaknesses are evident and significant should be considered for a Substandard classification.

(2) *Seasoned Loan Classification.* A loan which: (i) Has a remaining principal guaranteed loan balance of two thirds or less of the original aggregate of all existing B&I guaranteed loans made to that business.

(ii) Is in compliance with all loan conditions and B&I regulations.

(iii) Has been current on the B&I guaranteed loan(s) payments for 24 consecutive months.

(iv) Is secured by collateral which is determined to be adequate to ensure there will be no loss on the guaranteed loan.

(3) *Current Non-problem Classification*—Those loans that are current and are in compliance with all loan conditions and B&I regulations but do not meet all the criteria for a Seasoned Loan classification. All loans not classified as Seasoned or Current Non-problem will be reported on the quarterly status report with documentation of the details of the reason(s) for the assigned classification.

Administrative

Refer to RD Instruction 1980–E, Appendix G, Liquidation and Property Management Guide (available in any RD office) for advice on how to interact with the lender on liquidations and property management.

A. While the lender has the primary responsibility for loan servicing and protecting the collateral, the State Director is responsible for seeing that servicing as required by the Lender's Agreement and regulation is properly accomplished. Loan servicing is intended to be a preventive rather than a curative action. Prompt followup on delinquent accounts and early recognition of potential problems and pursuing a solution to them are keys to resolving many problem loan cases.

B. *Paragraph II of the Lender's Agreement.* 1. The Loan Note Guarantee is unenforceable by the lender to the extent any loss is occasioned by violation of usury laws, use of loan funds for unauthorized purposes, negligent servicing, or failure to obtain the required security regardless of the time at which

Rural Development acquires knowledge of the foregoing. As used herein, the phrase "use of loan funds for unauthorized purposes" refers to the situation in which the lender in fact agrees with the borrower that loan funds are to be so used and the phrase "unauthorized purposes" means any purpose not listed by the Lender in the completed application as approved by Rural Development.

2. With respect to the negligent servicing and use of loan funds for unauthorized purposes, the Loan Note Guarantee is unenforceable by the lender to the extent any loss is occasioned by negligent servicing and use of loan funds for unauthorized purposes regardless of the time Rural Development acquires knowledge of the negligent servicing or use of loan funds for unauthorized purposes by the lender. Only the amount of the loss caused by negligent servicing or use of loan funds for unauthorized purposes can be withheld from the final loss claim submitted by the lender. The dollar amount withheld from the final loss claim must be ascertainable. In order to determine the final loss amount, the guaranteed loan collateral and any collateral of the guarantor(s) must be liquidated and settled or a settlement with the guarantor(s) reached. In the event there is reason to suspect the lender of negligent servicing or use of loan funds for unauthorized purposes during the life of the loan, the lender should be notified in writing that (a) the acts of negligent servicing and/or use of loan funds for unauthorized purposes will cause the guarantee to be unenforceable by the lender to the extent these acts cause a loss; (b) any decision not to honor any part of the guarantee is not possible until the loan has been liquidated and a loss established; (c) if any loss occurs Rural Development will consider whether negligent acts of the lender caused a loss after the liquidation is complete; and (d) at the time Rural Development determines a loss has occurred as the result of negligent servicing the lender may appeal any adverse decision.

3. When facts or circumstances indicate that criminal violations may have been committed by an applicant, a borrower, or third party purchaser, the State Director will refer the case to the appropriate Regional Inspector General for Investigations, Office of Inspector General (OIG), USDA, in accordance with RD Instruction 2012–B (available in any Rural Development office) for criminal investigation. Any questions as to whether a matter should be referred will be resolved through consultation with OIG for Investigations and the State Director and confirmed in writing. In order to assure protection of the financial and other interest of the government, a duplicate of the notification will be sent to the Office of General Counsel (OGC). After OIG has accepted any matter for investigation, Rural Development staff must coordinate with OIG in advance regarding routine servicing actions on existing loans. A borrower or lender can be sued even though criminal fraud is present. If Rural Development has good reason to believe that, for example, a borrower or a lender made a false statement to obtain a loan or guarantee, or a lender submitted a loss claim to Rural Development which was false or fraudulent, it should promptly call the matter to the attention of OGC—even if no payment of the loss claim has occurred yet. (This would include those situations in which a borrower lied to the lender in order to get the loan, the lender believed the borrower and made the loan—which was guaranteed by Rural Development—and then the lender presented a loss claim to Rural Development for payment after the borrower defaulted on the loan.) Sometimes it might be necessary to ask OIG to do an investigation to establish all the aspects of the fraud. If at all possible, this should then be done prior to referral to OGC.

4. There are two methods the Government could use to seek relief for the fraud. One of the ways the Government could seek redress for the fraud is to sue under the False Claims Act (31 U.S.C. sections 3729–3731). If fraud is proven to have occurred, the False Claims Act provides for the recovery of double damages and a $2,000 penalty (and the costs of one civil suit) for each act involving, for example: (a) Knowingly submitting to a Government employee of false or fraudulent claim for payment or approval, (b) knowingly making or using a false record or statement to get a false or fraudulent claim paid or approved, or (c) conspiring to defraud the United States by getting a false or fraudulent claim allowed or paid. Suit under the False Claims Act must be filed within six years from the date of the commission of the act (e.g., presentation of the claim to Rural Development for payment). The double damage feature ought to be a good incentive to convince OIG to undertake necessary investigations to help establish the fraud.

5. In order to decide whether to file suit, the Department of Justice will need to know such things as: What was the amount of the loan or the loss paid to the lender or holder? How much did the scheme cost the Government? What is the difference in money between what the Government paid out and what it should have paid out? Does the borrower or lender have enough assets to make it worth suing? If Rural Development can answer these questions before referral to OGC—either on its own or by using OIG—than OGC can refer the matter that much more quickly to the Justice Department.

6. There is also a way to bring suit for civil fraud by alleging that "common law" fraud occurred. This would just involve proving that a borrower or a lender falsely represented by their words or actions, a matter of fact either by alleging something in a

false or misleading manner or by concealing something that should have been disclosed; and that Rural Development was deceived by this conduct, and relied on it to its detriment. Under "common law" fraud, only single damages could be recovered, and there would be no $2,000 penalty assessed. The action would generally have to be brought within three years from the date of the discovery of the fraud.

7. Neither the False Claims Act nor the right to bring a "common law" action for fraud precludes the Government from just suing to recover the money wrongfully or mistakenly paid by its employees. If the Justice Department decides not to pursue a civil frauds claim under the False Claims Act or "common law," it will return the matter to OGC. Depending on what stage the proceedings were in when the matter was first referred, Rural Development could then continue to negotiate with the lender or OGC could re-refer the case to Justice for any contract-based actions, including fraud or misrepresentation based on the terms of the guarantee.

C. *The State Director will assure that:* 1. [Reserved]

2. A timetable for routine site, borrower and lender visitations by Rural Development personnel is established before the Loan Note Guarantee is issued. As a guide, visits to newly established borrowers with the lender represented should be scheduled monthly. Visits to established, nonproblem borrowers must be made at least annually except for seasoned loans which will be visited at least bi-annually. Special attention problem accounts should be visited as frequently as the need demands. If possible, these visitations should be coordinated with the lender's visits.

3. During or in preparation for field visits, the following functions are to be performed:

(a) Current financial information is obtained in advance and analyzed for trends.

(b) Any issues revealed or problems not resolved from the last visitation are included in the agenda.

(c) Collateral is observed and its condition, maintenance, protection and utilization by the borrower appears to be satisfactory.

(d) A report of the visit is made on Form RD 449–39, "Field Visit Review (Business and Industrial Loans)," or otherwise documented and included in the loan file. The report should include an opinion of the borrower's status based upon observations made during the visit.

(e) Any instructions or directions to the lender should be confirmed by letter.

4. The Program Chief or Loan Specialist will conduct an annual meeting with each lender or its agent with whom a Loan Note Guarantee(s) or Contract of Guarantee(s) is outstanding. This cannot be redelegated. These meetings may be scheduled at the time Rural Development makes periodic field inspections to the borrower's place of business. At the meeting, a review will be made of the lender's performance in loan servicing, including enforcement of conditions and covenants in the loan agreements. The observations and results of the meeting will be documented. Form RD 449–39 may be used for this purpose. Servicing exceptions on the part of the lender which are noted by Rural Development will be confirmed by letter to the lender.

5. The lender performs an adequate analysis of borrower financial statements for Rural Development. Rural Development in turn will evaluate the lender's analysis and follow up with the lender on servicing action(s) required or negative observations not detected through the lender's analysis. The financial statement analysis of the lender, the financial statement and a memorandum reflecting Rural Development's analysis, including a comparison to previous and projected performance of the borrower, will be forwarded to the National Office, Attention: Business and Industry Division, only for the following loans:

(a) All loans within the first year of loan closing.

(b) Loans over one year old as determined by the State Director or a National Office assigned loan reviewer who is participating in a field review. In event of a disagreement between the State Director and an assigned loan reviewer as to which loans should be included, the assigned loan reviewer's decision will take precedence.

(c) All problem and delinquent loans.

(d) Loans that the State Director would like reviewed by the National Office.

6. Meetings are arranged between the lender, borrower and Rural Development to resolve any problems of late payment, etc.

D. *State Director authorities.* 1. The State Director may delegate authority for the conduct of all functions listed in § 1980.469 Administrative B., except item C. 4. in Administrative B.

2. The State Director may approve B&I guaranteed loan servicing actions as authorized in separate written approval authorities issued in accordance with Subpart A of Part 1901 of this chapter.

3. Servicing actions on loans which exceed the State Director's loan approval authority are to be referred together with the State Director's recommendations to the Director, Business and Industry Division, for prior review and concurrence.

[52 FR 6501, Mar. 4, 1987, as amended at 53 FR 40403, Oct. 17, 1988; 60 FR 26350, May 17, 1995; 61 FR 18495, Apr. 26, 1996]

§ 1980.470 Defaults by borrower.

[See § 1980.63 of subpart A, of this part.]

Administrative

Refer to Appendix G of FmHA or its successor agency under Public Law 103–354 Instruction 1980–E (available in any FmHA or its successor agency under Public Law 103–354 Office) for advice on how to interact with the lender on liquidations and property management.

A. In case of any monetary or significant non-monetary default under the loan agreement, the lender is responsible for arranging a meeting with the State Director, or its designee, and borrower to resolve the problem. A memorandum of the meeting, individuals who attend, a summary of the problem and proposed solution will be prepared by the Rural Development representative and retained in the loan file. When the State Director receives a notice of default on a loan, he/she will immediately notify the National Office in writing of the details and will subsequently report the problem loan to the National Office on the quarterly status report. The State Director will notify the lender and borrower of any decision reached by Rural Development.

B. In considering servicing options, some of which are identified in paragraph X. A of Form RD 449–35, the prospects for providing a permanent cure without adversely affecting the risks of the Rural Development and the lender must become the paramount objective. Within the State Director's authority temporary curative actions such as payment deferments, moratoriums on payments or collateral subordination, if approved, must strengthen the loan and be in the best interests of the lender and Rural Development. Some of these actions may require concurrence of the holder(s). A deferral, rescheduling, reamortization or moratorium is limited by the period of time authorized by this subpart for the purpose for which the loan(s) is made or the remaining useful life of the collateral securing the loan. For example, if the promissory note on a working captial loan is scheduled to mature in 2 years the loan could be rescheduled for 7 years or the remaining life of the collateral whichever is the lesser of the two.

C. Subsequent loan guarantee requests will be processed in accordance with provisions of §1980.473 of this subpart.

D. If the loan was closed with the multinote option, the lender may need to possess all notes to take some servicing actions. In these situations when Rural Development is holder of some of the notes, the State Director may endorse the notes back to the lender after the State Director has sought the advice and guidance of OGC, provided a proper receipt is received from the lender which defines the reason for the transfer. Under no circumstances will Rural Development endorse the original Form RD 449–34 to the lender.

E. The State Director's authority to approve servicing actions is defined in §1980.469, Administrative D.2.

F. Consultant services may be recommended by the State Director to assist Rural Development and the lender in determining which servicing action is appropriate. Requests for consultant services should be made by the State Director and addressed to the Administrator, Attn: Business and Industry Division. A full explanation of the loan history, an evaluation and scope of the proposed study and the need should be included in the request.

G. When the National Office determines it is necessary on individual cases, due to some special servicing requirements, it may, at its option, assume the servicing responsibility on individual cases.

H. The State Director will report all delinquent and problem loans quarterly to the Director, Business and Industry Division, by the 10th day of January, April, July and October.

I. The State Director will notify the Finance Office by memorandum of any change in payment terms such as reamortizations or interest rate adjustments and effective dates of any changes resulting from servicing actions.

[52 FR 6501, Mar. 4, 1987, as amended as 80 FR 9908, Feb. 24, 2015; 80 FR 9908, Feb. 24, 2015]

EDITORIAL NOTE: At 80 FR 9908, Feb. 24, 2015, §1980.470 was amended by removing "Refer to appendix G of this subpart (available in any FmHA or its successor agency under Public Law 103–354 Office)" from the introductory text and adding "Refer to RD Instruction 1980–E, Appendix G, Liquidation and Property Management Guide (available in any Rural Development office)" in its place; however the amendment could not be incorporated because the phrase to be replaced could not be found.

§1980.471 Liquidation.

(See §1980.64 of subpart A of this part.)

Refer to RD Instruction 1980–E, Appendix G, Liquidation and Property Management Guide (available in any Rural Development office) for advice on how to interact with the lender on liquidations and property management.

(a) Collateral acquired by the lender can only be released after a complete review of the proposal.

(1) There may be instances when the lender acquires the collateral of a business where the cost of liquidation exceeds the potential recovery value of the collection. Whenever this occurs the lender with the concurrence of

Rural Development on the collateral in lieu of liquidation.

(2) Sale of acquired collateral to the former borrower, former borrower's stockholder(s) or officer(s), the lender or lender's stockholder(s) or officer(s) must be based on an arm's length transaction with the concurrence of Rural Development.

Administrative

A. The State Director determines which Rural Development personnel will attend meetings with the lender.

B. Introduction to Paragraph XI and Paragraph XI B of the Lender's Agreement. Rural Development will exercise the option to liquidate only when there is reason to believe the lender is not likely to initiate liquidation efforts that will result in maximum recovery. When there is reason to believe the lender will not initiate efforts that will maximize recovery through liquidation, the State Director will forward the lender's liquidation plan, if available with appropriate recommendations, along with the State Director's exceptions to the lender's plan, if any, to the Director, Business and Industry Division, for evaluation and approval or rejection of the State Director's recommendation regarding liquidation. Only when compromise cannot be reached between Rural Development and the lender on the best means of liquidation will Rural Development consider conducting the liquidation. The State Director has no authority to exercise the option to liquidate without National Office approval. When Rural Development liquidates, reasonable liquidation expenses will be assessed against the proceeds derived from the sale of the collateral. In such instances the State Director will send to the Finance Office Form RD 1980-45, "Notice of Liquidation Responsibility."

C. State Directors are authorized to approve lender liquidation plans as authorized on separate written approval authorities issued in accordance with subpart A of part 1901 of this chapter. Within delegated authorities, the State Director may approve a written partial liquidation plan submitted by the lender covering collateral that must be immediately protected or cared for in order to preserve or maintain its value. Approval of the partial liquidation plan must be in the best interest of the government. The approved partial liquidation plan is only good for those actions necessary to immediately preserve and protect the collateral and must be followed by a complete liquidation plan prepared by the lender in accordance with the requirements of paragraph XII A of the Lender's Agreement.

D. Paragraph XI D. State Directors are responsible for review and acceptance of accounting reports as submitted by lenders and for submission of such reports to lenders when Rural Development is conducting liquidation, after they have been submitted with the State's recommendations to the Director, Business and Industry Division for prior review.

E. Paragraph XI E 2. State Directors are authorized to approve final reports of loss from the lender in separate written approval authorities issued in accordance with subpart A of part 1901 of this chapter. The State Director will submit to the Finance Office for payment any loss claims of the lender on Form RD 103-354 449-30, "Loan Note Guarantee Report of Loss." The Finance Office forwards loss payment checks to the State Director for delivery to lender. When a loss claim is involved on a particular loan guarantee, ordinarily one "Estimated Loss Report" will be authorized. Only one final "Report of Loss" will be authorized. A final Form RD 449-30 must be filed with the Finance Office at the completion of all liquidations. Finance Office will use this form to close out the account.

F. Paragraph XI E 3. Final loss payments will be made within the 60 days required but only after a review by Rural Development to assure that all collateral for the loan has been properly accounted for and liquidation expenses are reasonable and within approved limits. State Directors are responsible to see that such reviews are accomplished by the State within 30 days and final loss claims in excess of the State Director's approval authority are forwarded to be accepted or otherwise resolved by the Director, Business and Industry Division within the 60-day period. Any estimated loss payments made to the lender must be taken into consideration when paying a final loss on the Rural Development guaranteed loan. The estimated loss payment must be treated as a deduction from the principal amount of the loan and interest cannot be accrued on the principal amount of the loan that is equal to the estimated loss payment. Community and Business Program Chiefs (C&BP), Business and Industry Chiefs or Loan Specialists will conduct such reviews. The State Director may request National Office assistance in the conduct of any review. All reviews for final loss claim in excess of the State Director's approval authority (See subpart A of part 1901 of this Chapter) will be submitted to the National Office, Business and Industry Division, for concurrence prior to the State Director's approval of the claim. Close scrutiny of liquidation proceeds and their application in accordance with lien priorities is required. Before final loss payments are approved and to assist in the required review, the C&BP Chief, B&I Chief or Loan Specialist will prepare a narrative history of the guarantee transaction which will serve as the summary of occurrence which led to failure of the borrower and actions taken to maximize loan recovery. The original of this report will be filed in the loan case file. A copy of this report together with the review of the final loss claim will be included in the material sent to the Director, B&I Division, for review prior to approval of final loss payments.

§ 1980.472 Protective advances.

[See § 1980.65 subpart A of this part.]

Administrative

Refer to RD Instruction 1980–E, Appendix G, Liquidation and Property Management Guide (available in any Rural Development office) for advice on how to interact with the lender on liquidations and property management.

A. Protective advances will not be made in lieu of additional loans, in particular, working capital loans. Protective advances are advances made by the lender for the purpose of preserving and protecting the collateral where the debtor has failed to and will not or cannot meet its obligations. Ordinarily, protective advances are made when liquidation is contemplated or in process. A precise rule of when a protective advance should be made is impossible to state. A common, but by no means the only, period when protective advances might be needed is during liquidation. At this point, the borrower and success of the project are no longer of paramount importance, but preserving collateral for maximum recovery is of vital importance. Elements which should always be considered include how close the project is to liquidation or default, how much control the borrower will have over the funds, what danger is there that collateral may be destroyed and whether there will be a good chance of saving the collateral later if a protective advance in contemplation of liquidation is made immediately. A protective advance *must* be an indebtedness of the borrower.

B. The State Director must approve, in writing, all protective advances on loans within his/her loan approval authority which exceed a total commulative advance of $500 to the same borrower. Protective advances must be reasonable when associated with the value of collateral being preserved.

C. When considering protective advances, sound judgment must be exercised in determining that the additional funds advanced will actually preserve collateral interests and recovery is actually enhanced by making the advance.

§ 1980.473 Additional loans or advances.

(Refer to paragraph XIII of Form RD 449–35.)

Administrative

Only the State Director shall approve within his/her loan approval authority additional nonguaranteed loans or advances prior to or subsequent to the issuance of the Loan Note Guarantee. The State Director shall determine that there will be no adverse changes in the borrower's financial situation and that such loan or advance is not likely to adversely affect the collateral or the guaranteed loan.

§ 1980.474 [Reserved]

§ 1980.475 Bankruptcy.

(a) It is the lender's responsibility to protect the guaranteed loan debt and all the collateral securing it in bankruptcy proceedings. These responsibilities include but are not limited to the following:

(1) The lender will file a proof of claim where necessary and all the necessary papers and pleadings concerning the case.

(2) The lender will attend and where necessary participate in meetings of the creditors and all court proceedings.

(3) The lender, whose collateral is subject to being used by the trustee in bankruptcy, will immediately seek adequate protection of the collateral.

(4) Where appropriate, the lender should seek involuntary conversion of a pending Chapter 11 case to a liquidating proceeding under Chapter 7 or under Section 1123(b) (4) or seek dismissal of the proceedings.

(5) When permitted by the Bankruptcy Code, the lender will request modification of any plan of reorganization whenever it appears that additional recoveries are likely.

(6) Rural Development will be kept adequately and regularly informed in writing of all aspects of the proceedings.

(b) In a Chapter 11 reorganization, if an independent appraisal of collateral is necessary in Rural Development's opinion, Rural Development and the lender will share such appraisal fee equally.

(c) Expenses on Chapter 11 reorganization, liquidating Chapter 11 or Chapter 7 (unless the lender is directly handling the liquidation) cases are not to be deducted from the collateral proceeds.

(d) *Estimated loss payments.* See paragraph XVI of Form RD 449–35.

Administrative

Refer to appendix G of this subpart (available in any Rural Development office) for advice on how to interact with the lender on liquidation and property management.

A. It is the responsibility of the State Program Chief to see that Rural Development is being fully informed by the lender in all bankruptcy cases.

B. All bankruptcy cases should be reported immediately to the National Office by utilizing and completing a problem/delinquent status report. The Regional Attorney must be informed promptly of the proceedings.

C. Chapter 11 pertains to a reorganization of a business contemplating an ongoing business rather than a termination and dissolution of the business where legal protection is afforded to the business as defined under Chapter 11 of the Bankruptcy Code. Consequently, expenses incurred by the lender in a Chapter 11 reorganization can never be liquidation expenses unless the proceeding becomes a Liquidating 11. If the proceeding should become a Liquidating 11, reasonable and customary liquidation expenses may be deducted from proceeds of collateral provided the lender is doing the actual liquidation of the collateral as provided by the Lender's Agreement. Chapter 7 pertains to a liquidation of the borrower's assets. If and when liquidation of the borrower's assets under Chapter 7 is conducted by the bankruptcy trustee, the lender cannot claim expenses.

D. The State Director may approve the repurchase of the unpaid guaranteed portion of the loan from the holder(s) to reduce interest accruals during Chapter 7 proceedings or after a Chapter 11 proceeding becomes a liquidation proceeding. On loans in bankruptcy, any loss payment must be halted in accordance with the Lender's Agreement and carry the approval of the State Director.

E. The State Director must approve in advance and in writing the lender's estimated liquidation expenses on loans in liquidation bankruptcy. These expenses must be reasonable and customary and not in-house expenses of the lender.

F. The lender is responsible for advising Rural Development of the completion of the Chapter 11 reorganization plan; however, the Rural Development servicing office will monitor the lender's files to ensure timely notification of servicing actions.

G. If an estimated loss claim is paid during the operation of the reorganization plan, and the borrower repays in full the remaining balance of the loan as set forth in the plan without an additional loss sustained by the lender, a Final Report of Loss is not necessary. The Finance Office will close out the estimated loss account as a Final Loss at the time notification of payment in full is received.

H. If the bankruptcy court attempts to direct that loss payments will be applied to the account other than the unsecured principal first and then to unsecured accrued interest, the lender is responsible for notifying the Rural Development servicing office immediately. The Rural Development servicing office will then obtain advice from OGC on what actions Rural Development should take.

I. Protective Advances—Authorized protective advances may be included with the estimated loss payment associated with the Chapter 11 reorganization provided they were incurred in connection with liquidation of the account prior to the borrower filing bankruptcy.

J. Adequate Protection—The bankruptcy court can order protection of the collateral while the borrower is in a reorganization bankruptcy. The lender whose collateral is subject to being used by the trustee in bankruptcy should immediately seek adequate protection of the collateral, including petitioning for a super priority.

§ 1980.476 Transfer and assumptions.

(a) All transfers and assumptions will be approved in writing by Rural Development. Such transfers and assumptions will be to an eligible applicant.

(b) Transfers and assumptions will be considered without regard to § 1980.451 (d) of this subpart.

(c) The borrower will submit to Rural Development Form RD 449–4 for the required character evaluation prior to the execution of the Assumption Agreement.

(d) Available transfer and assumption options to eligible borrowers include the following:

(1) The total indebtedness may be transferred to another borrower on the same terms.

(2) The total indebtedness may be transferred to another borrower on different terms not to exceed those terms for which an initial loan can be made.

(3) Less than the total indebtedness may be transferred to another borrower on the same terms.

(4) Less than the total indebtedness may be transferred to another borrower on different terms.

(e) In any transfer and assumption case, the transferor, including any guarantor(s), may be released from liability by the lender with Rural Development written concurrence only when the value of the collateral being transferred is at least equal to the amount of the loan or part of the loan being assumed. If the transfer is for less than the entire debt:

(1) Rural Development must determine that the transferor and any guarantors have no reasonable debt-paying ability considering their assets and income at the time of transfer.

(2) The Rural Development County Committee must certify that the transferor has cooperated in good faith, used due diligence to maintain the collateral against loss, and has otherwise fulfilled all of the regulations of this subpart to the best of borrower's ability.

(f) Any proceeds received from the sale of secured property before a transfer and assumption will be credited on the transferor's guaranteed loan debt in inverse order of maturity before the transfer and assumption transaction is closed.

(g) When the transferee makes any cash downpayment in connection with the transfer and assumption:

(1) The lender will employ an independent appraiser, subject to concurrence of both the transferor and transferee, to make an appraisal to determine the fair market value of all the collateral securing the loan. Such appraisal report fee and any other costs related thereto will be paid by the transferor and the transferee as they mutually agree.

(2) The market value of the secured property being acquired by the transferee, plus any additional security the transferee proposes to give to secure the debt, will be adequate to secure the balance of the total guaranteed loan owed, plus any prior liens. If any cash downpayment is made, it may be paid directly to the transferor as payment for equity in the project provided:

(i) The lender recommends and Rural Development approves the case downpayment be released to the transferor. The lender and Rural Development may require that an amount be retained for an established period of time in escrow as a reserve account as security for use against any future default on the loan. Any interest accruing on such an escrow account may be paid periodically to the transferor.

(ii) Any payments that are to be made by the transferee to the transferor in respect to the downpayment do not suspend the transferee's obligation to continue to meet the guaranteed loan payments as they come due under the terms of the assumption.

(iii) The transferor will agree not to take any actions against the transferee in connection with such transfer in the future without first obtaining the written approval of Rural Development and the lender.

(iv) The lender determines that there is repayment ability for the guaranteed debt assumed and any other indebtedness of the transferee.

(h) The lender will make, in all cases, a complete credit analysis to determine viability of the project, subject to Rural Development review and approval, including any requirement for deposits in an escrow account as security to meet its determined equity requirements for the project.

(i) The lender will issue a statement to Rural Development that the transaction can be properly transferred and the conveyance instruments will be filed, registered, or recorded as appropriate and legally permissible.

(j) Rural Development will not guarantee any additional loans to provide equity funds for a transfer and assumption.

(k) The assumption will be made on the lender's form of assumption agreement.

(l) The assumption agreement will contain the Rural Development case number of the transferor and transferee.

(m) Loan terms cannot be changed by the Assumption agreement unless previously approved in writing by Rural Development, with the concurrence of any holder(s) and concurrence of the transferor (including guarantors) if they have not been released from personal liability. Any new loan terms cannot exceed those authorized in this subpart. The lender's request will be supported by:

(1) An explanation of the reasons for the proposed change in the loan terms.

(2) Certification that the lien position securing the guaranteed loan will

be maintained or improved, proper hazard insurance will be continued in effect and all applicable Truth in Lending requirements will be met.

(n) In the case of a transfer and assumption, it is the lender's responsibility to see that all such transfers and assumptions will be noted on all originals of the Loan Note Guarantee(s). The lender will provide Rural Development a copy of the transfer and assumption agreement. Notice must be given by the lender to Rural Development before any borrower or guarantor is released from liability.

(o) The holder(s), if any, need not be consulted on a transfer and assumption case unless there is a change in loan terms.

(p) If a loss should occur upon consummation of a complete transfer of assets and assumption for less than the full amount of the debt and the transferor-debtor (including personal guarantor) is released from personal liability, as provided in paragraph (e) of this section, the lender, if it holds the guaranteed portion, may file an estimated "report of Loss" on Form RD 449–30 to recover its pro rata share of the actual loss at that time. In completing Form RD 449–30, the amount of the debt assumed will be entered on Line 24 as Net Collateral (Recovery). Approved protective advances and accrued interest thereon made during the arrangement of a transfer and assumption, if not assumed by the transferee, will be entered on Form 449–30, lines 13 and 14.

Administrative

Refer to appendix G of this subpart (available in any Rural Development Office) for advice on how to interact with the lender on liquidations and property management.

A. The State Director may approve all transfer and assumption provisions if the guaranteed loan debt balance is within his/her individual loan approval authority including:

1. Consent in writing to the release of the transferor and guarantors from liability.

2. Any changes in loan terms.

NOTE: The assumption will be reviewed as if it were a new loan. The Loan Note Guarantee(s) will be endorsed in the space provided on the form(s).

B. A copy of the Assumption Agreement will be retained in the Rural Development file. The State Director will notify the Finance Office of all approved transfer and assumption cases on Form RD 1980–7, "Notice of Transfer and Assumption of a Guaranteed Loan," and submit Form RD 1980–50 for all new borrowers and Form RD 1980–51, "Add, Change, or Delete Guaranteed Loan Record," in order that Finance records may be adjusted accordingly.

C. Any transfer and assumption of less than the total indebtedness must be submitted to the Director, Business and Industry Division, for review and concurrence.

D. If the guaranteed loan debt balance is in excess of the State Director's loan approval authority, the State Director will forward the file, together with his/her recommendations, to the National Office for approval, ATTN: Business and Industry Division.

§§ 1980.477–1980.480   [Reserved]

§ 1980.481   Insured loans.

Applications from private parties for whom Rural Development and such borrowers agree that a guarantee lender is not available and from public bodies shall be processed as insured loans in accordance with the applicable provisions of this subpart and subpart A of part 1942 of this chapter, including the credit elsewhere requirement, except as provided in § 1980.488 of this subpart which provides for the guarantee of taxable bond issues of public bodies. Loans to public bodies will be used only to finance:

(a) Community facilities as defined in § 1980.402 of this subpart, and

(b) Constructing and equipping industrial plants for lease to private businesses (not including loans for operating such businesses) when the requesting loan is not available under subpart A of part 1942 of this chapter.

Administrative

A. Without specific written delegated authority, all insured loans require National Office concurrence prior to approval.

B. Applications from private parties for insured loans will not be encouraged.

C. Loan closings on insured loans will be in accordance with this subpart, the Regional Attorney and applicable provisions of subpart A of part 1942 of this chapter.

[52 FR 6501, Mar. 4, 1987, as amended at 53 FR 40403, Oct. 17, 1988]

§§ 1980.482–1980.487 [Reserved]

§ 1980.488 Guaranteed industrial development bond issues.

(a) Loans to public bodies will be guaranteed only in connection with the issuance of any class or series of industrial development bonds (as defined in section 103(c)(2) of the Internal Revenue Code of 1954, as amended (IRC)), the interest on which is included in gross income under IRC. No part of the loan guaranteed by Rural Development may extend to any class or series of industrial development bonds the interest on which is excludable from gross income under section 103(a)(1) of such Code. Before the execution of any Loan Note Guarantee, the lender will furnish Rural Development evidence regarding interest on bonds being taxable for Federal income tax purposes. Such evidence may be in the form of an unqualified opinion of a recognized bond counsel or a ruling from the Internal Revenue Service. Guaranteed loans to public bodies can only be used for constructing and equipping industrial plants for lease to private businesses engaged in industrial manufacturing and does not provide funds for debt refinancing, working capital and other miscellaneous fees, charges or services. The lessee will have to provide necessary capital and sufficient financial strength to provide for a sound project.

(b) If Rural Development and the applicant agree that a guaranteed lender is not available, the application may be considered for an insured loan under the provisions of § 1980.481 of this subpart.

Administrative

The lender is responsible for notifying the Rural Development of the taxability of the proposed bond issue.

§ 1980.489 [Reserved]

§ 1980.490 Business and industry buydown loans.

(a) *Introduction.* This section contains regulations for the Business and Industry Buydown (BIB) loan program. The purpose of this program is to provide loan guarantees with reduced interest rates to the borrowers, under the authority of Public Law 103–50 (107 Stat. 241). All provisions of Subparts A and E of this part apply to BIB loans except as provided in this section. All forms used in connection with a BIB loan will be those used with other B&I loans, except as provided in this section.

(b) *Location of applicants.* Businesses eligible for BIB loans shall be located within the area covered by the Presidential disaster declaration related to Hurricanes Andrew or Iniki or Typhoon Omar.

(c) *Interest rate.* (1) If the interest rate charged by the lender (note rate) on a BIB loan is a variable rate in accordance with § 1980.423 of this subpart, the base rate must be the prime rate as published in the Wall Street Journal and the note rate must not exceed the prime rate as published in the Wall Street Journal by more than 100 basis points. If the note rate is fixed, it must not exceed by more than 100 basis points the prime rate as published in the Wall Street Journal on the day the Loan Note Guarantee is issued.

(2) The note rate for a BIB loan must be the same for the entire loan, including both the guaranteed and unguaranteed portion.

(d) *Interest rate buydown.* (1) To be eligible for a BIB loan, the business must provide evidence and the lender and Rural Development must determine that, at least for the first year of the loan, the business will not have adequate cash flow to meet all of its financial obligations including the required payments on the proposed loan at the note rate, but that it can meet all obligations if the interest rate is reduced by 100 basis points.

(2) During the first year after a Loan Note Guarantee is issued for a BIB loan, Rural Development will pay one percentage point of interest on the loan directly to the lender, thereby reducing the interest due from the borrower by this amount. This interest payment shall be applied to both the guaranteed and unguaranteed portion of the loan pro ratably according to Rural Development regulations.

(3) Interest payments by Rural Development may continue in subsequent years if the borrower's cash flow is insufficient to pay all obligations including the required payments on the proposed loan at the note rate. On or about each yearly anniversary of the

promissory note the lender may submit a request to Rural Development for continued interest payments, along with current profit and loss and cash flow statements and cash flow projections to show that the continued payments are needed for another year. Rural Development will promptly review the material submitted, determine whether the continued interest payments by Rural Development are needed to provide for sufficient cash flow in the coming year, and notify the lender in writing of the determination. Once interest payments by Rural Development are terminated because the borrower's cash flow is determined to be sufficient to pay the note rate, such payments will not be made in subsequent years even if the cash flow decreases.

(4) This section does not authorize interest payments by Rural Development on B&I loans other than those approved under this section. To be eligible for interest payments by Rural Development, the loan must be designated as a BIB loan when approved and funded from funds authorized by Public Law 103–50.

(e) *Duration of BIB loan program.* No BIB loan will be obligated after September 30, 1994.

(f) *Administrative procedures.* (1) A lender that wants a B&I application considered under BIB authorities should so indicate by notation on Form RD 449–1 or by letter submitted with the Form RD 449–1.

(2) Rural Development will identify a loan as a BIB loan by notation in the top margin of Form RD 449–29 and by the "type of assistance" code listed on Form RD 1940–3, in accordance with the Forms Manual Insert.

(3) Rural Development will set out the interest buydown provisions in accordance with this section in the Conditional Commitment for Guarantee. When the Loan Note Guarantee is issued, the lender and Rural Development will execute Form RD 1980–48, "Business and Industry Interest Rate Buydown Agreement."

(4) The lender will request the interest payment from Rural Development by submitting Form RD 1980–23, "Request for Business and Industry Interest Buydown Payment," to the Rural

Development servicing office. Each request must cover exactly 1 year and be filed within 30 days after the anniversary date of the promissory note, except when interest buydown is terminated between anniversary dates. The Rural Development servicing office will review each request for consistency with Rural Development regulations and the Form RD 1980–48 and, if the claim is valid, will approve it and forward it to the Finance Office for issuance of the payment to the lender.

(g) *Termination of interest buydown.* When Rural Development purchases a portion of a loan, interest buydown will cease on the entire loan. Interest buydown will also cease upon termination of the Loan Note Guarantee or assumption/transfer of the loan. In the event of any action that causes the interest buydown to terminate, the lender will submit a claim on Form RD 1980–23 for interest buydown payments through the date of termination.

(h) *Loan purposes*—(1) *Refinancing.* Section 1980.452 *Administrative* C.1. (d) of this subpart does not apply to BIB loans if refinancing is needed as a direct consequence of the disaster. In such cases, the lender may be allowed to bring previously unguaranteed exposure under the guarantee. No loan will be refinanced unless the current market value of the collateral is at least equal to the amount of the loan to be refinanced plus any new loan amount.

(2) *Agriculture.* Section 1980.412 (e) of this subpart does not apply to BIB loans. BIB loans may be guaranteed for agriculture production, which means the cultivation, production (growing), and harvesting, either directly or through integrated operations, of agricultural products (crops, animals, birds, and marine life, either for fiber or food for human consumption), and disposal or marketing thereof, the raising, housing, feeding (including commercial custom feedlots), breeding, hatching, control and/or management of farm or domestic animals.

(3) *Other eligible businesses.* Eligible types of businesses also include:

(i) Commercial nurseries primarily engaged in the production of ornamental plants and trees and other nursery products such as bulbs, florists' greens, flowers, shrubbery, flower and vegetable seeds, sod, and the growing of vegetables from seed to the transplant stage.

(ii) *Forestry* which includes establishments primarily engaged in the operation of timber tracts, tree farms, forest nurseries, and related activities such as reforestation.

(iii) The growing of mushrooms or hydroponics.

(4) *Recreation and tourism.* Loans may be guaranteed for tourist or recreation facilities except for hotels, motels, bed and breakfasts, race tracks, gambling, or golf courses.

(5) *Meat processing facilities.* The provisions of §1980.411 (a)(8) of this subpart will not apply to BIB loans. Loans, including working capital or debt refinancing, may be guaranteed for businesses engaged in meat or poultry processing.

(i) *Small Business Administration.* Section 1980.451 (c) of this subpart will not apply to BIB loans. Applicants eligible for Small Business Administration assistance will be advised of the availability of that assistance.

(j) *Loan guarantee limits.* Notwithstanding the provisions of §1980.420 of this subpart, the guarantee percentage on any BIB loan will not exceed 80 percent.

(k) *Credit quality analysis.* In analyzing the credit quality of a proposed loan to a business that has lost assets to a natural disaster, primary emphasis will be placed on the operating history of the business, rather than its current financial condition. If the business has a sound, profitable and successful history prior to the disaster and there are reasonable projections to ensure it can operate successfully in the future, the proposed loan may be approved even if disaster losses have caused somewhat less equity and/or collateral than would normally be expected for a B&I loan guarantee. If the business appears to have had an unprofitable operation or inadequate cash flow prior to the disaster, the proposed loan guarantee will not be approved.

(l) *Equity requirements.* The equity requirements of §1980.441 of this subpart do not apply to BIB loans.

(m) *Collateral.* Section 1980.443 Administrative A. 2., 3., and 4. of this subpart will not apply to BIB loans. Collateral may be considered at its current market value without discount. Work-in-process inventory may be valued at the estimated market value of the finished product. All costs of producing the finished product must be included in the cash flow analysis.

(n) *Conditional approval.* A Form RD 449–14 may be issued prior to receipt of specific items needed to complete an application package provided:

(1) The lender and/or borrower demonstrates to the Government's satisfaction that it has a need for a prompt indication of the availability of the proposed loan guarantee and the conditions under which a guarantee are available;

(2) The specific items missing from the application package will take considerable time to obtain;

(3) The lender requests a commitment prior to providing the items;

(4) The attachment to Form RD 449–14 clearly states that the commitment is conditioned on satisfactory completion of the missing item(s) and a guarantee will not be issued unless all conditions of these regulations are met; and

(5) No Form RD 449–14 will be issued prior to the obligation date established with the Finance Office.

(o) *Financial statements.* All requirements of §1980.451(i)(13) of this subpart will apply except that for BIB loans minimum annual financial statements will be required as follows:

(1) For nonagricultural borrowers with a B&I indebtedness of $500,000 or less, an annual compilation by an independent certified public accountant or by an independent public accountant licensed and certified on or before December 31, 1970.

(2) For nonagricultural borrowers with a B&I indebtedness of $500,001 through $1 million, an annual review by an independent certified public accountant or by an independent public accountant licensed and certified on or before December 31, 1970.

(3) For nonagricultural borrowers with a B&I indebtedness of more than $1 million, an annual audited financial statement by an independent certified public accountant or by an independent public accountant licensed and certified on or before December 31, 1970.

(4) All agricultural loans will require annual financial statements per § 1980.113 of subpart B of this part.

(p) *Agriculture loans.* The following additional provisions apply to BIB loan guarantees for businesses engaged in agriculture production:

(1) *General policy.* Paragraph (p) of this section contains the regulations for making BIB loans to farmers for agricultural purposes. BIB loans made for agricultural purposes are subject to the provisions in subparts A and E of this part except as specified. In addition, certain sections of subpart B of this part referenced in this section are applicable subject to the limitations outlined in this section. Several key loan processing and loan servicing requirements stipulated in subpart B of this part do not apply to loans made to borrowers under this section.

(2) *Type of guarantee.* BIB loans will be processed under the Loan Note Guarantee option of § 1980.101 (e)(1) of subpart B of this part Only. No loan will be processed for a Contract of Guarantee (Line of Credit) under § 1980.101 (e)(2) of subpart B of this part.

(3) *Farm size.* Loan guarantees may be made under the BIB program without regard to the size of the farming operation.

(4) *Filing and processing preapplications and applications.* If the applicant has already developed material for an Rural Development Farmer Programs loan or if the financial and production information required by § 1980.113 of subpart B of this part is needed to document repayment ability or is required by the lender, § 1980.113 of subpart B of this part may apply with the following exceptions:

(i) Lines of credit will not be guaranteed.

(ii) If the application is submitted solely for a farm as defined in § 1980.106(b) of subpart B of this part, Form RD 1980-25, "Farmer Programs Application," or Form RD 449-1, will be used as an application for assistance.

(5) *Evaluation of applications.* If the application is developed and processed in accordance with § 1980.113 of subpart B of this part, the provisions outlined in § 1980.114 of subpart B of this part apply with the following exceptions:

(i) Timeframe requirements for the evaluation of applications and references to the Approved Lender Program are not applicable.

(ii) County Committee reviews of applications processed under this section will not be required. If the loan approval official finds the applicant is not eligible, the applicant will be notified in writing of the reasons for disapproval and his/her rights through inclusion of the Equal Credit Opportunity Act (ECOA) statement. An opportunity will be given for an appeal as set out in subpart B of part 1900 of this chapter.

(iii) When applied to BIB applications, references in § 1980.114 of this part to "County Office" shall normally be construed to mean "State Office." References to "County Supervisor" shall be construed to mean "Business and Industry Chief or Community and Business Programs Chief, or other appropriate FmHA or its successor agency under Public Law 103–354 official as designated by the State Director."

(6) *Terms of loan repayment.* (i) Principal and interest on the loan will be due and payable to coincide with the cash flow operating cycle of the business. Installments will be scheduled for payment as agreed upon by the lender and borrower on terms that reasonably assure repayment of the loan. The first installment to include a repayment of principal may be scheduled for payment after the project is operational and has begun to generate income. However, such installment will be due and payable within 6 years from the date of the debt instrument and at least annually thereafter. Interest will not be deferred and will be due at least annually from the date of the debt instrument. In granting a deferral of principal payment, the loan approval official must document based on pro forma financial statements and the nature of the crop that the deferral of payments is necessary.

(ii) The lender must ensure that loan repayment is scheduled to eliminate the possibility of a balloon payment at the end of the loan.

(7) *Agriculture BIB loan purposes.* Loans may be made only for the following purposes:

(i) Operating purposes as outlined in §1980.175 (c)(1) of subpart B of this part except for those stipulated in §1980.175(c)(1)(iv) and (vii).

(ii) Real estate purposes as outlined in §1980.180 (c) of subpart B of this part except for those stipulated in §1980.180 (c)(1) and (4).

(iii) Refinancing in accordance with paragraph (h)(1) of this section and §§1980.411 (a)(11), 1980.451 (i)(19), and 1980.452 Administrative C. (except §1980.452 Administrative C. 1. (d) of this subpart.

(8) *Sodbuster and swampbuster requirements.* The requirements found in 7 CFR part 1970 will apply to loans made to enterprises engaged in agricultural production.

[52 FR 6501, Mar. 4, 1987, as amended at 81 FR 11048, Mar. 2, 2016]

§§1980.491–1980.494 [Reserved]

§1980.495 RD forms and guides.

The following RD forms and guides, as applicable, are used in connection with processing B&I, D&D, and DARBE loan guarantees; they are incorporated in this subpart and made a part hereof:

(a) Form RD 449–1. "Application for Loan and Guarantee" is referred to as "Appendix A," or successor form,

(b) The "Certificate of Incumbency and Signature" or successor form,

(c) "Guidelines for Loan Guarantees for Alcohol Fuel Production Facilities" is referred to as "Appendix C. "

(d) "Alcohol Production Facilities Planning, Performing, Development and Project Control" is referred to as "Appendix D. "

(e) "Environmental Assessment Guidelines" is referred to as "Appendix E. "

(f) Form RD 449–14, "Conditional Commitment for Guarantee, " or successor form.

(g) "Liquidation and Property Management Guide" as found in RD Instruction 1980–E Appendix G.

(h) "Suggested Format for the Opinion of the Lender's Legal Counsel" is referred to as "Appendix H. "

(i) "Instructions for Loan Guarantees for Drought and Disaster Relief" and Forms RD 1980–68, "Lender's Agreement—Drought and Disaster Guaranteed Loans, " 1980–69, "Loan Note Guarantee—Drought and Disaster Guaranteed Loans, " and 1980–70, "Assignment Guarantee Agreement—Drought and Disaster Guaranteed Loans, or their successor forms.

(j) [Reserved]

(k) "Regulations for Loan Guarantees for Disaster Assistance for Rural Business Enterprises" and Forms RD 1980–71, "Lender's Agreement—Disaster Assistance for Rural Business Enterprises Guaranteed Loans," 1980–72 "Loan Note Guarantee—Disaster Assistance for Rural Business Enterprises Guaranteed Loans," and 1980–73 "Assignment Guarantee Agreement—Disaster Assistance for Rural Business Enterprises Guaranteed Loans" or their successor forms.

[80 FR 9910, Feb. 24, 2015]

§1980.496 Exception authority.

The Administrator may in individual cases grant an exception to any requirement or provision of this subpart which is not inconsistent with any applicable law or opinion of the Comptroller General, provided the Administrator determines that application of the requirement or provision would adversely affect the Government's interest. Requests for exceptions must be in writing by the State Director and submitted through the Assistant Administrator, Community and Business Programs. Requests must be supported with documentation to explain the adverse effect on the Government's interest, propose alternative courses of action, and show how the adverse effect will be eliminated or minimized if the exception is granted.

§1980.497 General administrative.

Refer to RD Instruction 1980–E, Appendix G, Liquidation and Property Management Guide (available in any Rural Development office) for advice on how to interact with the OGC on liquidations and property management.

(a) *Office of the General Counsel (OGC).* In performing the Rural Development functions with respect to B&I, D & D, and DARBE loans, the advice and assistance of OGC may be sought and followed on any legal matter. However, it is the responsibility of the lender to ascertain that all requirements for making, securing, and servicing the loan are duly met. If Rural Development has any questions concerning the lender's resolution of these matters, OGC should be consulted. Assistance of OGC will be requested on all loans as specified herein and all liquidations and workouts.

(b) *Contact with OGC.* Initial informal contact with OGC should be made as soon as possible. Rural Development State Directors should use the following format in formally requesting legal assistance on workouts.

(1) *Origination:* All written requests should come from the State Director.

(2) *Method:* Request should be made by referral memorandum to the Regional Attorney setting forth a brief statement of the facts, the reason assistance is requested, the extent of legal assistance sought, the date when Rural Development's response to the lender's liquidation plan (if any) is due and:

(i) *Projected losses on collateral: e.g.,* projected losses on collateral are expected to be significant.

(ii) *Unusual or complex nature of primary collateral: e.g.,* multi-state foreclosures or foreclosure of leases or general intangibles.

(iii) *Presence of other major creditors or of senior creditors: e.g.,* guaranteed loan collateral may be subject to a prior lien or other creditors may have rights in other assets of borrower, such as inventory and accounts receivable.

(iv) *Litigation is pending or threatened: e.g.,* bankruptcy, other foreclosure suits.

(3) *Materials to submit:* Referral memorandums will be accompanied by a copy of lender's liquidation plan together with a copy of Rural Development's planned response and principal loan papers, conditional commitment for guarantee, guarantee documents and any comments from the National Office. If lender refuses to prepare a plan, the State Director should so state. DO NOT SEND DOCKETS unless specifically requested by OGC.

(c) *Reviews prior to issuance of the loan note guarantee.* After the conditional commitment for guarantee has been issued and proposed with closing documents prepared by the lender and for-

warded to Rural Development with the lender's legal counsel's opinion in the suggested legal format of appendix H of this subpart, but prior to issuing the loan note guarantee, the State Director will forward the loan docket to the Regional Attorney for review. After an administrative review, the State Director will include with the docket a letter with recommendations and indicating any special items, documents or problems that need to be addressed specifically which may have a significant impact upon the loan or may be contrary to the regulation. The docket will be assembled for OGC review in accordance with § 1980.451 Administrative B 5 of this subpart and indexed and tabbed.

(d) *Please submit the following for OGC review.* Copies of:

(1) Letter from Rural Development National Office authorizing loan guarantee containing conditions (if applicable);

(2) Form RD 449-14, including any amendments;

(3) Loan Agreement;

(4) Promissory Notes;

(5) Security documents—Real Estate Mortgage, Security Agreement, Financing Statements, and Leases (if applicable);

(6) Personal or corporation guarantees with related security documents;

(7) Proposed Form RD 449-35.

(8) Proposed Form RD 449-34.

(9) Proposed Form RD 449-36, if any;

(10) Proposed Lender's Certification (§ 1980.60 of subpart A of this part); and

(11) Opinion of Lender's Counsel in form prescribed by OGC.

(e) *Do not submit for OGC review* feasibility studies, title information, or the original application unless specifically requested to do so.

(f) *OGC advice.* The Regional Attorney will review the docket and furnish advice to Rural Development on whether it may issue the LOAN NOTE GUARANTEE AFTER THE LOAN IS CLOSED. SUCH ADVICE IS FOR THE benefit of RURAL DEVELOPMENT ONLY AND DOES NOT RELIEVE THE LENDER OF ITS RESPONSIBILITIES UNDER RURAL DEVELOPMENT REGULATIONS. The Regional Attorney at his/her option may attend the loan closing. Upon receipt of the Regional Attorney's advice, the State Director will correct or cause to be corrected any noted deficiencies before issuing the Loan Note Guarantee.

(g) *Delegation of authority.* The State Director may delegate those administrative duties and responsibilities as authorized in the Administrative sections of this subpart, except those specifically reserved to the State Director.

§1980.498 **Business and Industry Disaster Loans.**

(a) *Introduction.* This section contains regulations for the Business and Industry Disaster (BID) loan program. The purpose of the program is to provide loan guarantees under the authority of the Dire Emergency Supplemental Appropriations Act, 1992, Public Law 102–368. These guaranteed loans cover costs arising from the consequences of natural disasters such as Hurricanes Andrew and Iniki and Typhoon Omar that occur after August 23, 1992, and receive a Presidential declaration. Also included are the costs to any producer of crops and livestock that are a consequence of at least a 40 percent loss to a crop, 25 percent loss to livestock, or damage to building structures from a microburst wind occurrence in calendar year 1992. No BID loan guarantee will be approved after September 30, 1993. All provisions of subparts A and E of part 1980 of this chapter apply to BID loans, except as provided in this section. All forms used in connection with a BID loan will be those used with other Business and Industry (B&I) loans, except as provided in paragraph (m) of this section.

(b) *Location of Applicants.* (1) Section 1980.405 of this subpart. "Rural area determinations," will not apply to BID loans. BID loans may be made in rural and nonrural areas.

(2) Eligible borrowers' businesses must be located within the area covered by the Presidential declaration except for those with qualifying losses from microburst wind in accordance with paragraph (a) of this section.

(c) *Loan Purposes.* Loans may be guaranteed for the purposes listed in §1980.411 of this subpart, "Loan Purposes," except as follows:

(1) *Relationship to disaster.* The purpose of any BID loan must be to cover costs that are a direct consequence of a natural disaster or microburst of wind in accordance with paragraph (a) of this section. The amount of the loan must not be greater than the amount needed as determined by the Rural Development Administration or its successor agency under Public Law 103–354 (RDA or its successor agency under Public Law 103–354) to cure problems caused by the natural disaster so that the business is reestablished on a successful basis. Facilities which were damaged or destroyed by the natural disaster may be repaired or replaced by modern facilities as necessary to ensure success. Replacement by modern facilities will not be made solely for the purpose of enlarging the business or increasing its production capacity. No loan for a change of purpose of the business will be guaranteed. Eligible refinancing or working capital loans should not exceed the amount needed to overcome the financial distress caused by the disaster. Losses that were adequately paid by insurance or by loans or grants from other sources will not be covered by BID loans. BID loans may be used to supplement insurance payments and/or assistance from other sources when the insurance coverage or other assistance is not sufficient.

(2) *Refinancing.* Section 1980.452, Administrative C.1.(d) of this subpart does not apply to BID loans. If refinancing is needed as a direct consequence of the disaster, the lender may be allowed to bring previously unguaranteed exposure under the guarantee. No loan will be refinanced unless the current market value of the collateral is at least equal to the amount of the loan to be refinanced plus any new loan amount.

(3) *Agriculture.* Section 1980.412(e) of this subpart does not apply to BID loans. BID loans may be guaranteed for agriculture production, which means the cultivation, production (growing), and harvesting, either directly or through integrated operations, of agricultural products (crops, animals, birds, and marine life, either for fiber or food for human consumption), and

disposal or marketing thereof, the raising, housing, feeding (including commercial custom feedlots), breeding, hatching, control and/or management of farm or domestic animals.

(4) *Other eligible businesses.* Eligible types of businesses also include:

(i) Commercial nurseries primarily engaged in the production of ornamental plants and trees and other nursery products such as bulbs, florists' greens, flowers, shrubbery, flower and vegetable seeds, sod, and the growing of vegetables from seed to the transplant stage.

(ii) Forestry which includes establishments primarily engaged in the operation of timber tracts, tree farms, forest nurseries, and related activities such as reforestation.

(iii) The growing of mushrooms or hydroponics.

(5) *Recreation and tourism.* Loans may be guaranteed for tourist or recreation facilities except for hotels, motels, bed and breakfasts, race tracks, gambling, or golf courses.

(6) *Meat processing facilities.* The provisions of § 1980.411(a)(8) of this subpart will not apply to BID loans. Loans, including working capital or debt refinancing, may be guaranteed for businesses engaged in meat or poultry processing.

(d) *Federal Emergency Management Agency (FEMA).* BID loans may be approved only to the extent that the assistance is not available from FEMA. The case file will be documented to show that FEMA assistance was not available or that FEMA assistance is not adequate to cover the costs as a consequence of the natural disaster.

(e) *Small Business Administration.* Section 1980.451 of this subpart will not apply to BID loans. Applicants eligible for Small Business Administration assistance will be advised of the availability of that assistance.

(f) *Loan guarantee limits.* Notwithstanding the provisions of § 1980.420 of this subpart, the guarantee percentage on any BID loan will not exceed 80 percent.

(g) *Credit quality analysis.* In analyzing the credit quality of a proposed loan to a business that has lost assets to a natural disaster, primary emphasis will be placed on the operating history of the business, rather than its current financial condition. If the business has a sound, profitable and successful history prior to the disaster and there are reasonable projections to ensure it can operate successfully in the future, the proposed loan may be approved even if disaster losses have caused somewhat less equity and/or collateral than would normally be expected for a B&I guarantee. If the business appears to have had an unprofitable operation or inadequate cash flow prior to the disaster, the proposed loan guarantee will not be approved.

(h) *Equity requirements.* The equity requirements of § 1980.441 of this subpart do not apply to BID loans.

(i) *Feasibility studies.* Feasibility studies as required by § 1980.442 of this subpart will not be required for BID loans if the business has a successful financial history that supports future plans and projections that indicate a successful operation with adequate repayment ability.

(j) *Collateral.* Section 1980.443, Administrative A. 2., 3., and 4. of this subpart will not apply to BID loans. Collateral may be considered at its current market value without discount. Work-in-process inventory may be valued at the estimated market value of the finished product. All costs of producing the finished product must be included in the cash flow analysis.

(k) *Conditional approval.* A Form RD 449-14, "Conditional Commitment for Guarantee," may be issued prior to receipt of specific items needed to complete an application package provided:

(1) The lender and/or borrower demonstrates to the Government's satisfaction that it has a need for a prompt indication of the availability of the proposed loan guarantee and the conditions under which a guarantee are available;

(2) The specific items missing from the application package will take considerable time to obtain;

(3) The lender requests a commitment prior to providing the items;

(4) The attachment to Form RD 449-14 clearly states that the commitment

is conditioned on satisfactory completion of the missing item(s) and a guarantee will not be issued unless all conditions of these regulations are met; and

(5) No Form RD 449–14 will be issued prior to the obligation date established with the Finance Office.

(1) *Financial statements.* All requirements of §1980.451(i)(13) of this subpart will apply except that it is modified for BID loans to require minimum annual financial statements as follows:

(1) For nonagricultural borrowers with a B&I indebtedness of $500,000 or less, an annual compilation by an independent certified public accountant or by an independent public accountant licensed and certified on or before December 31, 1970.

(2) For nonagricultural borrowers with a B&I indebtedness of $500,001 through $1,000,000, an annual review by an independent certified public accountant or by an independent public accountant licensed and certified on or before December 31, 1970.

(3) For nonagricultural borrowers with a B&I indebtedness of more than $1 million, an annual audited financial statement by an independent certified public accountant or by an independent public accountant licensed and certified on or before December 31, 1970.

(4) All agricultural loans will require annual financial statements per §1980.113 of subpart B of part 1980 of this chapter.

(m) *Agriculture loans.* The following additional provisions apply to BID loan guarantees for businesses engaged in agriculture production:

(1) *General policy.* This portion of this section contains the regulations for making BID loans to farmers for agricultural purposes. BID loans made for agricultural purposes are subject to the provisions in subparts A and E of part 1980 of this chapter except as specified. In addition, certain sections of subpart B of part 1980 of this chapter referenced in this section are applicable subject to the limitations outlined in this section. BID loans made for agricultural purposes are made under the Business and Industry authority of section 310B of the Consolidated Farm and Rural Development Act of 1972, as amended. In this regard, several key loan proc-

essing and loan servicing requirements stipulated in subpart B of part 1980 of this chapter do not apply to loans made to borrowers under this section. Only the material cross-referenced to subpart B of part 1980 of this chapter is to be utilized in lieu of or in addition to the requirements contained in subpart E of part 1980 of this chapter in processing loans under this section.

(2) *Type of guarantee.* See §1980.101(e)(1) of subpart B of part 1980 of this chapter. BID loans will be processed under the Loan Note Guarantee option ONLY. No loan will be processed for a Contract of Guarantee (Line of Credit) under this section.

(3) *Abbreviations and definitions.* (i) The abbreviations and definitions found in §1980.106 of subpart B of part 1980 of this chapter will apply to loans made under this section except for "family farm," "related by blood or marriage," and "subsequent loans."

(ii) Loan guarantees may be made under the BID program without regard to the size of the farming operation.

(4) *Loan eligibility requirements.* In addition to the requirements set forth in this subpart, the requirements in §1980.175(b) of subpart B of part 1980 of this chapter regarding controlled substances are applicable.

(5) *Filing and processing preapplications and applications.* If the applicant has already developed material for an Rural Development Farmer Programs loan or if the financial and production information required by §1980.113 of subpart B of part 1980 of this chapter is needed to document repayment ability or is required by the lender, §1980.113 of subpart B of part 1980 of this chapter may apply with the following exceptions:

(i) Lines of credit will not be guaranteed.

(ii) Timeframes for applicant/lender notification in §1980.113 of subpart B of part 1980 of this chapter do not apply.

(iii) If the application is submitted solely for a farm as defined in §1980.106(b) of subpart B of part 1980 of this chapter, Form RDal Development 449–1, "Application for Loan and Guarantee," will be used as an application for assistance.

(6) *Evaluation of applications.* If the application is developed and processed

275

in accordance with § 1980.113 of subpart B of part 1980 of this chapter, the provisions outlined in § 1980.114 of subpart B of part 1980 of this chapter applies with the following exceptions:

(i) Timeframe requirements for the evaluation of applications and references to the Approved Lender Program are not applicable.

(ii) County Committee reviews of applications processed under this section will not be required. If the loan approval official finds the applicant is not eligible, the applicant will be notified in writing of the reasons for disapproval and the opportunity given for an appeal as set out in subpart B of part 1900 of this chapter.

(7) *Terms of loan repayment.* (i) Principal and interest on the loan will be due and payable to coincide with the cash flow operating cycle of the business. Installments will be scheduled for payment as agreed upon by the lender and borrower on terms that reasonably assure repayment of the loan. The first installment to include a repayment of principal may be scheduled for payment after the project is operable and has begun to generate income. However, such installment will be due and payable within 6 years from the date of the debt instrument and at least annually thereafter. All accrued interest will be due at least annually from the date of the debt instrument. In no case will interest be deferred. In granting a deferral of principal payment, the loan approval official must document based on pro forma financial statements and the nature of the crop that the deferral of payments is necessary.

(ii) The lender must ensure that loan repayment is scheduled to eliminate the possibility of a balloon payment at the end of the loan.

(8) *BID agriculture loan purposes.* Loans may be made only for the following purposes:

(i) Operating purposes as outlined in § 1980.175(c)(1) of subpart B of part 1980 of this chapter except for those stipulated in paragraphs (c)(1) (iv) and (vii) of that section.

(ii) Real estate purposes as outlined in § 1980.180(c) of subpart B of part 1980 of this chapter except for those stipulated in paragraphs (c) (1) and (4) of that section.

(iii) Refinancing in accordance with paragraphs (c)(1) and (c)(2) of this section and §§ 1980.411(a)(11), 1980.451(i)(19) and 1980.452 ADMINISTRATIVE C [except 1980.452 ADMINISTRATIVE C 1(d)] of this subpart.

(9) *Sodbuster and swampbuster requirements.* The requirements found in 7 CFR part 1970 will apply to loans made to enterprises engaged in agricultural production.

[57 FR 45969, Oct. 5, 1992, as amended at 58 FR 34342, June 24, 1993; 58 FR 38952, July 21, 1993; 58 FR 41172, Aug. 3, 1993; 58 FR 48300, Sept. 15, 1993; 81 FR 11048, Mar. 2, 2016]

§ 1980.499 [Reserved]

§ 1980.500 OMB control number.

The reporting and recordkeeping requirements contained in this regulation have been approved by the Office of Management and Budget and have been assigned OMB control number 0575–0029. Public reporting burden for this collection of information is estimated to vary from 5 minutes to 58 hours per response, with an average of 4 hours per response including time for reviewing instructions, searching existing data sources, gathering and maintaining the data needed, and completing and reviewing the collection of information. Send comments regarding this burden estimate or any other aspect of this collection of information, including suggestions for reducing this burden, to the Department of Agriculture, Clearance Officer, OIRM, Room 404–W, Washington, DC 20250; and to the Office of Management and Budget, Paperwork Reduction Project (OMB# 0575–XXXX), Washington, DC 20503.

[55 FR 19245, May 8, 1990]

APPENDIXES A–B TO SUBPART E OF PART
1980 [RESERVED]

APPENDIX C TO SUBPART E OF PART
1980—GUIDELINES FOR LOAN GUAR-
ANTEES FOR ALCOHOL FUEL PRODUC-
TION FACILITIES

(1) *Alcohol production facility.* An alcohol
production facility is a facility in which alcohol, suitable for use by itself or in combination with other substances as a substitute for petroleum or petrochemical feed-
stocks and not suitable for beverage purposes, is manufactured from biomass.

(2) The alcohol production facility includes
all facilities necessary for the production
and storage of alcohol and the processing of
the by-products of alcohol production. The
intent is to limit the alcohol and by-products processing facilities to those facilities
which are necessary to yield marketable
products and necessary for the financial success of the project. Further refinements,
such as gasoline blending or the construction
of facilities which use the alcohol or by-
products in another manufacturing process,
are not considered part of the alcohol production facility.

(3) Application will be reviewed by both
B&I personnel and the State Office engineer
and forwarded to the National Office if approval is recommended.

(4) The applicant should have a startup
tangible book equity of 20–25 percent. (Appraisal surplus and subordinated debt are not
eligible equity items.)

(5) Loan maturity maximums will be as
follows:

Real Estate = 15–20 years
Machinery & Equipment = 10 years or less
depending on the estimated life of the
equipment involved
Working Capital = 3 years (It is assumed
that the additional equity required for
these projects will provide much of the
working capital needs.)

(6) Farmers Home Administration or its
successor agency under Public Law 103–354
will ordinarily only finance new facilities
and will not get involved in the refinancing
of existing ones.

(7) Priority consideration will be given to
the use of primary fuel other than petroleum
or natural gas.

(8) A positive energy balance must be indicated and supported by appropriate data; i.e.,
the energy content of the alcohol produced
at the alcohol production facility must be
greater than the energy used to produce the
alcohol and by-products.

(9) Plant location, in relation to feed-
stocks, primary fuel and markets for product
and by-products, will be an important consideration.

(10) Debt refinancing will only be considered in modest amounts and only when necessary to provide a satisfactory lien position.

(11) Feasibility studies are very important
and required and will be prepared by competent and knowledgeable independent parties.

(12) Participating lenders must either have
expertise or the availability of expertise in
this field.

(13) The proposed operating managers must
have experience in this or a related field.

(14) Alcohol Fuel Production Facilities are
eligible for assistance under the Drought and
Disaster (D&D) Guaranteed Loan and Disaster Assistance for Rural Business Enterprises (DARBE) programs described in this
subpart, and especially in appendix I and appendix K. Any such loan must meet the requirements for D&D and DARBE loans.

[52 FR 6522, Mar. 4, 1987, as amended at 53 FR
40403, Oct. 17, 1988; 54 FR 5, Jan. 3, 1989, and
54 FR 26946, June 27, 1989; 54 FR 42483, Oct. 17,
1989]

APPENDIX D TO SUBPART E OF PART
1980—ALCOHOL PRODUCTION FACILI-
TIES PLANNING, PERFORMING, DE-
VELOPMENT AND PROJECT CONTROL

(I) *Design Policy.* The borrower shall ensure
or cause to be ensured that:

(A) All project facilities are designed utilizing accepted engineering practices and are
conformed to applicable Federal, State and
local codes and requirements.

(B) Proven equipment and processes are
employed in all project facilities unless an
exception is granted by the Administrator or
designee of Rural Development ("Administrator") in accordance with paragraph (B)(2)
hereof and pilot equipment or processes are
used instead.

(1) Equipment and processes shall be considered "proven" if they have been successfully employed in other commercial facilities.

(2) Equipment and processes shall be considered pilot if they have not been used in a
commercial operation but have been operated on a scale such that all design and material problems have been identified and resolved and operations maintained to demonstrate that the equipment and process
may be successfully applied to the proposed
commercial operation. Pilot equipment and
processes may be considered for use in the
project subject to the following:

(a) The plans, specifications, and operational data for the applicable facilities are
reviewed by the Administrator or designee
and lender. If, in the opinion of Rural Development, the proposed processes or equipment
are insufficiently developed to assure reliable and successful operation of the project,

277

proven processes and equipment will be utilized.

(b) If pilot processes or equipment are used, the Administrator or designee will also require that:

(i) Reasonable provision is made in the project for conversion to proven equipment or processes; and

(ii) The borrower agrees to convert to proven equipment or processes if conversion is necessary to protect the interest of the Government in the project. A reserve account for this conversion may be required. This account will not be an eligible loan purpose.

(C) Facility and equipment design incorporates cost-effective primary fuel systems, energy recovery systems and conservation measures to the maximum extent that this is feasible and consistent with paragraphs (I), (A), and (B) of this appendix.

(II) *Technical Services.* (A) The borrower is responsible for selecting engineering consultants with suitable experience, training and professional competence in the design and construction of the project to assure that the completed project will operate at the prescribed levels of performance. In discharging its responsibility the borrower will obtain or cause to be obtained:

(1) Full engineering services for design and construction inspection for all project facilities. Resident inspection by qualified persons will be required.

(2) Agreements for engineering or design/build services which describe the project facilities in terms of the parameters critical to the successful operation of the project. The parameters shall include input quantities, conversion efficiency, rate of production and fuel consumption and product quality under normal operating conditions. The design parameters will be mutually agreed upon by the borrower, lender, the State Director and the project engineer, and may not be modified without the written concurrence of each of these parties. These agreements for engineering or design/build services will require, or the borrower will otherwise obtain, assurance satisfactory to the State Director that:

(a) The project engineer will maintain adequate insurance to protect the borrower, lender and the Government from incurring expenses resulting from errors and omissions of the engineer in performance of engineering services.

(b) The project engineer will certify that only proven equipment and processes will be utilized in the proposed development. The State Director may request evidence of successful operations of such proven equipment and process. If proven equipment or processes are not used in the project, the project engineer will identify these items and provide the information necessary for acceptance by the Administrator, borrower and lender in accordance with paragraph (I)(B)(2) of this appendix.

(c) If used equipment or existing facilities are incorporated into the project, they must be inspected by the project engineer or by another qualified engineer of the borrower. This engineer will prepare a report describing the proposed facilities or equipment and will comment on their suitability for use in the project. The report will also identify the modifications necessary for successful integration into the project. A cost estimate will also be included comparing new equipment and facilities to the proposed existing facilities or used equipment. Consideration must be given to the relative energy requirements of used and new facilities and their relative operation and maintenance costs.

(d) The project engineer or qualified individuals representing the manufacturer of principal equipment (or the designer/builder if the contractor has designed the plant) will visit the plant site at reasonable intervals for a period of one year after substantial completion of the project. Such personnel will be experienced in the proper operation and maintenance of applicable plant components. A report will be presented to the borrower within two weeks of each site visit advising the borrower of operation and maintenance deficiencies. A copy of each report will be forwarded to the State Director and lender by the borrower.

(e) The project engineer will prepare or supervise the preparation of a record drawing of all facilities. One copy will be submitted to the lender and the borrower.

(f) The project engineer or another group acceptable to the State Director and lender will prepare an operation and maintenance manual and assist the borrower in the start-up of the project. The operation and maintenance manual will describe the specific operation and maintenance procedures which must be performed for the project to operate at its rated capacity and efficiency and outline product testing, quality control, plant safety and emergency shut-down procedures.

(g) The project engineer will assist the borrower in determining acceptability of materials, equipment and construction during the construction period, review shop drawings, payment estimates and change orders, and assist in determining substantial completion of the project and final completion of individual contracts. (1) The project is substantially complete when:

(i) Construction is sufficiently completed in accordance with plans and specifications so that the project may be used for its intended purpose, and;

(ii) The project is producing products of the quantity and quality and at the conversion and energy efficiencies proposed in the completed application submitted by the lender and borrower and approved by the Rural Development.

(2) The State Director must concur that the project is substantially complete. The

following evidence, in form and substance satisfactory to the State Director and lender, must be submitted prior to such concurrence:

(i) A certificate from the project engineer stating that all facilities are substantially complete. Engineers who design specialized equipment or processes must also certify that construction/fabrication is acceptable in accordance with plans and specifications previously approved by them. The certification of the project engineer must be based upon a project start-up procedure where the complete project operates continuously to reach steady-state operating conditions. During this period contractors and engineers will identify and correct problems in operations, malfunctions in equipment, failure in materials and defects in workmanship. After this pre-startup, the certifying engineers will monitor project operations for a continuous period of at least 72 hours or 3 consecutive batch runs as appropriate to assure that all equipment is operating satisfactorily at rated capacity and efficiency.

(ii) Copies of system operation and performance data obtained during project start-up.

(iii) Exceptions to substantial completion and a list of nonsubstantial items which must be completed prior to release of any contractor's retainage.

(3) If the project is not producing products of the required quantity or quality at the prescribed conversion efficiencies, even though the project is otherwise physically complete in accordance with paragraph (1)(i) of this subparagraph, the project engineer will prepare a report identifying the corrective actions including an estimate of costs and additional time necessary to meet established performance criteria.

(4) The project must be certified to be substantially complete by an independent engineer if any portion of the project has been designed or constructed by the borrower or the project engineer has participated in any portion of the construction.

(B) Modification of plans and specifications will not be made without the written authorization of the project engineer.

(C) The Administrator, State Director or their representative's acceptance or concurrence in feasibility studies, preliminary engineering reports, plans, specifications, contract documents and payment estimates will not be construed as a representation of the adequacy of same, reliability of cost estimates or quality of construction, nor will such acceptance or concurrence be deemed a waiver of any of the Government's rights or remedies against any person or party. Reviews and construction inspections by the Administrator, State Director or their representatives are solely for the benefit of the Government and do not relieve the lender or

borrower of their obligation to conduct project reviews and inspections.

(III) *Project Construction.*

(A) Borrower will not award contracts for the construction of any project facilities unless and until:

(1) The borrower obtains applicable construction permits, right-of-ways, licenses and approvals of Federal, State and local authorities for the construction of such facilities.

(2) The State Director concurs in applicable plans, specifications and contract documents. Standard contract documents prescribed for use in Federally assisted projects may be used as a guide for determining the minimum standards for contract acceptability. These standard documents are contained in Guides 18 and 19 of subpart A of part 1942 of this chapter (available in any Rural Development office).

(B) The borrower has the responsibility, without recourse to the Government, for the settlement and satisfaction of all contractual and administrative issues arising out of procurements. This includes, but is not limited to, disputes, claims, protests of awards, or other matters of a contractual nature. Matters concerning violation of laws are to be referred to such local, State, or Federal authority as may have proper jurisdiction.

(C) The borrower's attorney will review executed contract documents including applicable performance and payment bonds and provide a certificate to the borrower and lender that they have been properly executed and that the persons executing these documents have been properly authorized to do so.

(D) In all contracts for construction or facility improvement awarded in excess of $100,000, the borrower will require bonds and a bank letter of credit or cash deposit in escrow, assuring performance and payment of 100 percent of the contract cost. The surety will normally be in the form of performance and payment bonds. Such assurance shall remain in full force and effect through any warranty period. Companies providing performance and payment bonds must hold a certificate of authority as an acceptable security on Federal bonds and eligible for listing in Treasury circular 510 as amended and be legally doing business in the State the project is located.

(E) Project Changes. Any change in the project which may affect collateral, its ultimate financial viability or compliance with the conditional commitment must have prior approval of the lender and Rural Development.

(1) Construction contracts will require that change orders receive prior approval from the lender when such changes:

(a) Increase or decrease contract price,

(b) Materially modify contract provisions,

(c) Increase or decrease time of completion,

(d) Affect project performance.

(2) All change orders will be recorded on a chronologically numbered contract change order as they occur. Change orders will not be included in payment estimates until approved by the borrower, project engineer, the lender and concurred in by Rural Development.

(F) *Warranty.*

(1) All major equipment must be guaranteed by the manufacturer to be free from defects in workmanship and materials for a period of one year after start-up of equipment.

(2) Equipment purchased by a construction contractor or design builder and all other work shall be further warranted to be free from defect in material and workmanship by the contractor or the design builder for a period of one year after substantial completion of the contract.

(3) Applicable provisions to this effect shall be included in equipment purchase orders or construction contracts.

(G) *Lease agreements.* Where the right of use or control of any property or equipment not owned by the borrower is essential to the successful operation of the project during the life of the loan, such right will be evidenced by written agreements or contracts between the owner(s) of the property or equipment and the borrower. Lease agreements shall not contain provisions for restricted use of the site or facility, forfeiture or similiar cancellation clauses and shall provide for the right to transfer and lease without restriction. Such lease contracts or agreements shall be approved by the lender and Rural Development.

(IV) *Project Control.*

(A) Lender will adopt project control procedures to assure that loan funds are applied for costs or expenses properly attributable to the project ("Eligible Project Costs") as proposed in the completed application submitted by the lender and borrower and approved by the Rural Development. A project monitoring account ("Project Monitoring Account") will be developed by lender for this purpose and concurred in by the State Director. This account will be divided into sufficient budget categories to permit adequate control of expenditures and identification of potential budget overruns.

(B) The first advance ("First Advance") of loan funds to the borrower will not commence from the Project Monitoring Account prior to lender's receipt of evidence that:

(1) The borrower has made adequate provisions for compliance with measures established by Rural Development to mitigate adverse historical and environmental impacts.

(2) Applicable engineering, design/build, construction management, inspection and plant start-up service agreements have been obtained and accepted by the State Director and lender.

(3) The project engineer has prepared a detailed cost estimate and construction schedule for all facilities related to the project. This estimate must indicate that the project can be completed with the funds available as shown on the Form RD 449–1, "Application for Loan and Guarantee." A reasonable contingency amount will be included in the estimate. This contingency shall be at least 20 percent of the estimated project costs for which firm bids have not been received plus 5 percent of project costs for which firm bids have been received. Construction interest and inspection costs will be based upon a reasonable contingency for unforeseen delays in project completion. The estimate shall include a listing with associated costs of any proposed leasing arrangements for property or equipment that is essential to the successful operation of the project.

(4) All funds necessary for construction of project facilities will be available when needed.

(5) The borrower has retained a project manager with sufficient experience and training to supervise project construction and engineering services on behalf of the borrower.

(C) After the first advance, future advances may be made from the Project Monitoring Account, in accordance with prudent lender practice, for all Eligible Project Costs established in the Project Monitoring Account, provided these payments are made in accordance with the terms of applicable contracts and are approved by the borrower and, when applicable, recommended by the project engineer.

(D) Payments for Eligible Project Costs incurred by the borrower prior to satisfaction of the conditions precedent to the first advance shall be made with borrower's funds or other nonguaranteed loan funds only. These payments however, may be reimbursed through the Project Monitoring Account as authorized by the State Director after compliance with Paragraph (IV)(B) hereof. The lender will not advance and the borrower will not be entitled to loan funds for reimbursement if such costs or expenses incurred by the borrower prior to the first advance, or at anytime thereafter, were for costs or expenses other than Eligible Project Costs. Costs and expenses accruing from but not limited to, interest charges imposed by construction, equipment, material or service contracts, penalty payments, damage claims, awards or settlements are not Eligible Project Costs unless specifically approved by the State Director.

(E) The lender will monitor the progress of construction and undertake the reviews and project inspections necessary to reasonably assure that funds are paid for Eligible Project Costs and that problems in project

280

development are expeditiously reported to the State Director.

(F) The lender will prepare a monthly report showing the expenditures made from each budget category of the Project Monitoring Account. This report will include a review of construction progress including proposed and approved contract change orders and, to the extend possible, identify problems or delays in construction or other matters which might affect successful startup of project. This report may be based upon information received from the project engineer and borrower and/or independent observations of the lender. The report will be initialed by the borrower and project engineer and submitted to the State Director.

(G) Transfer of loan funds between established or new categories of the Project Monitoring Account or any change in the total amount of funds committed to the project will be reported by the lender to the State Director as these changes occur.

APPENDIX E TO SUBPART E OF PART 1980—ENVIRONMENTAL ASSESSMENT GUIDELINES

In completing an assessment, it is important to understand the comprehensive nature of the impacts which must be analyzed. Consideration must be given to all potential impacts associated with the construction of the project and its operation and maintenance. The attainment of the project's major objectives often induces or supports changes in population densities, land uses, community services, transportation systems and resource consumption. The impacts of these activities must also be assessed.

The environmental reviewer should consult with appropriate experts from Federal, State and local agencies, universities and other organizations or groups whose views could be helpful in the assessment of potential impacts. In so doing, each discussion which is utilized in reaching a conclusion with respect to the degree of an impact should be summarized in the assessment as accurately as possible and include name, title, phone number, and organization of the individual contacted, plus the date of contact. Related correspondence should be attached to the assessment.

The Farmers Home Administration or its successor agency under Public Law 103–354 assessment should be prepared in the following format; it should address the listed items and questions and contain as attachments the indicated descriptive materials, as well as the environmental information submitted by the applicant.

These assessment guidelines have been designed to cover the wide variety of impacts which may be encountered. Consequently, not every issue or potential impact raised in these guidlines may be relevant to each

project. The purpose of the format is to give the preparer an understanding of a standard range of impacts, environmental factors and issues which may be encountered. In preparing an assessment, each topic heading identified by a roman numeral and each environmental factor listed under topic heading IV, such as air quality for example, must be addressed.

The amount of analysis and material that must be provided will depend upon the type and size of the project, the environment in which it is located and the range and complexity of the potential impacts. The amount of analysis and detail provided, therefore, must be commensurate with the magnitude of the expected impact. The analysis of each environmental factor (i.e., water quality) must be taken to the point that a conclusion can be reached and supported concerning the degree of the expected impact with respect to that factor.

(I) *Project description and need.* Identify the name, project number, location, and specific elements of the project along with their sizes, and, when applicable, their design capacities. Indicate the purpose of the project, Rural Development's position regarding the need for it, and the extent or area of land to be considered as the project site.

(II) *Primary beneficiaries and related activities.* Identify any existing businesses or major developments that will benefit from the project and those which will expand or locate in the area because of the project. Specify by name, product, service, and operations involved.

Identify any related activities which are defined as interdependent parts of an Rural Development action. Such undertakings are considered interdependent parts whenever they either make possible or support the Rural Development action or are themselves induced or supported by the Rural Development action or another related activity. These activities may have been completed in the very recent past and are now operational or they may reasonably be expected to be accomplished in the near future. Related activities may or may not be Federally permitted or assisted. When they are, identify the involved Federal agency(s).

In completing the remainder of the assessment, it must be remembered that the impacts to be addressed are those which stem from the project, the primary beneficiaries, and the related activities.

(III) *Description of project area.* Describe the project site and its present use. Describe the surrounding land uses; indicate the directions and distances involved. The extent of the surrounding land to be considered depends on the extent of the impacts of the project, its related activities, and the primary beneficiaries. Unique or sensitive areas must be pointed out. These include residential, schools, hospitals, recreational, historical sites, beaches, lakes, rivers, parks, floodplains, wetlands, dunes, estuaries, barrier islands, natural landmarks, unstable soils, steep slopes, aquifer recharge areas, important farmlands and forestlands, prime rangelands, endangered species habitats, or other delicate or rare ecosystems.

Attach adequate location maps of the project area, as well as (1) a U.S. Geological Survey "15 minute" ("7½ minute" if available) topographic map which clearly delineates the area and the location of the project elements, (2) the Department of Housing and Urban Development's floodplain map(s) for the project area, (3) site photos, (4) if completed, a standard soil survey for the project and, (5) if available, an aerial photograph of the site. When necessary for descriptive purposes or environmental analysis, include land use maps or other graphic information. All graphic materials shall be of high quality resolution.

(IV) *Environmental impact.* (1) Air Quality—Discuss, in terms of the amounts and types of emissions to be produced, all aspects of the project including beneficiaries' operations and known indirect effects (such as increased motor vehicle traffic) which will affect air quality. Indicate the existing air quality in the area. Indicate if topographical or meteorological conditions hinder or affect the dispersals of air emissions. Evaluate the impact on air quality given the types and amounts of projected emissions, the existing air quality and topographical and meteorological conditions. Discuss the project's consistency with the State's air quality implementation plan for the area, the classification of the air quality control region within which the project is located, and the status of compliance with air quality standards within that region. Cite any contacts with appropriate experts and agencies which must issue necessary permits.

(2) Water Quality—Discuss, in terms of amounts and types of effluents all aspects of the project, including primary beneficiaries' operations and known indirect effects which will affect water quality. Indicate the existing water quality of surface and/or underground water to be affected. Evaluate the impacts of the project on this existing water quality. Indicate if an aquifer recharge area is to be adversely affected. If the project lies within or will affect a sole source aquifer recharge area as designated by the Environmental Protection Agency (EPA), contact the appropriate EPA regional office to determine if its review is necessary. If it is, attach the results of its review.

Indicate the source and available supply of raw water and the extent to which the additional demand will affect the raw water supply. Describe the wastewater treatment system(s) to be used and indicate their capacity and their adequacy in terms of the degree of treatment provided. Discuss the characteristics and uses of the receiving waters for any sources of discharge. If the treatment systems are or will be inadequate or overloaded, describe the steps being taken for necessary improvements and their completion dates. Compare such dates to the completion date of the Rural Development project. Analyze the impacts on the receiving water during any estimated period of inadequate treatment.

Discuss the project's consistency with water quality planning for the area, such as EPA's Section 208 areawide waste treatment management plan. Describe how surface runoff is to be handled and the effect of erosion on streams.

Evaluate the extent to which the project may create shortages for or otherwise adversely affect the withdrawal capabilities of other present users of the raw water supply, particularly in terms of possible human health, safety, or welfare problems.

For projects utilizing a groundwater supply, evaluate the potential for the project to exceed the safe pumping rate for the aquifer to the extent that it would (1) adversely affect the pumping capability of present users, (2) increase the likelihood of brackish or saltwater intrusion, thereby decreasing water quality, or (3) substantially increase surface subsidence risks.

For projects utilizing a surface water supply, evaluate the potential for the project to (1) reduce flows below the minimum required for the protection of fish and wildlife or (2) reduce water quality standards below those established for the stream classification at the point of withdrawal or the adjacent downstream section.

Cite contacts with appropriate experts and agencies that must issue necessary permits.

(3) Solid Waste Management—Indicate all aspects of the project, including primary beneficiaries' operations, and known indirect effects which will necessitate the disposal of solid wastes. Indicate the kinds and expected quantities of solid wastes involved and the disposal techniques to be used. Evaluate the adequacy to these techniques especially in relationship to air and water quality. Indicate if recycling or resource recovery programs are or will be used. Cite any contacts with appropriate experts and agencies that must issue necessary permits.

(4) Land Use—Given the description of land uses as previously indicated, evaluate (a) the effect of changing the land use of the project site and (b) how this change in land use will affect the surrounding land uses and those within the project's area of environmental impact. Particularly address the potential impacts to the unique or sensitive areas discussed under Section III, Description of Project Area. Also address any changes in land use which may result from demand for

feedstock for the plant's operation. Describe the existing land use plan and zoning restrictions for the project area. Evaluate the consistency of the project and its impacts with these plans.

(5) Transportation—Describe available facilities such as highways and rail. Discuss whether the project will result in an increase in motor vehicle traffic and the existing roads' ability to safely accommodate this increase. Indicate if additional traffic control devices are to be installed. Describe new traffic patterns which will arise because of the project. Discuss how these new traffic patterns will affect the land uses described above, especially residential, hospitals, schools, and recreational. Describe the consistency of the project's transportation impacts with the transportation plans for the area and any air quality control plans. Cite any contact with appropriate experts.

(6) Natural Environment—Indicate all aspects of the project, including construction, beneficiaries' operations, and known indirect effects which will affect the natural environment including wildlife, their habitats, and unique natural features. Cite contacts with appropriate experts. If an area listed on the National Registry of Natural Landmarks may be affected, consult with the Department of Interior and document these consultations and any agreements reached regarding avoidance or mitigation of potential adverse impacts.

(7) Human Population—Indicate the number of people to be relocated and arrangements being made for this relocation. Discuss how impacts resulting from the project such as changes in land use, transportation changes, air emissions, noise, odor, etc., will effect nearby residents and their lifestyles or users of the project area and surrounding areas. Cite contacts with appropriate experts.

(8) Construction—Indicate the potential effects of construction of the project on air quality, water quality noise levels, solid waste disposal, soil erosion and siltation. Describe the measures that will be employed to limit adverse effects. Give particular consideration to erosion, stream siltation, and clearing operations.

(9) Energy Impacts—Indicate the project's and its primary beneficiaries' effects on the area's existing energy supplies. This discussion should address not only the direct energy utilization, but any major indirect utilization resulting from the siting of the project. Describe the availability of these supplies to the project site. Discuss whether the project will utilize a large share of the remaining capacity of an energy supply or will create a shortage of such supply. Discuss any steps to be taken to conserve energy.

(10) Discuss any of the following areas which may be relevant: noise, vibrations, safety, seismic conditions, fire prone loca-

tions, radiation, and aesthetic considerations. Cite any discussions with appropriate experts.

(V) *Coastal Zone Management Act.* Indicate if the project is within or will impact a coastal area defined as such by the state's approved Coastal Zone Management Program. If so, consult with the State agency responsible for the Program to determine the project's consistency with it. The results of this coordination shall be included in the assessment and considered in completing the environmental impact determination and environmental findings,

(VI) *Compliance with Advisory Council on Historic Preservation's regulations.* In this section, the environmental reviewer shall detail the steps taken to comply with the above regulations as specified in Subpart F of Part 1901 of this Chapter. First, indicate that the National Register of Historic Places, including its monthly supplements, has been reviewed and whether there are any listed properties located within the area to be affected by the project. Second, indicate the steps taken such as historical/archeological surveys to determine if there are any properties eligible for listing located within the affected area. Summarize the results of the consultation with the State Historic Preservation Officer (SHPO) and attach appropriate documentation of the SHPO's views. Discuss the views of any other experts contacted. Based upon the above review process and the views of the SHPO, state whether or not an eligible or listed property will be affected.

If there will be an effect, discuss all of the steps and protective measures taken to complete the Advisory Council's regulations. Describe the affected property and the nature of the effect. Attach to the asessment the results of the coordination process with the Advisory Council on Historic Preservation.

(VII) *Compliance with the Wild and Scenic Rivers Act.* Indicate whether the project will affect a river or portion of it which is either included in the National Wild and Scenic Rivers System or designated for potential addition to the System. This analysis shall be conducted through discussions with the appropriate regional office of the National Park Service or the Forest Service when its lands are involved, as well as the appropriate State agencies having implementation authorities. A summary of discussions held or any required formal coordination shall be included in the assessment.

(VIII) *Compliance with the Endangered Species Act.* Indicate whether the project will either (1) affect a listed endangered or threatened species or critical habitat or (2) adversely affect a proposed critical habitat for an endangered or threatened species or jeopardize the continued existence of a proposed endangered or threatened species. This analysis shall be conducted in consultation with the Fish and Wildlife Service and the National Marine Fisheries Service, when appropriate.

The results of any required coordination shall be included in the assessment along with any completed biological opinion and mitigation measures to be required for the project. These factors shall be considered in completing the environmental impact determination.

(IX) *Compliance with Executive Order 11988, Floodplain Management, and Executive Order 11990, Protection of Wetlands.* Indicate whether the project is either located within a 100-year floodplain (500-year floodplain for a critical action) or a wetland or will impact a floodplain or wetland. If so, determine if there is a practicable alternative project or location. If there is no such alternative, determine whether all practicable mitigation measures are included in the project and document as an attachment these determinations and the steps taken to inform the public, locate alternatives, and mitigate potential adverse impacts. See the U.S. Water Resource Council's *Floodplain Management Guidelines* for more specific guidance.

(X) *State Environmental Policy Act.* Indicate if the proposed project is subject to a State environmental policy act or similar regulation. Summarize the results of compliance with these requirements and attach available documentation.

(XI) *Consultation requirements.* Attach the comments of any State or local agency received through the implementation of Executive Order 12372, Intergovernmental Review of Federal Programs.

(XII) *Environmental analysis of participating Federal agency.* Indicate if another federal agency is participating in the project either through the provision of additional funds, a companion project, or a permit review authority. Summarize the results of the involved agency's environmental impact analysis and attach available documentation.

(XIII) *Reaction to project.* Discuss any negative comments or public views raised about the project and the consideration given to these comments. Indicate whether a public hearing or public information meeting has been held either by the applicant or Rural Development to include a summary of the results and any objections raised. Indicate any other examples of the community's awareness of the project, such as newspaper articles or public notifications.

(XIV) *Cumulative impacts.* Summarize the cumulative impacts of this project and the related activities. Give particular attention to land use changes and air and water quality impacts. Summarize the results of the environmental impact analysis done for any of these related activities and/or your discussion with the sponsoring agencies. Attach available documentation of the analysis.

(XV) *Adverse impact.* Summarize the potential adverse impacts of the proposal as pointed out in the above analysis.

(XVI) *Alternatives.* Discuss the feasibility of alternatives to the project and their environmental impacts. These alternatives should include (a) alternative location, (b) alternative designs, (c) alternative projects having similar benefits, and (d) no project.

(XVII) *Mitigation measures.* Describe any measures which will be taken or required by Rural Development to avoid or mitigate the identified adverse impacts. Such measures shall be included as special requirements or provisions to the offer of financial assistance.

APPENDIXES F–G TO SUBPART E OF PART 1980 [RESERVED]

APPENDIX H TO SUBPART E OF PART 1980—SUGGESTED FORMAT FOR THE OPINION OF THE LENDER'S LEGAL COUNSEL

(Legal Opinion to be Retyped on Lender's Counsel's Letterhead)

To: (Name of Lender).

I/We have acted as counsel to (Lender) _____ in connection with a $ (amount) _____ (type) _____ loan by the (Lender) _____ (hereinafter "the Lender" to (Borrower) _____ (hereinafter "Borrower"), the terms of which loans are set forth in a certain Loan Agreement (hereinafter "the Loan Agreement") executed by the Lender and Borrower on (date) _____.

In connection with this loan, I/we have examined:

1. The corporate records of Borrower, including its Articles of Incorporation, By-Laws and Resolutions of its Board of Directors.

2. The Loan Agreement between the Lender and Borrower.

3. The Security Agreement executed by Borrower on (date) _____.

4. The Guaranty (where applicable) executed on (date) _____ by (personal guarantors) _____.

5. Financing Statements executed by Borrower and the Lender.

6. Real Estate Mortgages dated _____ and executed by Borrower in favor of the Lender.

7. Real Estate Mortgages dated _____ and/or other security documents dated _____ executed by (personal guarantors) _____ in favor of the Bank.

8. The appropriate title and/or lien searches relating to Borrower's property.

9. The pledge of stock and instruments related thereto.

10. Such other materials, including relevant provisions of the laws of this state as I/we have deemed pertinent as a basis for rendering the opinion hereafter set forth.

In Some Circumstances

11. Lease(s) between Borrower and (lessor's name) _____ for the rental of (property being rented) _____, (if real property, give the address of the premises; if machinery equipment, etc., give brief, precise description of property for a (length of lease) _____ term commencing on (date) _____.

Based on the foregoing examinations, I am/ we are of the opinion and advise you that:

1. Borrower is a duly organized corporation in good standing under the laws of the Commonwealth/State of (State) _____.

2. Borrower has the necessary corporate power to authorize and has taken the necessary corporate action to authorize the Loan Agreement and to execute and deliver the Note, Security Agreement, Financing Statement, and Mortgage. Said instruments hereinafter collectively referred to as the "Loan Instruments."

3. The Loan Instruments were all duly authorized, executed, and delivered and constitute the valid and legally binding obligation of the Borrower and collectively create and valid (first) lien upon or valid security interest in favor of the Lender, in the security covered thereby, and are enforceable in accordance with their terms except to the extent that the enforceability (but not the validity) thereof may be limited by laws of bankruptcy, insolvency, or other laws generally affecting creditors' rights.

4. The execution and delivery of the Loan Instruments and compliance with the provisions thereof under the circumstances contemplated thereby did not, do not and will not in any material respect conflict with, constitute default under, or contravene any contract or agreement or other instrument to which the Borrower is a party or any existing law, regulation, court order, or consent decree or device to which the Borrower is subject.

5. All applicable Federal, State and local tax returns and reports as required have been duly filed by Borrower and all Federal, State and local taxes, assessments and other governmental charges imposed upon Borrower or its respective assets, which are due and payable, have been paid.

6. The guaranty has been duly executed by the Guarantors and is a legal, valid and binding joint and several obligations of the Guarantors, enforceable in accordance with its terms, except to the extent that the enforceability (but not the validity) thereof may be limited by laws of bankruptcy, insolvency, or other laws generally affecting creditors' rights.

7. All necessary consents, approvals, or authorizations of any governmental agency or regulatory authority or of stockholders which are necessary have been obtained. The improvements and the use of the property

comply in all respects with all Federal, State, and local laws applicable thereto.

8. (In cases involving subordinate or other than first lien position) That the mortgage/ deed of trust on Borrower's real estate and (fixtures, e.g., machinery and equipment) and the security interest on (type of collateral, e.g., machinery and equipment, accounts, receivables and inventory) both given as security to the Lender for the Loan, will be subordinate to (first mortgagee) _____ given as security for a loan in the amount of $_____ and the security interest in Borrower's (type of collateral, e.g., accounts inventory) _____ given to (secured creditor) _____ as security for a loan (state type of loan, i.e., revolving line of credit, _____ if known) in the amount of $_____.

9. That there are no liens, as of the date hereof, on record with respect to the property of Borrower other than those set forth above.

10. There are no actions, suits or proceedings pending or, to the best of our knowledge, threatened before any court or administrative agency against Borrower which could materially adversely affect the financial condition and operations of Borrower.

11. Borrower has good and marketable title to the real estate security free and clear of all liens and encumbrances other than those set forth above. I/we have no knowledge of any defect in the title of the Borrower to the property described in the Loan Instruments.

12. Borrower is the absolute owner of all property given to secure the repayment of the loan, free and clear of all liens, encumbrances, and security interests.

13. Duly executed and valid functioning statements have been filed in all offices in which it is necessary to file financing statements to fully perfect the security interests granted in the Loan Instruments.

14. Duly executed real estate mortgages/ deeds of trust have been recorded in all offices in which it is necessary to record to fully perfect the security interests granted in the Loan Instruments.

15. (IN SOME OTHER CIRCUMSTANCES) The Indemnification Agreement has been duly executed by the Indemnitors and is a legal, valid and binding joint and several obligation of the Indemnitors, enforceable in accordance with its terms, except to the extent that the enforceability (but not the validity) thereof may be limited by laws of bankruptcy, insolvency, or other laws generally affecting creditors' rights.

16. That the lease contains a valid and enforceable right of assignment and right of reassignment, enforceable in accordance with its terms, except to the extent the enforceability (but not the validity) thereof may be limited by laws of bankruptcy, insolvency, or other laws generally affecting creditors' rights.

17. The Lender's lien has been duly noted on all motor vehicle titles, stock certificates or other instruments where such notations are required for proper perfection of security interests therein.

18. That a valid pledge of the outstanding and unissued stock and/or shares of Borrower has been obtained and the Lender has a validly perfected and enforceable security interest in the shares/stock of Borrower, except to the extent the enforceability thereof may be limited by laws of bankruptcy, insolvency, or other laws generally affecting creditors rights.

[52 FR 6522, Mar. 4, 1987]

APPENDIX I TO SUBPART E OF PART 1980—INSTRUCTIONS FOR LOAN GUARANTEES FOR DROUGHT AND DISASTER RELIEF

A. *In general.* Drought and Disaster (D&D) guaranteed loans are authorized by section 331 ("Disaster Assistance for Rural Business Enterprises") of the Disaster Assistance Act of 1988, which provides for guarantees of up to 90 percent of the unpaid principal amount of qualifying loans. Interest and protective advances are not covered by the guarantee. Drought and Disaster Guaranteed Loans may be either to assist in alleviating financial distress caused to rural business entities, directly or indirectly, by drought, hail, excessive moisture, or related conditions occurring in 1988, or to assist such entities that refinance or restructure debt as a result of losses incurred, directly or indirectly, because of such natural disasters. Where used in this appendix, the term "natural disaster(s)" refers only to drought, hail, excessive moisture, and related conditions occurring in 1988. All provisions of Subparts A and E of Part 1980 of this chapter apply to D&D loans, except as provided in this appendix. All forms used in connection with a D&D loan will be those used in connection with a B&I guaranteed loan, except for the following three forms that are incorporated in this Appendix I of this Subpart E, made a part hereof:

(1) Form FmHA or its successor agency under Public Law 103–354 1980–68, "Lender's Agreement—Drought and Disaster Guaranteed Loans," or successor form will be used instead of Form FmHA or its successor agency under Public Law 103–354 449–35, "Lender's Agreement."

(2) Form FmHA or its successor agency under Public Law 103–354 1980–69, "Loan Note Guarantee—Drought and Disaster Guaranteed Loans," or successor form will be used instead of Form FmHA or its successor agency under Public Law 103–354 449–34, "Loan Note Guarantee."

(3) Form FmHA or its successor agency under Public Law 103–354 1980–70, "Assignment Guarantee Agreement—Drought and Disaster Guaranteed Loans," or successor form will be used instead of Form FmHA or its successor agency under Public Law 103–354 449–36, "Assignment Guarantee Agreement."

B. *Loan purpose.* Except for §§1980.411(a)(11), 1980.412, and section C., below, loan proceeds may be used for purposes described in §1980.411(a) if such use of loan proceeds will assist in alleviating financial distress caused, directly or indirectly, by drought, hail, excessive moisture, or related conditions which occurred in 1988. In lieu of the debt refinancing requirements in §1980.411(a)(11), the following refinancing requirements apply to D&D loans. Loan proceeds to be used for refinancing must be used solely for refinancing or restructuring of debts as a result of losses incurred, directly or indirectly, as a result of drought, hail, excessive moisture, or related condition occurring in 1988, and such refinancing or restructuring of debt(s) must be essential for the borrower to meet its financial obligations in a timely fashion. In addition, D&D loan proceeds may be used for hotels, motels, tourist or recreation facilities which meet the eligibility requirements for D&D guaranteed loans.

C. *Ineligible loan purposes.* See §1980.412. Except for hotels, motels, tourist and recreation facilities mentioned in section B of this appendix, purposes listed as ineligible B&I loan purposes are ineligible D&D loan purposes. In addition, D&D guaranteed loans may not be used for:

(1) Business expansion, acquisition of real estate, machinery, equipment, inventory, other goods or services, or for any other purpose unless related directly to the financial distress or loss that is the basis for the D&D guaranteed loan.

(2) Any eligible agricultural production purpose if annual tillage of the soil is involved.

(3) Refinancing or restructuring debt(s) which are or were in payment default more than 60 consecutive days during the 12 months preceding the date of the adverse financial effect of the natural disaster of 1988 upon the borrower.

D. *Transactions which will not be guaranteed.* In addition to transactions listed in §1980.413, Rural Development will not guarantee:

(1) D&D guaranteed loan(s) to any borrower if the total cumulative principal amount of D&D guaranteed loan(s) to that borrower would exceed $500,000, or

(2) Any D&D guaranteed loan if the completed application is not received by Rural Developmenton or before September 30, 1991.

E. *Borrower equity requirements.* See §1980.441. In lieu of the borrower equity requirements in §1980.441, paragraphs (a) and (b), the following applies to D&D loans.

Tangibles balance sheet equity must be positive when the Loan Note Guarantee is issued. Equity must be such that, when considered with other credit factors, repayment of the loan and the continued success of the business operation are reasonably assured. Requirements of § 1980.441(c) apply to D&D guaranteed loans.

F. *Filing and processing preapplications and applications.* See § 1980.451. All requirements of § 1980.451 remain in effect. But, in addition to the information required as part of a preapplication under § 1980.451(f), and unless previously submitted, as a part of an application under § 1980.451(i) evidence is required which demonstrates:

(1) The causal relationship between a 1988 natural disaster and the financial distress or loss upon which the preapplication or application is based; and,

(2) That the amount of the loan requested is not greater than the amount necessary for curing the problems caused by the natural disaster. Financial distress or loss shall be determined on the basis of a comparison of financial data for comparable periods of time and need not necessarily be based on data at the year's end. Evidence submitted may include, but is not limited to, the following:

(a) Evidence of financial loss or distress (including loss or distress caused by business interruption) resulting from physical damage caused by natural disaster, or

(b) Evidence that the financial loss and/or distress of the business is the direct or indirect result of loss of sales, business interruption, loss of markets, shortage of raw materials, or decline in patronage or customers caused by a natural disaster. It must be shown that business operations were damaged as a result of such natural disaster.

G. *Loan guarantee limit.* See § 1980.20 of Subpart A. The maximum loss covered by the Loan Note Guarantee, Form FmHA or its successor agency under Public Law 103–354 1980–69, can never exceed the percentage of guarantee multiplied by the unpaid principal amount of the loan as evidenced by the note(s) or by assumption agreement(s). Interest, capitalized interest, and protective advances are not covered by the guarantee of a D&D loan.

H. *Percentage of guarantee.* See § 1980.420. The maximum percentage of guarantee on a D&D loan is 90 percent of the unpaid principal.

I. *Lender's existing unguaranteed exposure.* The provisions of § 1980.452 Administrative C. 1(d) do not apply.

J. *No direct or "insured" loans.* Sections 1980.423(b), 1980.488(b), 1980.481, 1980.411(b), and other provisions of this subpart dealing with "insured" or direct loans do not apply to D&D loans. All D&D loans are Rural Development guaranteed loans. Rural Develop-

ment has no authority to make D&D loans directly to borrowers.

[54 FR 5, Jan. 3, 1989, as amended at 54 FR 14792, Apr. 13, 1989; 54 FR 26946, June 27, 1989; 80 FR 9911, Feb. 24, 2015]

EDITOR'S NOTE: At 80 FR 9911, Feb. 24, 2015, appendix I was amended by removing "Form FmHA or its successor agency under Public Law 103–354" in paragraphs X.D and X.F and adding "Form RD" in its place; however the amendment could not be incorporated because the paragraphs did not exist. Additionally the appendix was amended by removing "will be used" in paragraphs IV., and E.1 and 2, and adding "or successor form will be used" in its place; however, the amendments could not be incorporated because the paragraphs did not exist.

APPENDIX J TO SUBPART E OF PART 1980
[RESERVED]

APPENDIX K TO SUBPART E OF PART 1980—REGULATIONS FOR LOAN GUARANTEES FOR DISASTER ASSISTANCE FOR RURAL BUSINESS ENTERPRISES

A. In general

Disaster Assistance for Rural Business Enterprises (DARBE) guaranteed loans are authorized by Section 401 of the Disaster Assistance Act of 1989, which provides for guarantees of up to 90 percent of the unpaid principal and interest amount of qualifying loans, or $2,500,000 whichever is less, to any one borrower. DARBE guaranteed loans may be either to assist in alleviating financial distress caused to rural business entities, directly or indirectly, by drought, freeze, storm, excessive moisture, earthquake, or related conditions occurring in 1988 or 1989, or to assist such entities that refinance or restructure debt as a result of losses incurred, directly or indirectly, because of such natural disasters. Where used in this appendix, the term "natural disaster(s)" refers only to drought, freeze, storm, excessive moisture, earthquake, and related conditions occurring in 1988 or 1989. All provisions of subparts A and E of part 1980 of this chapter apply to DARBE loans, except as provided in this appendix. All forms used in connection with a DARBE loan will be those used in connection with a Business and Industrial (B&I) guaranteed loan, except for the following three forms that are incorporated in this appendix K of this subpart E, made a part hereof:

(1) Form RD 1980–71, "Lender's Agreement—Disaster Assistance for Rural Business Enterprise Guaranteed Loans," or successor form will be used instead of Form RD 449–35, "Lender's Agreement."

287

(2) Form RD 1980-72, "Loan Note Guarantee—Disaster Assistance for Rural Business Enterprise Guaranteed Loans," or successor form will be used instead of Form RD 449-34, "Loan Note Guarantee."

(3) Form RD 1980-73, "Assignment Guarantee Agreement—Disaster Assistance for Rural Business Enterprise Guaranteed Loans," or successor form will be used instead of Form RD 449-36, "Assignment Guarantee Agreement."

B. Loan purposes

Loan proceeds may be used for purposes described in §1980.411(a), except in lieu of the debt refinancing requirements in §1980.411(a)(11), the following refinancing requirements apply to DARBE loans. Loan proceeds to be used for refinancing must be used solely for refinancing or restructuring of debts as a result of losses incurred, directly or indirectly, as a result of drought, freeze, storm, excessive moisture, earthquake, or related conditions occurring in 1988 or 1989, and such refinancing or restructuring of debt(s) must be essential for the borrower to meet its financial obligations in a timely fashion. DARBE loan proceeds may be used for hotels, motels, tourist, or recreation facilities which meet the eligibility requirements of DARBE guaranteed loans in addition to the eligible loan purposes as stated in RD Instruction 1980-E. In addition, DARBE loan proceeds may be used for business enterprises engaged in agricultural production (production agriculture) which means the cultivation, production (growing), and harvesting, either directly or through integrated operations, of agricultural products (crops, animals, birds, and marine life, either for fibers or food for human consumption), and disposal or marketing thereof, the raising, housing, feeding (including commercial custom feedlots), breeding, hatching, control and/or management of farm and domestic animals. Other eligible uses of loan proceeds under agricultural production include:

(1) Commercial nurseries primarily engaged in the production of ornamental plants and trees and other nursery products such as bulbs, florists' greens, flowers, shrubbery, flower and vegetable seeds, sod, and the growing of vegetables from seed to the transplant stage.

(2) Forestry which includes establishments primarily engaged in the operation of timber tracts, tree farms, forest nurseries, and related activities such as reforestation.

(3) Loans for livestock and poultry processing as identified under eligible purposes.

(4) The growing of mushrooms or hydroponics.

In addition, those business enterprises which qualify for assistance as agricultural production must be ineligible entities for FmHA or its successor agency under Public Law 103-354 farmer program loans because the entity exceeds the definition of a family-size farm as defined by RD Instruction 1941-A, §1941.4(d).

C. Ineligible loan purposes

RD Instruction 1980-E, §1980.412 are ineligible purposes for DARBE guaranteed loans except for hotels, motels, tourist, recreation facilities and agricultural production (production agriculture) as defined in §1980.412(e), DARBE guaranteed loans may not be used for:

(1) Business expansion, acquisition of real estate, machinery, equipment, inventory, other goods or services, or for any other purpose unless related directly to the financial distress or loss that is the basis for the DARBE guaranteed loan.

(2) Alleviating financial distress of entities engaged in agricultural production that are eligible for other Rural Development -type farm loan programs.

D. Transactions which will not be guaranteed

In addition to transactions listed in RD Instruction 1980-E, §1980.413, except for §1980.413(a)(3), Rural Development will not make DARBE guaranteed loans if the completed application is not received by Rural Development on or before September 30, 1991, nor will Rural Development make subsequent DARBE guarantee loans.

E. Borrower equity requirements

See RD Instruction 1980-E, §1980.441. In lieu of the borrower equity requirements in §1980.441, paragraphs (a) and (b), the following applies to DARBE loans. Tangible balance sheet equity must be positive when the Loan Note Guarantee is issued. Equity must be such that, when considered with other credit factors, repayment of the loan and the continued success of the business operation are reasonably assured. Requirements of §1980.441(c) apply to DARBE guaranteed loans.

F. Filing and processing preapplications and applications

See RD Instruction 1980-E, §1980.451. All requirements of §1980.451 remain in effect. In addition to the information required as part of a preapplication under §1980.451(f), and unless previously submitted as a part of an application under §1980.451(i) evidence is required which demonstrates to Rural Development 's satisfaction:

(1) The causal relationship between a 1988 or 1989 natural disaster and the financial distress or loss upon which the preapplication or application is based; and,

(2) That the amount of the loan requested is not greater than the amount necessary for curing the problems caused by the natural disaster. Financial distress or loss shall be determined on the basis of a comparison of

financial data for comparable periods of time and need not necessarily be based on data at the year's end. Evidence submitted may include, but is not limited to, the following:

(a) Evidence of financial loss or distress (including loss or distress caused by business interruption) resulting from physical damage caused by natural disaster, or

(b) Evidence that the financial loss and/or distress of the business is the direct or indirect result of loss of sales, business interruption, loss of markets, shortage of raw materials, or decline in patronage or customers caused by a nautral disaster. It must be shown that business operations were damaged as a result of such natural disaster.

(3) Evidence of compliance with Sodbuster and Swampbuster requirements as referenced in paragraph K below.

G. Loan guarantee limit. The total principal amount of DARBE guaranteed loans to any one borrower cannot exceed $10,000,000. The maximum loss covered by Form RD 1980–72, "Loan Note Guarantee DARBE," (or successor form) issued on any one borrower can never exceed the percentage of guarantee multiplied by the unpaid principal and accrued interest on the loan as evidenced by the note(s) or by assumption agreement(s), and protective advances, or $2,500,000, whichever is the lesser amount.

H. Percentage of guarantee. The provisions of RD instruction 1980–E, § 1980.420 will not apply to DARBE. For loans in excess of $2,000,000, the percentage of guarantee will be calculated so that the guaranteed portion of the principal amount of the loan cannot exceed $2,000,000. For loans of $2,000,000 or less the maximum percentage of guarantee will be 90 percent. For example, a loan of $10,000,000 would not exceed a 20 percent guarantee; a $5,000,000 loan would not exceed a 40 percent guarantee.

I. Lender's existing unguaranteed exposure

The provisions of § 1980.452 ADMINISTRATIVE C. 1(d) do not apply.

J. No direct or insured loans

RD Instruction 1980–E, §§ 1980.423(b), 1980.488(b), 1980.481, 1980.411(b), and other provisions of this subpart dealing with insured or direct loans do not apply to DARBE loans. All DARBE loans are Rural Development guaranteed loans. Rural Development has no authority to make DARBE loans directly to borrowers.

K. Sodbuster and Swampbuster requirements

The provisions of 7 CFR part 1970 will apply to loans made to rural business enterprises engaged in agricultural production.

[54 FR 42483, Oct. 17, 1989, as amended at 55 FR 137, Jan. 3, 1990; 55 FR 19245, May 8, 1990; 80 FR 9911, Feb. 24, 2015; 81 FR 11048, Mar. 2, 2016]

Exhibit G to Subpart E of Part 1980

NOTE: The Exhibit is not published in the Code of Federal Regulations. It is available in any Rural Development office.

[54 FR 1599, Jan. 13, 1989, as amended at 80 FR 9911, Feb. 24, 2015]

Subparts F–I [Reserved]

Subpart K—Strategic Economic and Community Development

SOURCE: 80 FR 28816, May 20, 2015, unless otherwise noted.

§ 1980.1001 Purpose.

The purpose of this subpart is to give priority to Projects that support implementation of strategic economic development and community development plans on a Multi-jurisdictional basis for applications submitted for the programs identified in § 1980.1002.

§ 1980.1002 Programs.

The Agency may elect to reserve funds from one or more of the programs listed in paragraphs (a) through (h) of this section.

(a) Community Facility Loans (7 CFR part 1942, subpart A).

(b) Fire and Rescue and Other Small Community Facilities Projects (7 CFR part 1942, subpart C).

(c) Community Facilities Grant Program (7 CFR part 3570, subpart B).

(d) Community Programs Guaranteed Loans (7 CFR part 3575, subpart A).

(e) Water and Waste Disposal Programs Guaranteed Loans (7 CFR part 1779).

(f) Water and Waste Loans and Grants (7 CFR part 1780, subparts A, B, C, and D).

(g) Business and Industry Guaranteed Loanmaking and Servicing (7 CFR part 4279, subparts A and B; 7 CFR part 4287, subpart B).

(h) Rural Business Development Grants (7 CFR part 4280, subpart E).

§ 1980.1003 Applicability of Program Regulations.

Except as supplemented by this subpart, the provisions of the programs identified in § 1980.1002 are incorporated into this subpart.

§ 1980.1004 Funding.

Unless the Agency publishes a notice that indicates otherwise, the Agency will reserve funds according to the procedures specified in paragraphs (a) through (c) of this section for each of the programs identified in § 1980.1002 each fiscal year.

(a) *Individual program basis.* The Agency will reserve funds on an individual program basis.

(b) *Percentage of funds.* The Agency will reserve 10 percent of the funds made available in a fiscal year to each program identified in § 1980.1002 unless the Agency specifies a different percentage. If the Agency specifies a different percentage, the Agency will publish a notice indicating the percentage. The Agency may reserve the same or different percentages for each program in a single fiscal year.

(c) *Unobligated funds.* If a program's funds reserved under this subpart remain unobligated as of June 30 of the fiscal year in which the funds are reserved, the Agency will return such remaining funds to that program's regular funding account for obligation for all eligible Projects in that program.

§ 1980.1005 Definitions.

In addition to the definitions found in the regulations for the programs identified in § 1980.1002, the following definitions apply to this subpart. If the same term is defined in any of the regulations for the programs identified in § 1980.1002, for purposes of this subpart, that term will have the meaning identified in this subpart.

Adopted means that a Plan has been officially approved for implementation by the appropriate entity or entities in the Jurisdiction(s) affected by the Plan (for example, a State, Indian Tribe, county, city, township, town, borough, etc.).

Agency means the Rural Business-Cooperative Service, the Rural Housing Service, or the Rural Utilities Service, or their successor agencies.

Carried Out Solely in a rural area means either:

(1) The Project is physically located in a rural area; or

(2) All of the beneficiaries of the services provided by the Project either reside in a rural area (for individuals)

or are located in a rural area (for businesses).

Investment means either monetary or non-monetary contributions to the implementation of the Plan's objectives.

Jurisdiction means a unit of government or other entity with similar powers. Examples include, but are not limited to: City, county, district, special purpose district, township, town, borough, parish, village, State, and Indian tribe.

Multi-Jurisdictional means at least two Jurisdictions.

Philanthropic organization means an entity whose mission is to provide monetary, technical assistance, or other items of value for religious, charitable, scientific, literary, or educational purposes.

Plan means a comprehensive economic development or community development strategy that outlines a region's vision for shaping its economy, and includes, as appropriate and necessary, consideration of such aspects as natural resources, land use, transportation, and housing. Such Plans bring together key community stakeholders to create a roadmap to diversify and strengthen their communities and to build a foundation to create the environment for regional economic prosperity. To be acceptable under this subpart, the Plan must be vetted and supported by the Jurisdictions affected by the Plan and must contain at a minimum the following:

(1) A summary of the economic conditions of the region;

(2) An in-depth analysis of the economic and community strengths, weaknesses, opportunities, and threats for the region, to include consideration of such aspects as the environmental and social conditions;

(3) Strategies and implementation Plan to build upon the region's strengths and opportunities and to resolve the weaknesses and threats facing the region;

(4) Performance measures that evaluate the successful implementation of the Plan's objectives; and

(5) Support of key community stakeholders.

Project means the eligible proposed use(s) for which funds are requested as described in the application material submitted to the Agency for funding under the underlying program.

§§ 1980.1006–1980.1009 [Reserved]

§ 1980.1010 Project eligibility.

In order to be eligible to receive funds under this subpart, the Project must meet the following:

(a) The Project must meet the Project eligibility criteria of the applicable program identified in § 1980.1002;

(b) The Project must be Carried Out Solely in a rural area; and

(c) The Project must support the implementation of a Plan on a Multi-Jurisdictional basis.

§§ 1980.1011–1980.1014 [Reserved]

§ 1980.1015 Applications.

In addition to the application material specific to the applicable program identified in § 1980.1002, each applicant seeking funding under this subpart must provide the information specified in paragraphs (a) through (d) of this section.

(a) *Applicant.* The applicant must submit:

(1) Name of the applicant;

(2) Telephone number of the applicant;

(3) Email address of the applicant; and

(4) A statement indicating whether or not the applicant is or includes one of the following:

(i) State government;

(ii) County government;

(iii) Municipal government; or

(iv) Tribal government.

(b) *Plan.* Each application must include the following information:

(1) The name of the Plan the Project supports;

(2) The date the Plan became effective;

(3) The dates the Plan is to remain in effect;

(4) Contact information for the entity(ies) approving the Plan, including name(s), telephone number(s), and email address(es);

(5) As found in the most current version of the Plan, the name and description of each objective that the Project will directly support;

(6) A description of the service area of the Plan;

(7) Documentation that the Plan was developed through the collaboration of multiple stakeholders in the service area of the Plan, including the participation of combinations of stakeholders;

(8) Documentation that the Plan demonstrates an understanding of the applicable region's assets that could support the Plan;

(9) Documentation indicating whether or not the Plan includes monetary or non-monetary contributions from ·Federal agencies other than the U.S. Department of Agriculture;

(10) Documentation indicating whether or not the Plan includes monetary or non-monetary contributions from one or more Philanthropic organizations.

(11) Documentation that the Plan contains:

(i) Clear objectives and

(ii) The ability to establish measurable performance measures and to track progress towards meeting the Plan's objectives; and

(12) If available, a Web site address link to the Plan.

(c) *Project.* Each application must include the following information:

(1) The name of the Project;

(2) Sufficient detail to allow the Agency to determine that the Project has been Carried Out Solely in a rural area as defined in § 1980.1005;

(3) A detailed description of how the Project directly supports each objective identified under paragraph (b)(5) of this section; and

(4) If the application is from an applicant that includes a State, county, municipal, or tribal government, a letter from the appropriate entity(ies) indicating that:

(i) The Project is consistent with the Plan and

(ii) The Plan has been Adopted.

(d) *Agency coordination.* To help ensure coordination among the programs included in this subpart, the Agency is requiring applicants provide the Agency the information in paragraphs (d)(1) through (3) of this section.

(1) *Program areas.* Identify the program area(s) (*i.e.,* Community Facilities, Water and Waste, Rural Business and Cooperative Development) from which funds are being sought.

(2) *Multiple applications.* If the applicant is submitting in the same fiscal year more than one application for funding under this subpart, identify in each application the other application(s) by providing:

(i) The name(s) of the Project(s);

(ii) The program area(s) for which funds are being sought; and

(iii) The date that each application was submitted to the Agency.

(3) *Previous applicants.* If the applicant has previously submitted one or more applications for funding under this subpart, the applicant must provide in the current application the following information for each previous application:

(i) The date the application was submitted;

(ii) The name of the Project;

(iii) The program area(s) from which funds were sought;

(iv) Whether or not the Project was selected for funding; and

(v) If the Project was selected for funding,

(A) The name(s) of the specific program(s) that provided the funding;

(B) The date and amount of the award; and

(C) Whether any of the funding came from the funds reserved under this subpart.

§§ 1980.1016–1980.1019 [Reserved]

§ 1980.1020 Scoring.

The Agency will score each eligible application seeking funding under this subpart as described in this section.

(a) *Underlying program scoring.* The Agency will score each application using the criteria for the applicable program identified in § 1980.1002. The maximum number of points an application can receive under this paragraph is based on the scoring criteria for the applicable underlying program, including any discretionary points that may be awarded.

(b) *Section 6025 scoring.* The Agency will score each application using the criteria identified in paragraphs (b)(1) and (2) of this section. The maximum number of points an application can receive under this paragraph is 20 points.

(1) *Project's direct support of a Plan's objectives.* The Agency will score each application on the basis of the number of a Plan's objectives the Project directly supports. The maximum score under this paragraph is 10 points.

(i) If the Project directly supports implementation of 3 of the Plan's objectives, 10 points will be awarded.

(ii) If the Project directly supports implementation of 2 of the Plan's objectives, 5 points will be awarded.

(iii) If the Project directly supports implementation of less than 2 of the Plan's objectives, no points will be awarded.

(2) *Characteristics of a Plan.* The Agency will score the Plan associated with a project based upon the characteristics of the Plan, which are identified in paragraphs (b)(2)(i) through (v) of this section. Applicants must supply sufficient documentation that demonstrates to the Agency the criteria identified in paragraphs (b)(2)(i) through (v) of this section. The maximum score under this paragraph is 10 points.

(i) *Collaboration.* If the Plan was developed through the collaboration of multiple stakeholders in the service area of the Plan, including the participation of combinations of stakeholders, such as State, local, and tribal governments, nonprofit institutions, institutions of higher education, and private entities, two points will be awarded.

(ii) *Resources.* If the Plan demonstrates an understanding of the applicable regional assets that could support the Plan, including natural resources, human resources, infrastructure, and financial resources, two points will be awarded.

(iii) *Other Federal Agency Investments.* If the Plan includes Investments from Federal agencies other than the U.S. Department of Agriculture, two points will be awarded.

(iv) *Philanthropic organization Investments.* If the Plan includes Investments from Philanthropic organizations, two points will be awarded.

(v) *Objectives and performance measures.* If the Plan contains clear objectives and the ability to establish measurable performance measures and to track progress toward meeting the objectives, two points will be awarded.

(c) *Total score.* The Agency will sum the scores each application receives under paragraphs (a) and (b) of this section in order to rank applications.

§§ 1980.1021–1980.1024 [Reserved]

§ 1980.1025 Award process.

(a) Unless RD indicates otherwise in a notice, the award process for the applicable underlying program will be used to determine which Projects receive funding under this subpart.

(b) In years when funding is made available under this subpart, Projects not receiving funding under this subpart are eligible to compete for funding under the applicable underlying program. The scores for such Projects when competing for underlying program funding will not include the score assigned to the application under § 1980.1020(b).

(c) In years when funding is not made available under this subpart, Projects are eligible to compete for funding for the applicable underlying program. The scores for such Projects when competing for underlying program funding will include the score assigned the application § 1980.1020(b) as described in a notice published in the FEDERAL REGISTER.

§ 1980.1026 Evaluation of Project information.

To assist the Agency in evaluating the effectiveness of this subpart, each applicant that receives funding under this subpart must submit to the Agency all measures, metrics, and outcomes of the Project that are reported to the entity(ies) who are monitoring Plan implementation. This information will be submitted for as long as the Plan is in effect.

§§ 1980.1027–1980.1100 [Reserved]

PARTS 1981–1999 [RESERVED]

FINDING AIDS

A list of CFR titles, subtitles, chapters, subchapters and parts and an alphabetical list of agencies publishing in the CFR are included in the CFR Index and Finding Aids volume to the Code of Federal Regulations which is published separately and revised annually.

Table of CFR Titles and Chapters
Alphabetical List of Agencies Appearing in the CFR
List of CFR Sections Affected

Table of CFR Titles and Chapters

(Revised as of January 1, 2019)

Title 1—General Provisions

I   Administrative Committee of the Federal Register (Parts 1—49)
II   Office of the Federal Register (Parts 50—299)
III   Administrative Conference of the United States (Parts 300—399)
IV   Miscellaneous Agencies (Parts 400—599)
VI   National Capital Planning Commission (Parts 600—699)

Title 2—Grants and Agreements

    SUBTITLE A—OFFICE OF MANAGEMENT AND BUDGET GUIDANCE FOR GRANTS AND AGREEMENTS
I   Office of Management and Budget Governmentwide Guidance for Grants and Agreements (Parts 2—199)
II   Office of Management and Budget Guidance (Parts 200—299)
    SUBTITLE B—FEDERAL AGENCY REGULATIONS FOR GRANTS AND AGREEMENTS
III   Department of Health and Human Services (Parts 300—399)
IV   Department of Agriculture (Parts 400—499)
VI   Department of State (Parts 600—699)
VII   Agency for International Development (Parts 700—799)
VIII   Department of Veterans Affairs (Parts 800—899)
IX   Department of Energy (Parts 900—999)
X   Department of the Treasury (Parts 1000—1099)
XI   Department of Defense (Parts 1100—1199)
XII   Department of Transportation (Parts 1200—1299)
XIII   Department of Commerce (Parts 1300—1399)
XIV   Department of the Interior (Parts 1400—1499)
XV   Environmental Protection Agency (Parts 1500—1599)
XVIII   National Aeronautics and Space Administration (Parts 1800—1899)
XX   United States Nuclear Regulatory Commission (Parts 2000—2099)
XXII   Corporation for National and Community Service (Parts 2200—2299)
XXIII   Social Security Administration (Parts 2300—2399)
XXIV   Department of Housing and Urban Development (Parts 2400—2499)
XXV   National Science Foundation (Parts 2500—2599)
XXVI   National Archives and Records Administration (Parts 2600—2699)

Title 2—Grants and Agreements—Continued

Title 3—The President

Title 4—Accounts

Title 5—Administrative Personnel

Title 5—Administrative Personnel—Continued

Title 6—Domestic Security

Title 7—Agriculture

Title 12—Banks and Banking—Continued

Title 13—Business Credit and Assistance

Title 14—Aeronautics and Space

Title 15—Commerce and Foreign Trade

Title 15—Commerce and Foreign Trade—Continued

Title 16—Commercial Practices

Title 17—Commodity and Securities Exchanges

Title 18—Conservation of Power and Water Resources

Title 19—Customs Duties

Title 20—Employees' Benefits

Title 23—Highways—Continued

Title 24—Housing and Urban Development

Title 25—Indians

Title 26—Internal Revenue

Title 27—Alcohol, Tobacco Products and Firearms

Title 28—Judicial Administration

Title 29—Labor

Title 29—Labor—Continued

Title 30—Mineral Resources

Title 31—Money and Finance: Treasury

309

Title 34—Education—Continued

Title 35 [Reserved]

Title 36—Parks, Forests, and Public Property

Title 37—Patents, Trademarks, and Copyrights

Title 38—Pensions, Bonuses, and Veterans' Relief

Title 45—Public Welfare—Continued

Title 46—Shipping

Title 47—Telecommunication

Title 48—Federal Acquisition Regulations System

Title 49—Transportation

Title 50—Wildlife and Fisheries

Alphabetical List of Agencies Appearing in the CFR
(Revised as of January 1, 2019)

Agency	CFR Title, Subtitle or Chapter
Administrative Conference of the United States	1, III
Advisory Council on Historic Preservation	36, VIII
Advocacy and Outreach, Office of	7, XXV
Afghanistan Reconstruction, Special Inspector General for	5, LXXXIII
African Development Foundation	22, XV
Federal Acquisition Regulation	48, 57
Agency for International Development	2, VII; 22, II
Federal Acquisition Regulation	48, 7
Agricultural Marketing Service	7, I, IX, X, XI
Agricultural Research Service	7, V
Agriculture, Department of	2, IV; 5, LXXIII
Advocacy and Outreach, Office of	7, XXV
Agricultural Marketing Service	7, I, IX, X, XI
Agricultural Research Service	7, V
Animal and Plant Health Inspection Service	7, III; 9, I
Chief Financial Officer, Office of	7, XXX
Commodity Credit Corporation	7, XIV
Economic Research Service	7, XXXVII
Energy Policy and New Uses, Office of	2, IX; 7, XXIX
Environmental Quality, Office of	7, XXXI
Farm Service Agency	7, VII, XVIII
Federal Acquisition Regulation	48, 4
Federal Crop Insurance Corporation	7, IV
Food and Nutrition Service	7, II
Food Safety and Inspection Service	9, III
Foreign Agricultural Service	7, XV
Forest Service	36, II
Grain Inspection, Packers and Stockyards Administration	7, VIII; 9, II
Information Resources Management, Office of	7, XXVII
Inspector General, Office of	7, XXVI
National Agricultural Library	7, XLI
National Agricultural Statistics Service	7, XXXVI
National Institute of Food and Agriculture	7, XXXIV
Natural Resources Conservation Service	7, VI
Operations, Office of	7, XXVIII
Procurement and Property Management, Office of	7, XXXII
Rural Business-Cooperative Service	7, XVIII, XLII
Rural Development Administration	7, XLII
Rural Housing Service	7, XVIII, XXXV
Rural Telephone Bank	7, XVI
Rural Utilities Service	7, XVII, XVIII, XLII
Secretary of Agriculture, Office of	7, Subtitle A
Transportation, Office of	7, XXXIII
World Agricultural Outlook Board	7, XXXVIII
Air Force, Department of	32, VII
Federal Acquisition Regulation Supplement	48, 53
Air Transportation Stabilization Board	14, VI
Alcohol and Tobacco Tax and Trade Bureau	27, I
Alcohol, Tobacco, Firearms, and Explosives, Bureau of	27, II
AMTRAK	49, VII
American Battle Monuments Commission	36, IV
American Indians, Office of the Special Trustee	25, VII
Animal and Plant Health Inspection Service	7, III; 9, I

318

319

320

List of CFR Sections Affected

All changes in this volume of the Code of Federal Regulations (CFR) that were made by documents published in the FEDERAL REGISTER since January 1, 2014 are enumerated in the following list. Entries indicate the nature of the changes effected. Page numbers refer to FEDERAL REGISTER pages. The user should consult the entries for chapters, parts and subparts as well as sections for revisions.

For changes to this volume of the CFR prior to this listing, consult the annual edition of the monthly List of CFR Sections Affected (LSA). The LSA is available at *www.govinfo.gov*. For changes to this volume of the CFR prior to 2001, see the "List of CFR Sections Affected, 1949–1963, 1964–1972, 1973–1985, and 1986–2000" published in 11 separate volumes. The "List of CFR Sections Affected 1986–2000" is available at *www.govinfo.gov*.

7 CFR—Continued

7 CFR—Continued

7 CFR—Continued

7 CFR—Continued

7 CFR—Continued

7 CFR—Continued

7 CFR—Continued

2016

7 CFR

7 CFR—Continued

2017

7 CFR

2018

7 CFR